Introduction to
Oceanography

David A. Ross

Woods Hole Oceanographic Institution

Introduction to Oceanography

Third edition

Prentice-Hall, Inc., Englewood Cliffs, New Jersey 07632

Library of Congress Cataloging in Publication Data

ROSS, DAVID A., (date)
  Introduction to oceanography.

  Bibliography: p.
  Includes index.
    1. Oceanography.   I. Title.
  GC16.R6 1982   551.46   81-17702
  ISBN 0-13-491357-4   AACR2

Editorial/production supervision: Kathleen M. Lafferty-Lyddy
Cover and chapter opening designs: Maureen Olsen
Manufacturing buyer: John B. Hall

Printed in the United States of America

10 9 8 7 6 5 4 3

ISBN 0-13-491357-4

Prentice-Hall International, Inc., *London*
Prentice-Hall of Australia Pty. Limited, *Sydney*
Prentice-Hall of Canada, Ltd., *Toronto*
Prentice-Hall of India Private Limited, *New Delhi*
Prentice-Hall of Japan, Inc., *Tokyo*
Prentice-Hall of Southeast Asia Pte. Ltd., *Singapore*
Whitehall Books Limited, *Wellington, New Zealand*

Description and Credits for Photographs Used without Legends

title page   "Sea smoke" due to a large difference in the temperature between the air and the ocean. (Photograph courtesy of Jim Broda.)

chapter 1   Working at sea. (Photograph courtesy of Woods Hole Oceanographic Institution.)

2   The research vessel *Westward* in Labrador waters. The ship is used by the Sea Education Association of Woods Hole for training of students. (Photograph courtesy of Peter Beamish.)

3   Instrument anchor being lowered from research vessel *Knorr*. (Photograph courtesy of Dick Nowak.)

4   A cluster of giant sea worms found in the hot spring vents in the Galapagos Rift along the East Pacific Rise by scientists from Woods Hole Oceanographic Institution. The worms have brilliant red tips and live in one-inch diameter tubes attached to rocks on the ocean bottom. An eight-foot tube and a worm more than five feet long were recovered during a recent expedition. Clams and mussels are in the foreground; some measure nearly one foot. (Photograph courtesy of Holger Jannasch.)

5   A box corer being readied for lowering. (Photograph courtesy of Allan Driscoll.)

6   Variety of pillow lava observed along the East Pacific Rise. (Photograph courtesy of Woods Hole Oceanographic Institution.)

7   Lowering of an array of water sampling equipment. (Photograph courtesy of Susan Kadar.)

8   Retrieval of large bongo nets. (Photograph courtesy of Frank A. Bailey, National Marine Fisheries Service.)

9   Launching of an acoustic device. (Photograph courtesy of Jim Doutt.)

10   Penguins on an iceberg. (Photograph courtesy of Woods Hole Oceanographic Institution.)

To my parents

Contents

I AM CONTINUALLY SURPRISED at the changes that have occurred in the field of oceanography over the last five years since the second edition of *Introduction to Oceanography*. Clearly, the ocean is being visualized more and more for its use rather than just as an abstract curiosity. The mineral and biological resources of the ocean, especially anything having energy potential, have been a special area of interest. Equally important have been the increased awareness of pollution problems and the importance and use of the coastal zone.

An increased awareness and understanding of how the ocean interacts with the atmosphere and how it influences climate have developed recently. The potential for better prediction or even control of climate is an especially exciting field. The concept of sea-floor spreading and plate tectonics has continued to gain acceptance. Especially important has been the finding of unique bottom fauna along ocean ridges that owe their existence to the processes of sea-floor spreading. The ongoing United Nations Law of the Sea Conference has entered its second decade and may be close to a conclusion, although conflicts still exist over mineral exploitation. Because of all these developments, the third edition includes the chapters: Climate and the Ocean, The Coastal Zone, Innovative Uses of the Ocean, and a separate chapter on Sea-Floor Spreading. All other chapters have been updated to include discoveries of recent years. Over 130 new figures, a summary for each chapter, and new references, especially some from *Scientific American* and *Oceanus*, have been added.

As in previous editions, I have tried to explain and describe oceanography in a manner understandable to nonscientists. I have stressed the role of the separate scientific disciplines in their applications to oceanography. Social sciences, such as law and economics, are also included.

I have been very fortunate to discuss many aspects of this edition with my colleagues in the Woods Hole community and elsewhere. Many people have given me photographs and other material, and I have acknowledged this in the appropriate places within the text. However, I wish to thank especially several who have provided particularly large amounts of material: David Aubrey, Robert Ballard, Vicky Cullen, K. O. Emery, George Grice, Shelley Lauzon, William MacLeisch, Paul Ryan, Donald Souza, and Ivan Valiela. Special appreciation is due Ellen Gately who assisted in most stages of preparation. I would like to also thank Logan Campbell, Kathleen Lafferty-Lyddy, and Maureen Olsen of Prentice-Hall for their encouragement and help.

I would like to thank my wife, Edith, who gave me help and encouragement and privacy when they were needed. Finally, I have dedicated this work to my father, who didn't live to see the effects that his hard work had on his son, and to my mother.

DAVID A. ROSS

Woods Hole, Massachusetts

Introduction to Oceanography

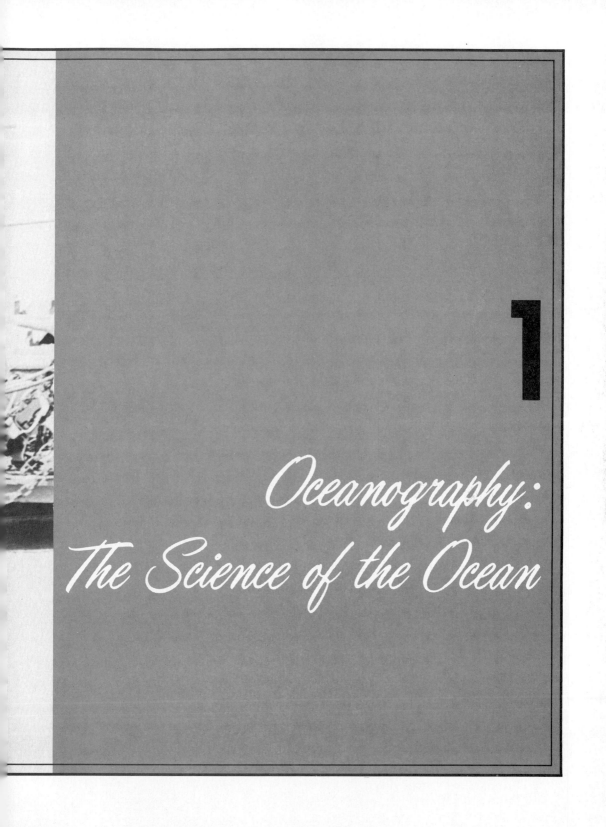

1

Oceanography: The Science of the Ocean

THE OCEAN IS one of the last frontiers available to the human race. Even as we are slowly learning more and more scientifically about the marine environment, we are, according to many, polluting parts of it beyond a level from which it may recover. In the political arena, the countries of the world are dividing up the ocean to increase their shares of its biologic and mineral resources. The food and oil shortages of recent years have caused people to look toward the ocean for at least a partial solution to these problems, and many new and imaginative ideas have been proposed for ocean use. The answers to the problems and the implications of all the above ocean uses and ideas are not completely known. One objective of this book is to describe what we know about the ocean and what we can do.

In recent years the relationships between events on land and events in the ocean have become better understood and more appreciated. Some of these relationships are obvious. For example, debris from weathering and erosion on land is carried into the ocean by rivers and streams. Likewise the ocean is the principal source of the water for the rain and snow that feed the streams. Even more important has been the developing awareness of the influence of the ocean on climate and of the possibility of adjusting or controlling the world's climate. The ocean is critically important as a source of food, especially animal protein, and energy. The latter can come from conventional sources, such as hydrocarbon deposits found buried within the sediments of the continental margin, as well as from unconventional methods using the physical, chemical, and biological processes active within the ocean (see Chapter 14). Mineral deposits, such as manganese nodules and phosphorite, and possibilities of obtaining fresh water and different elements by desalination may change some economic and industrial aspects on land.

The ocean also profoundly influences land in other, perhaps less obvious, ways. For example, most of the world's commerce travels by ship and much of our communication goes through cables laid on the ocean floor. The military use of the ocean by the superpowers is so evenly balanced that it may provide a moderating influence on the arms race. Finally, the ongoing series of U.N. Conferences to develop a comprehensive law of the sea may provide a unifying legal concept for ocean usage.

What is oceanography? Many definitions are possible; a simple one is *the application of all science to the phenomena of the ocean.* The key word in this definition is *all,* for to truly understand the ocean and how it works, one should know something about almost all fields of science and their relationships to the marine environment. Thus oceanography is not a single science but rather a combination of various sciences. Most oceanographers divide oceanography into four main parts: (1) chemical oceanography; (2) biological oceanography; (3) physical oceanography; and (4) geological oceanography, which includes marine geophysics. In recent years fifth and sixth components of oceanography have evolved. The fifth division is ocean engineering, and the sixth is sometimes called marine policy.

Components of Oceanography

The chemical oceanographer is concerned with chemical reactions that occur both in the ocean and on the sea floor. Many of these reactions are biologically influenced, especially those in the upper layers of the ocean. Biological oceanography involves the study of the distribution and environmental aspects of life in the ocean. Physical reactions, such as changes and motion of seawater, are included in the realm of the physical oceanographer. The marine geologist, or geological oceanographer, studies the sediments and topography of the ocean floor. The deeper structure of the ocean floor and its physical properties are the domain of the marine geophysicist. Both marine geologists and marine geophysicists are especially interested in the origin and evolution of the earth and its ocean basins. The ocean engineer is mainly concerned with the development of technology for oceanographic research and exploitation. The field of marine policy is not well defined but is generally considered as the application of social and political sciences such as economics, law, and policy toward the use and management of the ocean.

Although these divisions seem to break up oceanography into neat little niches, in practice it is otherwise. For example, a marine geologist taking a sample of the sediment under the equatorial Pacific would obtain material composed mainly of the shells of dead microscopic organisms. Most of these organisms lived in surface waters more than 3.2 kilometers (2 miles) above the bottom. If samples were taken of the ocean bottom 160 kilometers (100 miles) north or south of the equator, the number of shells obtained would be considerably less. This is because unique physical oceanographic conditions exist in the region of the equator, where the right combination of currents and winds keeps the water well mixed. The mixing, in turn, influences the chemistry of the water; nutrients that are necessary for the life cycle of the organisms are brought from depth to the surface, where they can be used. Thus these geologic deposits on the sea floor are intimately influenced by the chemistry, physics, and biology of the water above.

This example suggests that perhaps the different divisions of oceanography are somewhat artificial and unnecessary since an oceanographer should be versed in all of these fields. Oceanography has advanced so rapidly, however, that it is essentially impossible for a scientist to be expert in all its aspects. Consequently, most oceanographers specialize in one or two of the divisions listed earlier. You should always remember that the divisions are not rigid and that the different fields are closely related.

In this book biological oceanography, chemical oceanography, physical oceanography, and marine geology and geophysics are treated in separate chapters. A detailed discussion of ocean engineering is beyond the scope of this book, but I have included a chapter on important oceanographic instrumentation. Marine policy is discussed in the sections on mineral resources, pollution, law of the sea, and the coastal zone. Separate chapters are devoted to sea-floor spreading (the origin of the ocean basins) and innovative uses of the ocean.

Why Study the Ocean?

Now that we have defined oceanography, we can consider some general aspects, such as why a person studies oceanography. Clearly the ocean is a hostile and not readily accessible environment that does not easily yield its secrets. This very secrecy, this lure and romance of the sea, has drawn many persons to oceanography. Equally enticing is the knowledge that water covers about 72 percent of the world, and people have always been interested in their environment. The depths of the ocean are no exception.

To the trained marine scientist the ocean may have the answer to some of people's important scientific questions and problems. Within the sediment layers of the ocean floor are recorded the geologic history of the earth and, in fossils, its biological history. By studying ocean sediments we can learn about ancient climate and how it changed and thus better understand our present climate. Evidence indicates that life began in the ocean a few billion years ago, and since then evolution has produced the vast quantities and varieties of life now found in the ocean. This abundance of life is an important food source and holds the promise of solving some of today's food problems. Other biological products of the sea, such as pearls or the shells of dead organisms, have varied uses—some shells, for example, are valuable as building materials. The ocean is also an important source of some commercially valuable chemical resources, including iodine, bromine, potassium, magnesium, manganese, and other elements. Desalination of ocean water is yielding increasingly important amounts of freshwater in arid areas of the world. Sea-floor mineral accumulations like phosphorite, manganese nodules, heavy-metal-rich muds, sand, and gravel are valuable commodities that in some instances are already being exploited. Accumulations of oil and gas below the sea floor are important natural, but nonrenewable, resources already supplying about

20 percent of the world's needs. Much of the solar energy that reaches the earth is stored in the ocean (and perhaps could be used—see Chapter 14) and helps power oceanic and atmospheric circulation. In this manner the ocean plays an important, but incompletely understood, role in influencing the weather and climatic patterns of the earth.

The ocean is necessary to commerce, communication, and national defense. Over 90 percent of the trade between countries is carried by ships; beneath the ocean, cables link the communication networks of many of the world's countries. The seas have been a battlefield for much of our history, and some of today's oceanic research is concerned with national defense. Finally, the ocean is important for recreation; the sports of fishing, boating, water skiing, scuba diving, and swimming attract ever greater numbers of persons each year. This latter aspect underscores one of the major problems facing humans—pollution, which if not controlled can make some of these activities become things of the past.

Oceanography in the 1980s

Oceanography in the 1980s will probably be different than in the past. First, the problems resulting from the Law of the Sea Conference (see Chapter 15) will certainly result in a more restrictive environment for oceanographic research. National interest, especially among developing countries, will undoubtedly stress the exploration and exploitation of marine resources rather than encourage pure marine scientific research. One of the more important problems that oceanographers will focus on in the 1980s will be climate. Recent investigations have shown that variable climatic conditions existed in the past and that some of the variations were related to oceanographic phenomena. An understanding of the causes of climatic changes could lead to techniques that could modify climate and weather (like hurricane control). This interesting aspect is discussed further in Chapter 10. Another area of increased interest will be in obtaining energy from the ocean via instrumentation like the Ocean Thermal Energy Conversion system (OTEC) (Figure 1-1) or by new ideas (see Chapter 14).

Several members of the U.S. Congress have suggested that the 1980s be considered as the Decade of Ocean Resource Use and Management. Eight major uses have been identified, including fisheries and other forms of food, ocean energy, marine transportation, defense, nonfuel minerals, waste disposal, recreation, and environmental protection. In some instances one use can conflict with another, creating a need for adequate management techniques.

Oceanography as a Career

The modern oceanographer generally receives training in one of two ways: either through formal graduate school training in oceanography or from an education in an associated scientific field. Because oceanography is the application of all

FIGURE 1-1 Model of an OTEC, or Ocean Thermal Energy Conversion system, that can produce energy by using the temperature differences between the surface and deep waters of the ocean (see Chapter 14 for more details). The system shown is one proposed by TRW and Global Marine. (Photography courtesy of TRW.)

science to the phenomena of the ocean, an aspiring research oceanographer would probably do best to obtain a sound training in the basic sciences as an undergraduate and then to specialize in a specific science or oceanography in graduate school. Probably the best approach for a person interested in becoming a marine technician would be to study in a specific scientific field, such as biology, geology, or engineering. Those desiring further information about a career in oceanography should read an article by Dr. Robert B. Abel (see reference list at end of this chapter) and consult the various sources of information that he mentions.

College graduates who have a bachelor's degree in one of the associated sciences and want employment in marine science usually work first as laboratory or research assistants. Scientists with more advanced training or experience may have teaching or research positions. Whatever their training, most marine scientists spend part of their time at sea. Generally oceanographic cruises last from a few days to several months. Most of the time at sea is spent acquiring data, sometimes under adverse conditions.

There are about 20 universities and research laboratories where good advanced training in oceanography is available, and over 150 universities and colleges offer at least a few courses in marine science or related fields. Among the larger research institutions are Scripps Institution of Oceanography (Figure 1-2), Woods Hole Oceanographic Institution (Figure 2-7), School of Marine and Atmospheric Science of the University of Miami, and Lamont–Doherty Geological Observatory. These large research institutions and other excellent schools, such as those associated with the University of Rhode Island, Texas A & M, University of Southern California, Oregon State University, University of Hawaii, and University of Washington, often emphasize deep-sea research. However, in recent years these institutions and other smaller, equally competent laboratories have increased their interest in more local, nearshore problems.

Oceanographers are employed by universities, research laboratories, state agencies, environmental groups, industry, and the federal government. In the United States the largest federal agency concerned with the ocean is NOAA, or the National Oceanic and Atmospheric Administration. This organization, with a broad charter, has over 14,000 employees, 25 ships (Figure 1-3), and a budget

FIGURE 1-2 Scripps Institution of Oceanography, La Jolla, California. In the background is the San Diego Campus of the University of California. (Photograph courtesy of Glasheen Graphics, La Jolla, Calif.)

FIGURE 1-3 *Researcher*, a 84.7-m (278-ft) long ocean research vessel that is part of the scientific fleet of NOAA, the National Oceanic and Atmospheric Administration. The vessel is highly automated and can conduct all types of marine research. *Researcher* can accommodate 76 officers, scientists, and crew and can stay at sea for more than a month.

of about $800 million. Included within NOAA are the National Ocean Survey, National Marine Fisheries, and Sea Grant. Other government agencies involved in marine science are the Office of Naval Research, National Science Foundation, U.S. Geological Survey, Department of Energy, Environmental Protection Agency, the Coast Guard, and the State Department. The outlook for future employment in the field of oceanography is considered to be good, especially in environmental and energy-related work.

The federal government has recognized the importance of the ocean and has established or supported several long-range programs such as the Deep Sea Drilling Project (see pages 47–49), International Decade of Ocean Exploration (from 1970 to 1980), and Sea Grant. The programs associated with the International Decade of Ocean Exploration encouraged the participation of scientists from foreign countries in studies of worldwide oceanographic phenomena. The Sea Grant Program, a part of NOAA, is a combined effort involving education, research, and advisory services that emphasizes marine resources. Participation in this program is not limited to marine scientists, but also involves lawyers, teachers, economists, and the like. The Sea Grant Program is similar in concept to the Land Grant Program, initiated about 100 years ago, that developed our national agricultural and engineering capabilities.

There are perhaps 3,000 people in the United States who would qualify as trained marine scientists, or about 1 for every 80,000 of our population—a very small percentage indeed, especially when one considers that 72 percent of the earth's surface is covered by water.

Before turning in following chapters to the origin of the ocean, the history of oceanography, and oceanographic instrumentation, I shall discuss some terms and statistics commonly used in oceanography and some basic characteristics of the ocean.

Terms and Statistics

Oceanographers, for various reasons, use a confusing mixture of terms when discussing the ocean. The metric system, adopted by most scientists, is also generally used in oceanography. This system is based on multiples of 10. The smallest unit with which we shall be concerned is a micron (μ); 1,000 μ equal 1 millimeter (mm), 10 mm equal 1 centimeter (cm), 100 cm equal 1 meter (m), and 1,000 m equal 1 kilometer (km). A kilometer is about 0.6 mile (mi) (Tables 1-1 and 1-2).

Depth is measured either in fathoms or in meters. A fathom is 6 feet (ft), or approximately the length of a line a person can hold between outspread hands; 100 fathoms equal 183 m.

Velocity is usually measured in knots (kn); 1 kn equals 1 nautical mile (nmi) (6,080 ft) per hour. The commonly used metric equivalent of 1 kn is approximately

TABLE 1-1   Metric–English Equivalents

Unit	METRIC			ENGLISH		
	Centimeters	Meters	Kilometers	Inches	Feet	Miles
cm	1	1/100	1/100,000	0.3937	—	—
m	100	1	1/1,000	39.37	3.28	—
km	100,000	1,000	1	—	3,280	0.624
in.	2.54	—	—	1	1/12	—
ft	30.48	0.3048	—	12	1	1/5,280
mi	—	1,609	1.609	—	5,280	1
	Grams		**Kilograms**	**Ounces**		**Pounds**
g	1		1/1,000	0.035		—
kg	1,000		1	—		2.20
oz	28.35		—	1		1/16
lb	453.54		0.453	16		1

TABLE 1-2 Conversion of Various Units Used in Oceanography

To convert	Into	Multiply by
centimeters	inches	0.3937
meters	feet	3.28
meters	centimeters	100.0
meters	fathoms	0.546
kilometers	miles	0.624
kilometers	meters	1000.0
grams	ounces (avdp)	0.035
kilograms	pounds	2.2
degrees Celsius	degrees Fahrenheit	$(°C \times \frac{9}{5}) + 32$

50 cm per second. The term *knot* was derived from a device called a Dutchman's log, which is a piece of wood or log attached to a string with knots tied in it at equal units. The log was thrown overboard and, as the ship moved away, the string played out. The number of knots in the line passing overboard in a certain duration of time were counted, and thus the speed of the ship was measured in knots.

Temperature is measured in degrees Celsius (°C); 0°C equals 32°F (Fahrenheit) (the freezing point of water), 20°C equals 68°F (about room temperature), and 100°C equals 212°F (the boiling point of water).

General Characteristics of the Ocean

The average area and volume of different parts of the ocean have been calculated (Table 1-3). The total volume of the ocean is about $1,350 \times 10^6$ (1,350 million) cubic kilometers (km³) or about 318×10^6 mi³. The average depth of the ocean is 3,729 m, which is equal to 2,036 fathoms, 12,216 ft, or 2.3 mi.

The distribution of elevation over the world is shown in Figure 1-4. This type of representation is called the *hypsographic,* or *hypsometric,* curve. The curve shows the amount of the earth's surface above any given elevation or depth. Two important facts are evident from Figure 1-4:

1. Two elevations dominate: one at about 100 m, the other at about −5,000 m.
2. There is a sharp intermediate zone between these elevations.

The two elevations clearly indicate two different parts of the earth's crust: the deep-sea floor and the portion of land just at or near sea level. Note that if all the water were removed from the ocean, this major difference would still exist. The intermediate zone corresponds to the transition area between the continental and the oceanic regions. This part of the curve is represented in the ocean by the continental slope.

TABLE 1-3   Area, Volume, and Mean Depth of the Ocean

Ocean and adjacent seas	Area (millions of km²)	Volume (millions of km³)	Mean depth (m)
Pacific	181.344	714,410	3,940
Atlantic	94.314	337,210	3,575
Indian	74.118	284,608	3,840
Arctic	12.257	13,702	1,117
Totals and mean depth	362.033	1,349,929	3,729

Source: Data from Menard and Smith, 1966.

The continents and the ocean basins have an uneven distribution on the earth's surface. A continental area usually has an oceanic area opposite it on the other side of the earth, and most of the land is north of the equator (Figures 1-5 and 1-6). Less than 35 percent of the total land is found south of the equator, and between latitudes 50° and 65° S there is essentially no land at all. In this area the oceans are essentially connected.

Since most of the earth is covered by water, perhaps our planet should have been called Water.

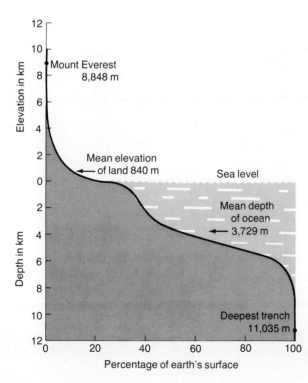

FIGURE 1-4   Hypsographic curve showing the percentage area of the earth's surface above a given elevation or depth.

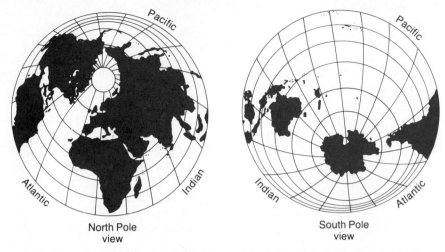

FIGURE 1-5 North and south polar views of the earth. Note that water dominates the southern part of the earth, whereas land prevails in the northern part. Note also that the major oceans are connected around the Antarctic continent.

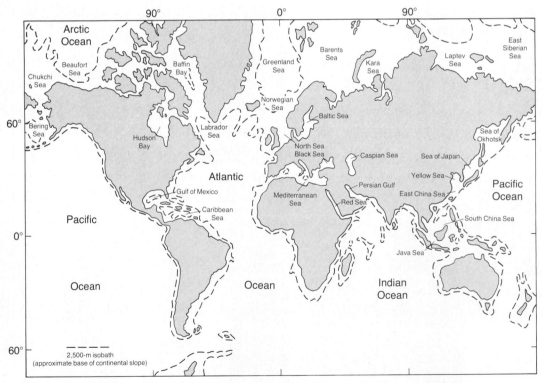

FIGURE 1-6 The major oceans and seas of the world. Also shown is the approximate position of the 2,500 m (8,202 ft) isobath, or bottom contour. (Adapted from U.S. Geological Survey Circular 694, 1974.)

14

Specific Characteristics of the Ocean Floor

The ocean floor can be subdivided into two major divisions called the *continental margin* and the *ocean basin,* or *deep sea* (Figure 5-17). The continental margin (about 21 percent of the total ocean) is where the oceans and continents merge and includes the generally broad continental shelf, relatively steep and narrow continental slope, and deep and wide continental rise (see Chapter 5 for more detailed discussion). The deep sea is dominated by the broad and relatively flat abyssal plains and the impressive oceanic ridges.

The Pacific Ocean is the largest ocean, having an area and volume greater than the Atlantic and Indian Oceans combined (Table 1-3). It is also the deepest ocean. The Pacific is essentially circular in shape and has numerous deep trenches and associated islands around its perimeter (Figure 1-7). This perimeter is also the site of frequent earthquakes and active volcanoes (Mount St. Helens sits along this zone). Several large marginal seas (bodies of water, usually shallow, some-what isolated by land or by shallow submerged regions, called *sills*, from the main ocean) flank the western part of the Pacific Ocean. These include the Sea of

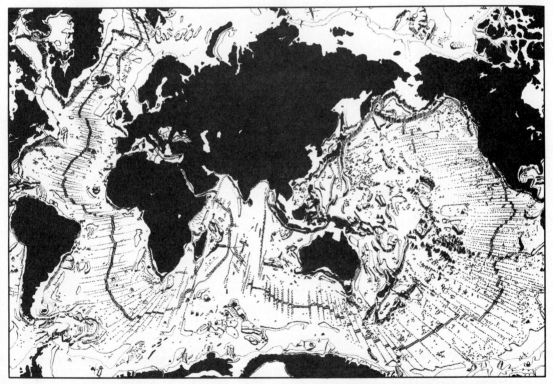

FIGURE 1-7 The general bottom characteristics of the oceans based in part on the excellent charts made by Dr. Bruce C. Heezen and Miss Marie Tharp of the Lamont–Doherty Geological Observatory.

Okhotsk, Sea of Japan, Yellow Sea, and East and South China Seas. Many marginal seas are sites of known or probable hydrocarbon accumulation.

The Atlantic Ocean is a relatively narrow ocean. The surrounding continents—Africa, South America, Europe, and North America—appear as if they would fit together like pieces of a jigsaw puzzle if the water of the ocean were removed and they were pushed together. This symmetry has led to the concept of sea-floor spreading, which has shown that these continents were indeed once all connected and have recently separated to form the ocean basins. This exciting concept will be discussed in more detail in Chapter 6.

Several large but shallow seas are adjacent to and actually are part of the Atlantic Ocean. These include the Mediterranean, Baltic, North, Norwegian, and Caribbean Seas and the Gulf of Mexico. Numerous large rivers like the Amazon, Congo, and Mississippi discharge large quantities of sediment and freshwater into the Atlantic.

The Indian Ocean, somewhat triangular in shape, is situated mainly south of the equator. Although one of the first oceans to be explored and sailed, it is still one of the least understood of the major oceans. Three large rivers, the Ganges, Indus, and Brahmaputra, discharge into the Indian Ocean. The Persian Gulf and the Red Sea are two important marginal seas in the northwestern part of the Indian Ocean.

The Arctic Ocean is somewhat circular in shape and is considerably shallower than the other oceans. It should not, according to some, be classified as a separate ocean but rather as an almost landlocked arm of the Atlantic.

A distinguishing feature of all these oceans is a mid-ocean ridge that essentially almost encircles the globe (Figures 1-7 and 5-33). This ridge goes by several different names, depending on which ocean it is in; that is, Mid-Atlantic Ridge, East Pacific Rise, or Mid-Indian Ridge. These ridges are probably the most distinctive topographic features of the earth, but their existence and extent was confirmed only a few decades ago. The origin of these features and other aspects of the ocean are described in Chapter 6.

Summary

The ocean is one of the last frontiers available and is an extremely important one. In the coming years increasing use will be made of the ocean including its biological and mineral resources and possibly for weather modification.

Oceanography can be defined as the application of *all* science to the phenomena of the ocean. In general, oceanography is divided into four main parts: chemical oceanography, biological oceanography, physical oceanography, and geological (and geophysical) oceanography. Ocean engineering and marine policy could be considered as additional fields. In much oceanographic work a knowledge of all fields is necessary.

Oceanography in the 1980s will clearly be changed because of the Law of the Sea Conference and a trend toward more applied uses of the ocean, such as increased energy from the ocean. Climate and its relationship to the marine environment will also be an important aspect for study.

Although oceanography is taught at many institutions there really are only a small number of oceanographers employed in the United States. Many U.S. government agencies are concerned with marine problems. The largest is NOAA, or the National Oceanic and Atmospheric Administration.

The ocean has a volume of 1,350 million km^3 (318 million mi^3) and an average depth of 3,729 m (12,216 ft). Oceans tend to dominate in the Southern Hemisphere while land is predominant in the Northern Hemisphere. Seventy-two percent of the earth is covered by water.

The sea floor can be divided into two major features: the continental margin and the deep-sea floor. The Pacific is the largest and deepest ocean. An area of high earthquake and volcanic activity rings much of the Pacific. A mid-ocean ridge runs through the Pacific, Indian, and Atlantic Oceans and is perhaps the most distinctive topographic feature on the earth's surface.

Suggested Further Readings

ABEL, R. B., "Careers in Oceanography," *Sea Technology,* **19** (1978), pp. 25–27.

BORGESE, E. B., and N. GINSBURG, *Ocean Yearbook,* **1**. Chicago: University of Chicago Press, 1978.

GROSS, M. G., *Oceanography: A View of the Earth*, 3rd. ed. Englewood Cliffs, N.J.: Prentice-Hall, Inc., 1982.

ROSS, D. A., *Opportunities and Uses of the Ocean*. New York: Springer-Verlag, 1980.

SKINNER, B. J., and K. K. TUREKIAN, *Man and the Ocean*. Englewood Cliffs, N.J.: Prentice-Hall, Inc., 1973.

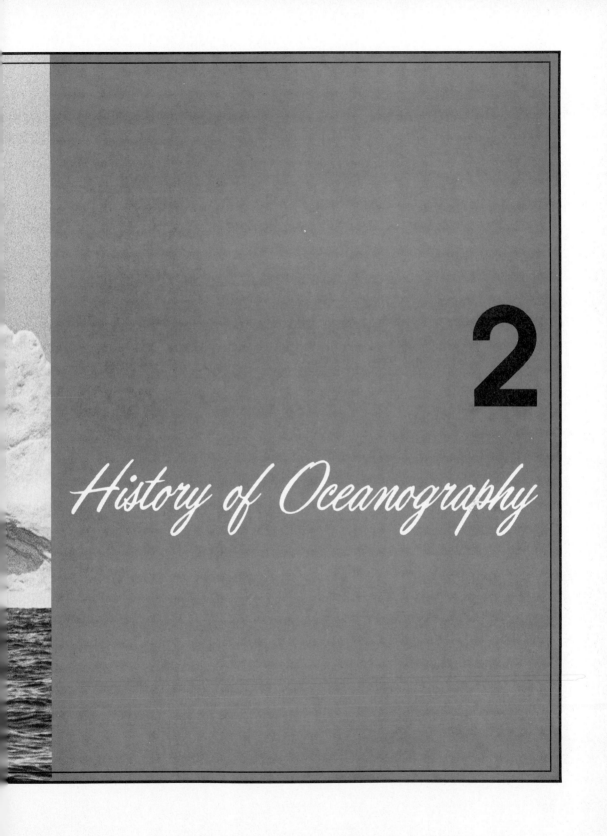

2

History of Oceanography

Early History

THE EARLY STUDY of the sea was usually motivated by a practical rather than an abstract curiosity about the ocean. One simply wanted to go from one place to another in as short a time as possible. Therefore, the beginning of oceanography is closely connected with people's early thoughts about geography and the development of trade. By 800 B.C., voyages had been made around Africa. In 500 B.C., Parmenides stated that the earth was round, and by 400 B.C., the rise and fall of tides had been related to the phases of the moon. The Greek mathematician Eratosthenes in 250 B.C. determined the circumference of the earth with remarkable precision and made a fairly accurate chart of the world as it looked to people at that time (Figure 2-1).

There developed, among ancient scholars, two hypotheses about the distribution of land and water. Eratosthenes and Strabo believed that the continents of the world formed a single island surrounded by the ocean. Ptolemy, who lived in the middle of the second century A.D., believed that the Atlantic and Indian Oceans were enclosed seas like the Mediterranean. He also held that the eastern and western points of the world approached each other very closely and that by sailing west one could reach the eastern extremity. It was this idea that led Columbus to sail west, where he discovered America instead of gaining his objective, India.

The Indian Ocean was the first ocean used for trade but the last to be explored in detail. A condition exists in the Indian Ocean that makes it uniquely suitable for sailing vessels. During the summer monsoon season, the wind blows from the southwest; during the winter monsoon season, it blows from the northeast. Thus vessels with the simplest square-rigged sails could travel across the entire ocean in one season and return in the next. After the fall of the Roman Empire, trade decreased, and during the Dark Ages most of the knowledge men had acquired about the sea was lost.

While the theological problems of southern Europe were restraining explo-

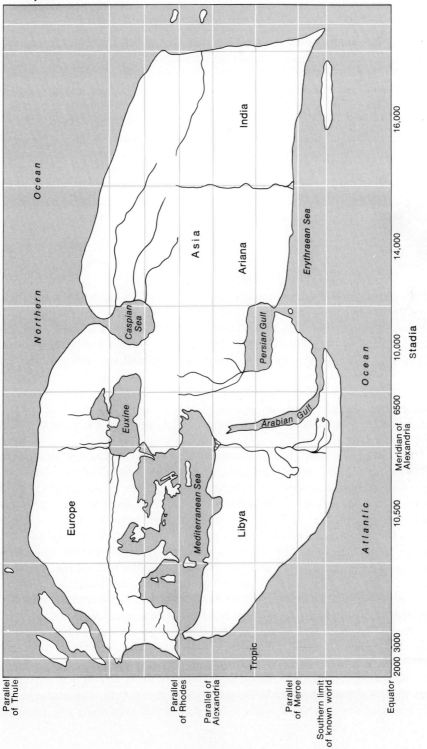

FIGURE 2-1 Eratosthenes' map of the world produced at about 250 B.C. (From M. R. Cohen and I. E. Drabkin, *A Source Book in Greek Science*, Cambridge, Mass.: Harvard University Press, 1948.)

ration by the sailors of these countries, the Vikings of northern Europe sailed to and established colonies in Iceland and Greenland prior to A.D. 1000. The Vikings were helped in their travels by relatively temperate climatic conditions that reduced ice and made sailing easier. By A.D. 982 Eric the Red had sailed to Baffin Island and later his son Leif Ericsson reached Newfoundland. It is probable that the Vikings also established colonies in the United States; however, by A.D. 1200 the climate had so deteriorated that the colonies were isolated and had to be abandoned. Also, around this time Arabs were sailing over and around the Indian Ocean and had even reached China. In China the magnetic lodestone was found and eventually adapted as a compass. It was not until the late 1400s that the people of southern Europe again explored the sea. Some of these early explorers, Diaz, Vasco da Gama, and Columbus, are known to everyone.

After the Dark Ages

By the early 1500s, one had to accept the fact that the earth was round, not flat. Magellan's voyage around the world (1519–1522) proved this without a doubt. He also may have been the first to attempt a sounding in the deep sea. Magellan used a sounding line of only 100 or 200 fathoms (182 to 364 m) in length in the Pacific and did not reach bottom. He thus concluded that this was the deepest part of the ocean. Actually a more successful sounding attempt may have been made 16 centuries before Magellan. Posidonius, who was born about 135 B.C., claimed that the sea near Sardinia had been sounded to a depth of about 1,000 fathoms (1,828 m). Unfortunately, there is little information about the methods he used.

Some of the basic characteristics of the ocean, such as the fact that deeper waters are generally cooler than surface waters and that salinity is fairly constant, were known by the end of the sixteenth century. One of the most successful of the early explorers was Captain James Cook, who early in his career became known as an excellent sailor, astronomer, and mathematician. Because of these talents he was chosen in 1768 to lead an expedition aboard the *Endeavour* to the South Pacific for an astronomical study. On this expedition he discovered the Society Islands and charted much of New Zealand and eastern Australia. Later expeditions took him south of the Antarctic Circle, to Easter Island, to South America and North America, and to the Bering Sea. Cook was the first explorer who had instruments to measure accurately latitude and longitude. After his last voyage (1776–1779) the broad general outlines of the oceans were known and only Antarctica remained to be discovered. However, the depth and character of the sea floor were still generally unknown.

Attempts at sounding the ocean depths by using a long line were made by Ellis in 1749, by Mulgrave in 1773, and by Soresby in 1817. The first real success, however, was by Sir John Ross in 1818; he obtained a sounding and a mud sample from a depth of 1,050 fathoms (1,919 m) in Baffin Bay west of Greenland. Sir Clark Ross (Sir John's nephew), during an Antarctic expedition (1839–1843), obtained soundings of 2,425 fathoms (4,433 m) in the South Atlantic and 2,677 fathoms (4,893 m) off the Cape of Good Hope. The art of sounding was advanced by a device built by Midshipman Brooke of the U.S. Navy in 1854; at the end of the sounding line he attached a detachable weight that dropped off when the line hit the bottom. The weight dropping off the line made it easier to detect when the sounding line reached the bottom. The introduction of steel cable in 1870 was another important advancement. But it was not until 1925, with the Meteor Expedition, that soundings were routinely and continuously made across the ocean. The *Meteor* measured depth electronically and did not have to stop for each measurement.

Early Oceanographic Research in the United States

Oceanographic research in the United States can be considered to have started in 1770 when Benjamin Franklin (helped by Timothy Folger, a Nantucket ship captain) published his map of the Gulf Stream. Franklin had correctly concluded that a northeastward-flowing current along the east coast of the United States must be responsible for the relatively faster travel time of ships sailing to Europe than returning from Europe. Most pictures of this chart actually are ones that Franklin published in 1786, since copies of the originals had been "lost" for close to 200 years. A copy was found in 1978 (Figure 2-2) in the Bibliothèque Nationale in Paris by Philip Richardson, a physical oceanographer.

A significant early contributor to American oceanography was Matthew Fontaine Maury of the U.S. Navy. Maury used data from the log books of ships that had crossed the Atlantic to establish the relationship between currents and oceanic weather. He published his findings in 1855 in a book called *The Physical Geography of the Sea*, one of the first English-written books about oceanography. Maury also accumulated records of deep-sea soundings, and in 1854 he published the first bathymetric map (a map that shows the bottom topography) of the North Atlantic Ocean (Figure 2-3). Another American, William Ferrel, intrigued by Maury's book, was the first to explain scientifically the motion of the surface waters of the ocean as being due to the winds.

An early large expedition was the U.S. Exploring Expedition that began in 1838. Its objective was to collect all the information it could from both land and sea to help increase commercial prospects for the United States. This expedition could be considered as the first systematic scientific expedition made by the United States. In 1836 the U.S. Congress appropriated $300,000 for a surveying and exploring expedition to the Pacific and South Seas, but the expedition was delayed until 1838 because of conflicts over how big a role the U.S. Navy should play in the program.

Eventually Lieutenant Charles Wilkes was given the command of the ex-

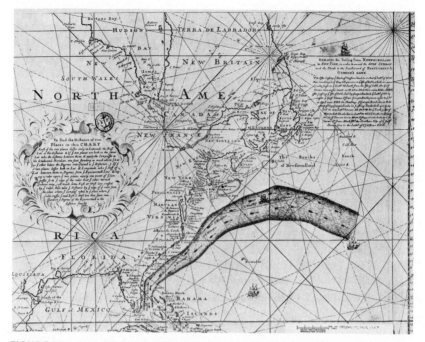

FIGURE 2-2 A portion of the Franklin–Folger chart of the Gulf Stream. (Photograph courtesy of Dr. Philip Richardson, adapted from his 1980 article in *Science* magazine; photocopy of chart provided by Bibliotheque Nationale.)

pedition and most of the scientific operations were done by navy officers, while the nine nonnavy scientists were under strong navy control.

The expedition started with three warships and three support ships. By the time they reached the Pacific, three of the ships had been disabled by storms. Probably the most famous aspect of this expedition occurred when Wilkes sailed south and claimed to have discovered Antarctica—a claim that rightly belongs to Sir Clark Ross. The expedition ended in New York in 1842 after sailing between 150,000 to 165,000 km (or about 80,000 to 90,000 mi). However, there was little public interest in the expedition; in addition, much of the scientific collection had been lost or improperly handled. There were also insufficient funds for publication of the reports of the expedition. Eventually only 100 copies each of 20 volumes were published; the final copy appeared 32 years after the expedition. Only 2 volumes were directly related to oceanography (hydrography and meteorology), while 4 others were on marine animals. The expedition was not generally considered to be scientifically successful but did establish U.S. interest in the Pacific Ocean.

Following the Civil War, the U.S. Coast and Geodetic Survey (now the National Ocean Survey, part of NOAA) started several areas of important research. The survey charted much of the coast of the United States, measured

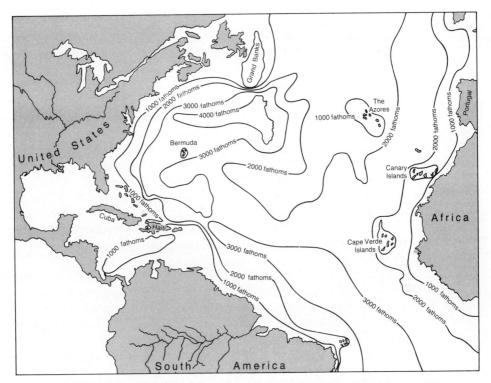

FIGURE 2-3 Maury's 1854 bathymetric map of the North Atlantic Ocean. (After Murray and Hjort, 1912.)

and studied the Gulf Stream, and studied the U.S. continental shelf. Using the ship *Blake* it conducted several dredging and sampling cruises along the U.S. east coast in the 1870s and 1880s.

Early Contributions from Other Nations

An important, although not strictly oceanographic, expedition was the voyage of the *Beagle* from 1831 to 1836, with the young naturalist Charles Darwin aboard. Darwin's findings about evolution and other aspects of the natural world stimulated other scientists to explore the ocean further.

Another important contributor to the developing field of oceanography was the English biologist Edward Forbes. Forbes is generally considered to be the initiator of the field of biological oceanography. Unfortunately, however, he is best known for one of his errors. In the 1850s he suggested that the ocean was divided into two main zones based on its content of life. He felt that the upper zone, from the surface to about a depth of 548 m or 300 fathoms, contained essentially all the life to be found in the ocean. The lower, or azoic zone, below

300 fathoms, he thought to be devoid of life. The reasoning was generally based on the assumption that light would not penetrate below this depth. This hypothesis had several difficulties including the fact that living organisms had already been dredged up from the azoic zone. Nevertheless, Forbes's prestige and the support of his students were sufficient to keep his hypothesis alive until the late 1870s, even though he died in 1854.

At about the same time Forbes's work was being debated, Thomas Huxley, a close friend of Charles Darwin's, was suggesting two more interesting ideas about the ocean. The first came from his study of deep-sea calcareous oozes that he noted to be very similar to chalk cliffs found on land. Indeed, both were composed of the small calcareous shells of planktonic animals and plants. Huxley then concluded that the chalk cliffs found on land had originally formed as oozes in the deep-ocean floor and were eventually compacted and uplifted to their present position. This idea is not completely wrong as most chalks are originally formed in the marine environment, but not necessarily in the deep sea as visualized by Huxley.

Huxley's second idea was much more spectacular; it came from the studies of samples of deep-sea mud that had been preserved for several years in "spirits." He noticed a thin jellylike layer in the sample bottle overlying the mud; after examining this layer under a microscope, he concluded that it contained protoplasm. Huxley called this material *Bathybius haeckelii,* which was considered by some, including Huxley and the German naturalist Ernst Haeckel after whom it was named, to be a basic form of life from which all other life forms had originated. Huxley thought that this material covered much of the sea floor. Many years later it was found that this miraculous material was nothing but a precipitate of calcium sulfate resulting from the mixing of alcohol and seawater.

The various hypotheses proposed by Forbes and Huxley, although eventually found to be incorrect, created considerable interest in the ocean and were directly responsible for many later expeditions. Actually, numerous oceanographic expeditions were taking place in the 1850 to 1870 interval but the beginning of deep-sea research is generally thought to have started with the Challenger Expedition (1872–1876). This expedition, under the direction of Sir Charles Wyville Thomson, circumnavigated the world. The *Challenger*, a 68.8 m (226 ft), 2,300-ton steam corvette (a fast vessel, somewhat smaller than a destroyer) made observations of many aspects of oceanography (Figure 2-4). The vessel had a crew of 243 and a scientific party of 6 and covered 127,000 km (68,890 mi), made 492 deep soundings and 133 dredgings, and obtained data from 362 oceanographic stations (Figure 2-5). At these oceanographic stations, data were collected on weather, currents, water temperature, water composition, marine organisms, and bottom sediments. More than 4,700 new species of marine life were discovered (an amazing average of about 5 new species for each day spent at sea). A deep-sea sounding of 8,180 m (4,475 fathoms or 26,850 ft), the deepest that had been made at that time, was made in the Marianas Trench. This area is now called the Challenger Deep.

FIGURE 2-4 The research vessel *Challenger*.

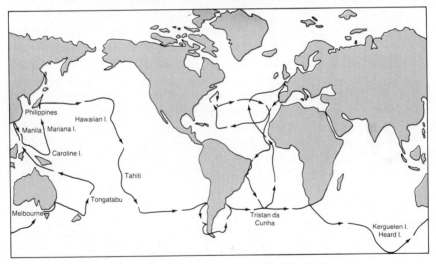

FIGURE 2-5 Route traversed by *Challenger*.

At each of the 362 stations the scientists usually attempted to make the following observations:

1. Measure the depth of the water.
2. Obtain a sample of the bottom with a tube in the sounding weight.
3. Sample the animal and plant life in the surface and intermediate waters by towing nets.

4. Obtain a sample of the fauna living on the bottom by dredging.

5. Measure the water temperature at the surface and at various depths below the surface.

6. Obtain samples of the water near the bottom and measure its temperature before it is brought to the surface.

7. Obtain samples of seawater from various depths.

8. Measure surface currents as to their direction and speed and occasionally attempt to measure subsurface currents.

9. Make atmospheric and meteorological observations.

The amount of data collected by the Challenger Expedition was immense. The reports of the expedition filled 29,500 pages of 50 volumes and took 23 years to complete. One of the great achievements of scientific exploration, the expedition also showed how little anyone knew about the sea.

After the Challenger Expedition, interest in oceanography increased. Many countries wanted to have worldwide expeditions. The German ship *Gazelle* circumnavigated the world (1874–1876), as did the Russian steamer *Vitiaz* (1886–1889). During this period other smaller-scale expeditions were also making significant contributions to oceanography. Especially noteworthy were the Austrian ship *Pola*, which worked in the Red and Mediterranean Seas from 1890 to 1898; the U.S. ship *Blake*, under the direction of Alexander Agassiz, which worked the Caribbean region from 1877 to 1880; and the Norwegian expedition of the *Fram* under the direction of Fridtjof Nansen from 1893 to 1896. The *Fram* was a wooden ship so constructed that she could be frozen into the ice of the Arctic Ocean. One objective of this cruise was to drift across the North Pole. Although the attempt failed, it was established that the Arctic was not a shallow sea, as had been thought, but a deep ocean. An attempt to recreate the Fram Expedition was completed in 1979, although in this case a field station on the ice was used rather than a vessel. The expedition (*Fram I*) objectives, which were successful, were to collect geophysical information from the Arctic Ocean north of Greenland.

Modern Oceanography

It was not until the early twentieth century that a good general picture of the oceans was obtained. Slowly the general topography of the ocean floor was mapped (Figure 2-6). A new era of oceanography started with the Meteor Expedition (1925–1927), which made one of the first detailed studies of a particular part of the ocean. Previously, most oceanographic expeditions had made only isolated, nonsystematic observations. The *Meteor* made 14 crossings of the South Atlantic in a 25-month interval. Collecting data day and night through all weather and seasons, *Meteor* was one of the first to use an electronic echo sounder to measure ocean depths, gathering more than 70,000 soundings of the ocean. The results of these soundings clearly revealed the ruggedness of the ocean floor.

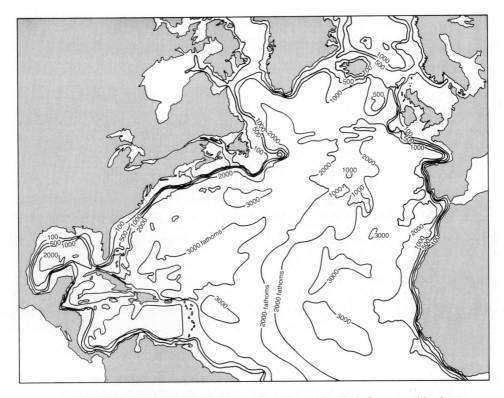

FIGURE 2-6 The bathymetry of the ocean as known in 1911. Compare this chart with Figures 1-7 and 2-3. (After Murray and Hjort, 1912.)

Actually the first continuous echo-sounding profile across an ocean basin was made from the U.S.S. *Stewart* in 1922. A total of 900 soundings were made using a sound source and accurately measuring the total time it took for sound to travel from the ship to the ocean floor and reflect to the ship.

During the time of the Meteor Expedition and in following years oceanography became an accepted science. It was also during this period that many large U.S. marine laboratories were established. Scripps Institution of Oceanography (Figure 1-2) in La Jolla, California, was formed as part of the University of California in 1912. Since that time it has grown to become the largest oceanographic institution in the United States. In 1871 the United States established a Fish Commission that was headed by Spencer Fullerton Baird. Baird established a small summer laboratory at Woods Hole, Massachusetts, on Cape Cod. This area eventually developed into one of the largest centers of marine science, containing both the Woods Hole Oceanographic Institution and the Marine Biological Laboratory (Figure 2-7).

During and after World War II, interest in oceanography expanded rapidly. Advances in technology, the military use of the ocean, and national disasters

FIGURE 2-7 Woods Hole Oceanographic Institution (center) and Marine Biological Laboratory (to the right) in Woods Hole, Massachusetts. The research vessels *Chain*, *Knorr*, and *Atlantis II* can be seen tied to the Oceanographic's dock, while a small collecting boat, the *Verill*, is tied at the Marine Biological Laboratory dock.

spurred research. The dive of the bathyscaph *Trieste* to a depth greater than 10,850 m (more than 35,600 ft) and the voyage of the nuclear submarine *Nautilus* under the North Pole showed that all parts of the ocean could be explored. However, the loss of the submarines and entire crews of *Thresher* in 1963 and *Scorpion* in 1968 sadly showed that the ocean is still a formidable adversary.

Research in the 1960s and 1970s was often done on an international scale, and scientists from many institutions and countries participated in joint programs, such as the International Geophysical Year studies of the Indian Ocean in the early 1960s. Several large and expensive programs were developed by oceanographers and many captured the imagination of the public with results reported almost daily in the newspapers. The first was Project Mohole, an attempt to drill through the earth's crust. Eventually abandoned, the project nevertheless gave impetus to the JOIDES (Joint Oceanographic Institutions Deep Earth Sampling) program. The objective of this project was to drill and sample just the upper layers of the earth's crust. An initial stage, off Florida in 1965, was successful and prompted a new program in 1968, the Deep Sea Drilling Project that has drilled over 850 holes in the deep parts of the Atlantic, Pacific, and Indian Oceans (Figures 3-11, 3-12, and 3-13). This program, using the drilling vessel *Glomar*

Challenger, has been extremely successful and perhaps as scientifically important as the first Challenger Expedition almost 100 years earlier. Studies of the sediments and rocks collected have helped establish and verify the concept of sea-floor spreading and have shown how the earth's climate has varied during the past (see Chapter 6). This project, which is mainly funded by the U.S. National Science Foundation, also receives financial support from several foreign countries. Representatives from these countries and some other nations plan and organize the program, giving it a true international flavor. In the 1980s the drilling program will expand and emphasize deep drilling on the continental margin. A different vessel, *Glomar Explorer*, will be used for this work. This vessel (Figure 12-14) was the one initially built by Howard Hughes and financed by the CIA in an attempt to raise a sunken Soviet submarine (see pages 415–416 for further discussion).

The new drilling program, called the ocean-margin drilling program, will be a 10-year effort that may cost over $600 million. It is anticipated that drilling and oil companies and foreign countries will contribute to part of the cost, but the principal funding will come from the National Science Foundation. Included in the design will be blowout preventers in case hydrocarbon deposits are encountered. Initial plans are for deep drilling on continental margins to better understand their development. This program, however, may be eliminated or delayed due to budget cuts.

The 1970s were designated as the International Decade of Ocean Exploration, and the National Science Foundation established an IDOE program. This program funded several large-scale programs. One was the GEOSECS program; GEOSECS stands for Geochemical Ocean Sections. Its principal objective was the study of mixing and circulation processes in the ocean; to reach this goal there were expeditions by the United States, West Germany, and Japan, and land-based studies in Canada, India, Italy, and France. Large volumes of water were collected from all the oceans and stored in a "water library," where samples are available to all scientists. The data and water collected are also valuable in establishing base lines for the determination of pollution effects.

Another large IDOE program was MODE (Mid-Ocean Dynamics Experiment) in which an international group of physical oceanographers studied water motions in the deep ocean. A later phase, POLYMODE, is focusing on the western Atlantic.

The possibility of living and working on the ocean floor has always fascinated scientists and nonscientists alike. With the promise of obtaining so many resources from the sea floor, the need for direct observation of the sea floor has become critical. Divers, in simulated conditions, have been able to work at pressures of about 610 m (2,000 ft), but the average depth of the ocean is 3,729 m (12,216 ft). To this day, however, it has been technically easier to walk on the moon than to walk on the deep-ocean floor.

In the last few years research submersibles have become extremely valuable research tools. Submersibles (Figures 3-6, 3-7, 3-8, and 3-9) have made extensive studies of the Mid-Atlantic Ridge and the East Pacific Rise. These studies have discovered extensive areas of mineralization and active volcanic activity (see the discussion in Chapters 5 and 6).

The 1980s should be a period in which large-scale studies will continue, but emphasis may move toward applied research. Problems such as climate, energy from the sea, coastal zone use, increasing biological productivity, and environmental concerns will be among the dominant themes. Satellite use will certainly become more important (Figures 3-17 and 11-14), especially as the cost of fuel increases makes ship use more expensive. Results from the Law of the Sea Conference (see Chapter 15) will restrict access of marine scientists to other countries, which will probably not be beneficial to our understanding of the ocean since marine problems rarely observe national boundaries.

Summary

The early study of the ocean was motivated by practical reasons—one wanted to reach, by sailing, a location as quickly as possible. Nevertheless, much was learned about the ocean before the time of Christ. However, much was lost in the Dark Ages. In the years following the early 1500s, explorers learned much about the general geography of the earth and its oceans. Early U.S. contributions included Benjamin Franklin's map of the Gulf Stream and Matthew Fontaine Maury's study of ocean currents.

European contributions included those of Charles Darwin on the *Beagle*, Edward Forbes, and Thomas Huxley (the latter two are remembered more for their mistakes). However, it was the world-encircling voyage of the *Challenger* (1872–1876) that gave the major thrust to the start of oceanography as a science. The period of modern oceanography is thought to have begun in 1925 with the work of *Meteor* in the southern Atlantic. This was a comprehensive study of a specific area of the ocean using, among other things, a continuous echo-sounding device. It was also during this time that the Scripps Institution of Oceanography in California and the Woods Hole Oceanographic Institution were founded and started to develop.

Oceanography expanded rapidly following World War II, partly as a result of advances in technology, military uses of the ocean, and some disasters. By the 1960s much oceanographic research was being done on an international scale with the participation of many countries. Projects like the Deep Sea Drilling Project, GEOSECS, and MODE helped contribute to major advances in our knowledge of the ocean. The future will probably see these large-scale projects continue but emphasis may be more on applied problems, like climate, energy from the sea, and environmental concerns.

Suggested Further Readings

DARWIN, C., *The Voyage of the Beagle*, ed. MILLICENT E. SELSAM. New York: Harper and Row, Pub., 1959.

DEACON, G. E. R., *Seas, Maps, and Men: An Atlas—History of Man's Exploration of the Oceans*. Garden City, N.Y.: Doubleday, 1962.

DEACON, MARGARET, *Scientists and the Sea, 1650–1900: A Study of Marine Science*. London: Academic Press, 1971.

LINKLATER, ERIC, *The Voyage of the Challenger*. London: John Murray (Publishers) Ltd., 1972.

MACLEISCH, W. H., ed., "A Decade of Big Ocean Science," *Oceanus,* **23,** no. 1 (1980).

MURRAY, J., and J. HJORT, *The Depths of the Ocean,* London: Macmillan, 1912.

RICHARDSON, P. L., "Benjamin Franklin and Timothy Folger's First Printed Chart of the Gulf Stream," *Science,* **207** (1980), pp. 643–45.

SCHLEE, SUSAN, *The Edge of an Unfamiliar World: A History of Oceanography*. New York: Dutton, 1973.

SCHLEE, SUSAN, *History of Oceanography*. London: Robert Hale Ltd., 1975.

SEARS, MARY, and DANIEL MERRIMAN, eds., *Oceanography: The Past*. New York: Springer-Verlag, 1980.

STANTON, WILLIAM, *The Great United States Exploring Expedition of 1938–1942*. Berkeley: University of California Press, 1975.

THOMSON, C. W., *The Voyage of the Challenger*. New York: Harper, 1878.

WENK, E., "Genesis of a Marine Policy—The IDOE," *Oceanus,* **23,** no. 1 (1980), pp. 2–11.

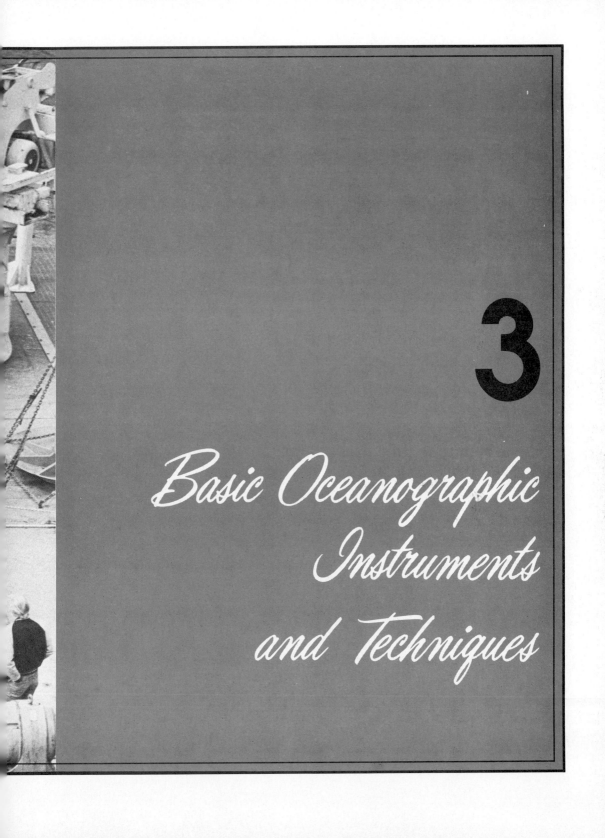

3

Basic Oceanographic Instruments and Techniques

THE INSTRUMENTS AND SHIPS of oceanographers are their means of getting to and into the ocean to see, sample, and touch it. Without such equipment oceanographers would have learned little about the details of the marine environment. In actuality, many of the major marine scientific breakthroughs have resulted from a technological development—some new technique or instrument that permitted the critical observation or measurement to be made. Ocean engineers have been responsible for the development of many of these instruments.

This chapter considers oceanographic research vessels and some of those instruments and techniques common to more than one field of oceanography. Instruments unique to an individual field of oceanography are discussed in the appropriate chapter.

Oceanographic Research Vessels

Probably the most important oceanographic instrument is the research ship—without ships oceanographers would have no platforms on which to carry and use their other instruments at sea. These ships can range in size from large ocean-going vessels to small vessels for estuary or coastal-zone work. Until recently many of the large oceanographic ships were not originally built for research but were converted from other uses into research ships. These vessels have performed admirably even though they had certain limitations imposed by their original nonresearch design. Modern offshore oceanographic research requires ships that can perform numerous different tasks, that can accommodate many kinds of equipment, and that are seaworthy (Figures 3-1, 3-2, and 3-3). The following arc some of the important considerations in the acquisition and operation of a research ship:

1. *Costs:* initial, operating, and maintenance.
2. *Agility:* the ability to maneuver at low speeds or when stopped.

(a)

(b)

FIGURE 3-1 Research vessels *Knorr* (75 m or 246 ft) (a) and *Oceanus* (54 m or 177 ft) (b). See also Figure 3-2. (a) Research vessel *Knorr* is named in honor of Ernest R. Knorr, a distinguished early hydrographic engineer and cartographer. With a full complement of 25 officers and crew and 24 scientists the vessel has an endurance time of 60 days, a cruising speed of 20.3 km/hr or 11 kn, and a range of 18,530 km or 10,000 n mi. (b) Research vessel *Oceanus,* which normally works in the Atlantic Ocean, has a complement of 12 officers and crew and 12 scientists with an endurance time of 30 days and a cruising speed of 25.9 km/hr or 14 kn. (Photograph of *Knorr* courtesy of Woods Hole Oceanographic Institution; photograph of *Oceanus* by Frank Medeiros.)

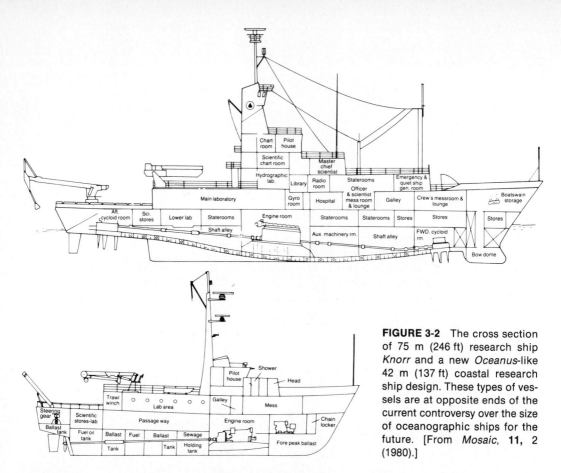

Chart room
Pilot house
Scientific chart room
Master chief scientist
Hydrographic lab.
Library
Radio room
Staterooms
Emergency & quiet ship gen. room
Main laboratory
Gyro room
Hospital
Officer & scientist mess room & lounge
Galley
Crew's messroom & lounge
Boatswain storage
Aft. cycloid room
Sci. stores
Lower lab
Staterooms
Engine room
Staterooms
Staterooms
Stores
Stores
Shaft alley
Aux. machinery rm.
Shaft alley
FWD. cycloid rm.
Bow dome

FIGURE 3-2 The cross section of 75 m (246 ft) research ship *Knorr* and a new *Oceanus*-like 42 m (137 ft) coastal research ship design. These types of vessels are at opposite ends of the current controversy over the size of oceanographic ships for the future. [From *Mosaic*, **11**, 2 (1980).]

Shower
Pilot house
Head
Trawl winch
Lab area
Galley
Mess
Steering gear
Scientific stores-lab
Passage way
Engine room
Chain locker
Ballast tank
Fuel oil tank
Ballast
Fuel
Ballast
Sewage
Fore peak ballast
Tank
Tank
Holding tank

FIGURE 3-3 *Meteor*, a German oceanographic research ship built in 1964 and operated by the Deutsches Hydrographisches Institute. The ship is 88 m long (about 269 ft) and displaces 3,085 tons. It has 13 laboratories, several aquariums, and numerous winches. It has a crew of 55 and can accommodate 24 scientists. The ship has a maximum cruising speed of 25 km/hr (about 13.5 kn) and can stay at sea for 40 days at a time. This ship is named after the original *Meteor*, which worked in the South Atlantic in 1925.

3. *Seaworthiness:* stability in rough seas.
4. *Laboratory facilities.*
5. *Quietness of electronic equipment used in operating the ship:* no interference with electronic equipment or computers used in collecting oceanographic data (Figure 3-4).
6. *Winch capabilities:* ability to lower instruments to the bottom of the deep sea.
7. *Endurance:* ability to stay at sea for at least a month.
8. *Recreation and living facilities sufficient for officers, crew, and scientists.*
9. *Dependability:* Use in science cruises planned as much as a year in advance in areas where repairs may be difficult.

Research ships designed today incorporate the above requirements and also have facilities for new and unique instrumentation and techniques. Some new ships are able to handle deep submergence vehicles, drilling rigs, and large oceanic buoys.

Most modern oceanographic ships have shipboard computer systems, as shown in Figure 3-4. Among the numerous advantages in having a computer at

FIGURE 3-4   Photograph of a shipboard computer system (IBM 1710). (Photograph courtesy of Woods Hole Oceanographic Institution.)

sea, one stands out: By presenting data in a form that is immediately usable and understandable, the computer allows the formation and testing of hypotheses while the scientist is still at sea. Computers on a research ship must withstand a wide range of conditions, especially those of high humidity and the rapid up-and-down movements of the ship.

The changing economic (such as inflation and the increase in fuel costs) and legal conditions (Law of the Sea, see Chapter 15) of the past few years have had an effect on research vessels. For example, only about $14 million was needed in 1970 to maintain 35 vessels, whereas $23 million was not enough in 1979 for 28 ships (see Ludwigson, 1980). More than half the U.S. academic research fleet is more than 10 years old, and even fewer are fuel efficient (Table 3-1). Recent fuel price increases have resulted in some vessels being laid up. In general, the larger the ship the more it costs to run. Other pressures on large research vessels include the possibility of performing some aspects of oceanography by satellites (see later section) and an increasing tendency toward coastal research, using smaller ships (Figure 3-2). The trend toward coastal work may be influenced by Law of the Sea restrictions, fuel costs, and emphasis on more applied research, especially in getting energy from the sea. On the other hand, the complexity of many studies, coastal or elsewhere, requires large numbers of technicians, scientists, and equipment. Thus in many instances, a large vessel may still be the best one. However, most of the recent vessels built for the academic oceanographic community have been of the coastal variety (that is, the *Oceanus*-type ships of Figures 3-1 and 3-2).

Research Submersibles

Research submersibles are an important research tool for oceanographers. The sinking of the submarines *Thresher* and *Scorpion* and the U.S. Navy's interest in deep submergence programs have encouraged the development of numerous research submersibles. Submersibles come in a variety of shapes, sizes, and capabilities and can perform different and sometimes very specialized tasks.

The use of underwater vehicles is not a new idea; the first recorded device was powered by hand and used in 1620. It was built by a Dutchman, Cornelius Dnebbel, who was King James I of England's personal friend. The device was a kind of rowboat enclosed with leather that used 12 oarsmen for propulsion. The vehicle apparently could stay underwater for several hours at a depth of 4 to 5 m (13 to 16 ft). It is rumored that the king made a trip in it under the Thames River.

The first use of one of these devices as a military vessel was made by David Bushnell in 1776. His ship, called *Turtle*, was a one-person, hand-powered submersible that was supposed to attach explosives to unsuspecting British ships (Figure 3-5). Most of the technical aspects worked except the mechanism to attach the explosives. A few decades later Robert Fulton devised an improved vessel

TABLE 3-1 General Characteristics of the U.S. Academic Fleet

Vessel	Length (m/ft)	Crew/ scientists	Year built or converted	Operator[a]	Owner[a]/capital source
63 m and up					
Robert D. Conrad	63/208	25/18	1962	LDGO	USN
Thomas G. Thompson	64/209	22/19	1965	U. Wash	USN
Thomas Washington	64/209	19/23	1965	SIO	USN
Atlantis II	64/210	25/25	1963	WHOI	WHOI/NSF
Knorr	75/245	25/25	1969	WHOI	USN
Melville	75/245	20/30	1970	SIO	USN
48 to 60 m					
Vema	60/197	21/14	1923	LDGO	LDGO
Kana Keoki	48/156	12/16	1967	UNIHI	UNIHI
Moana Wave	53/174	11/9–11	1973	UNIHI for Navelex	USN
Gyre	55/179	11/18	1973	TAMU	USN
Columbus Iselin	52/170	12/13	1972	RSMAS	U. Miami/NSF
Oceanus	54/177	12/12	1975	WHOI	NSF
Wecoma	54/177	12/16	1975	OSU	NSF
Endeavor	54/177	12/16	1976	URI	NSF
New Horizon	52/170	12/13	1978	SIO	U. Calif.
32 to 41 m					
Ridgely Warfield	32/106	8/10	1967	CBI	JHU/NSF
Velero IV	36/110	11/12	1948	USC	USC
Eastward	36/118	13/13	1964	Duke U.	Duke U./NSF
Alpha Helix	41/133	12/12	1965	U. Alaska	NSF
Less than 30 m					
Calanus	19/63	2/6	1970	RSMAS	U. Miami/NSF
Hoh	20/65	2/6	1943	U. Wash.	USN
Onar	20/65	2/6	1954 (1963)	U. Wash.	USN
Blue Fin	22/72	4/8	1972	Skidaway	Skidaway
Cayuse	24/80	7/8	1968	MLML	OSU/1/2 NSF, 1/2 Oreg.
Longhorn	24/80	5/10	1971	U. Texas	U. Texas
Ellen B. Scripps	29/95	5/8	1965	SIO	U. Calif

Source: From *Mosaic*, 1980.

[a] CBI = Chesapeake Bay Institute, The Johns Hopkins University
LDGO = Lamont-Doherty Geological Observatory, Columbia University
MLML = Moss Landing Marine Laboratories
OSU = Oregon State University
RSMAS = Rosenstiel School of Marine and Atmospheric Sciences, University of Miami
SIO = Scripps Institution of Oceanography, University of California at San Diego
Skidaway = Skidaway Institute of Oceanography, University of Georgia System
USC = University of Southern California, Institute for Marine and Coastal Studies
TAMU = Texas A & M University
URI = University of Rhode Island
USN = United States Navy
UNIHI = University of Hawaii
WHOI = Woods Hole Oceanographic Institution
Navelex = U.S. Navy Electronic Systems Command

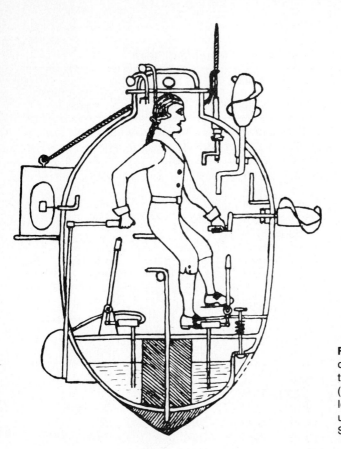

FIGURE 3-5 *Turtle*—note the pointed corkscrewlike device on the top right that is used to attach a powder charge (contained in the watertight box on the left) to the hull of an enemy ship. (Figure courtesy of New York Historical Society.)

that used compressed air for breathing and actually blew up a surplus vessel. It was, however, unsuccessful in real operations.

By 1886 the French had built a 16.7-m (55-ft) long vessel that used batteries to develop electrical power for propulsion. Before the beginning of the twentieth century the English, Italians, Spaniards, and Americans had built submarines that also used electrical power, but they all were limited in that their batteries could not be charged at sea. Gasoline-driven engines (usable only when the ship was on the surface and could charge the batteries) were introduced in the 1890s by two American engineers. By World War I such submarines were a common part of life.

One of the first uses of submersibles for science was by William Beebe in the early 1930s. He and his colleagues built a "bathysphere" (a strong sphere with viewing ports) that was lowered by cable from a surface support ship. His record dive to 923 m (3,028 ft) in 1934 did much to excite interest in the ocean.

The next important phase was led by the Swiss physicist Auguste Piccard. Piccard's device was not attached to a surface ship, and after several versions were built he eventually dove it to 4,175 m (13,700 ft) in 1954 in the Mediterranean. After this, Piccard, with his son Jacques, built the *Trieste*, which the U.S. Navy purchased and eventually used to make a record dive of 10,910 m (35,795 ft) in the Challenger Deep near Guam. This exploit clearly showed that scientific submersibles could be used almost anywhere.

Many new devices were built, and by 1970 about 60 submersibles were operational although most were used for oil well and pipeline construction and the like, rather than for scientific studies. In the last few years, however, research submersibles have made some remarkable discoveries, especially along the mid-ocean ridges (see pages 174–176). *Alvin*, an American submarine, (Figures 3-6 and 3-7) and several French vessels (Figure 3-8) have been especially prominent in this work. These exciting vessels are just starting to reach their full scientific potential.

A submersible possesses numerous advantages over a surface vessel. The most important is direct observation, permitting a scientist to see and photograph what he or she is trying to measure or sample (Figure 5-34, for example). Other

FIGURE 3-6 *Alvin*, a research submersible supported by the National Science Foundation, the Office of Naval Research, and the National Oceanic and Atmospheric Administration. The mechanical arm is visible in the front of the vessel; immediately in front of it is an instrument package. (Photograph by John Porteous, Woods Hole Oceanographic Institution.)

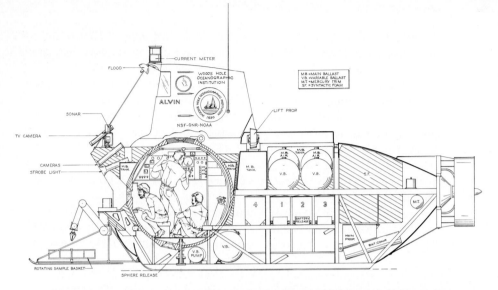

FIGURE 3-7 Cutaway view of *Alvin*; its total length is 6.4 m or 21 ft. The submersible uses a variable ballast system consisting of interconnected pressure-proof aluminum spheres (M.B. on figure) and collapsible rubber bags (V.B.) partially filled with oil. As oil is pumped from the spheres into the bags, the amount of seawater displaced by the vehicle is increased (thus increasing the buoyancy and making the submersible lighter) while the weight of the vehicle remains the same. (Photograph courtesy of Woods Hole Oceanographic Institution.)

FIGURE 3-8 The French submersible *Archimède* having its batteries recharged after a dive on the Mid-Atlantic Ridge during Project FAMOUS. *Archimède* is capable of dives to a depth of about 11,000 m or 36,000 ft. (Centre National pour l'Exploration des Océans photograph released in the United States by the National Oceanographic and Atmospheric Agency.)

advantages of submersibles are their ability to operate independently of the sea surface and to explore small features that may not even be detected from a surface vessel. By using a submersible, a scientist can make measurements in the same place for several hours, whereas an unanchored ship might drift away from a given area. Another advantage is that a submersible can return to the same place on the bottom, while a surface vessel is limited by navigational problems in finding any given area (see section on navigation).

A disadvantage of submersibles includes their dependence on surface ships to tow or carry them to their dive site. Difficulties in launching and retrieving often limit the use of a submersible to periods of mild sea conditions. Because submersibles are often very small, observers must remain in an uncomfortable position. Other disadvantages are that the cost of each dive usually is very high and that submersibles can generally explore only a small area. But despite these disadvantages, submersibles are one of the most important research tools of the oceanographer.

Some of the adventures of these submersibles are amazing. *Aluminaut* (an aluminum submersible not now in use) and *Alvin* in 1966 participated in the

FIGURE 3-9 Photograph of swordfish that attacked *Alvin* and got stuck in the external hull of the submersible. The fish was about 2.4 m long (about 8 ft) and weighed about 90 kg or 200 lb. It ended up making a good meal for the crew of *Alvin* and its support ship. (Photograph courtesy of Woods Hole Oceanographic Institution.)

successful search and recovery of an H-bomb lost off the Spanish coast. *Alvin*, while on a dive on the Blake Plateau off Florida, was attacked by an apparently nearsighted swordfish (Figure 3-9). *Alvin* also once sank at sea, fortunately without the loss of any life, and about a year later was recovered from the ocean bottom. The recovery was accomplished using *Aluminaut* and *Mizar* (a surface vessel); ironically, these three ships were previously teamed in the H-bomb recovery.

One successful scientific expedition using research submersibles was the recently completed Project FAMOUS (French–American Mid-Ocean Undersea Study), which made numerous dives on the Mid-Atlantic Ridge, about 320 km or 200 mi southwest of the Azores. Three submersibles were used: the American *Alvin*, the French *Archimède* (Figure 3-8), and *Cyana*. These submersibles dove as deep as 2,700 m (about 9,000 ft) in the extremely rugged terrain along the ridge. Numerous examples of recent volcanic activity resulting from sea-floor spreading were observed and photographed (Figure 3-10). Also participating in the FAMOUS project were numerous surface ships that made extremely detailed surveys of the ridge. This research program was followed by an even more exciting one made in the Pacific in the Galapagos region (see pages 174–176). These recent

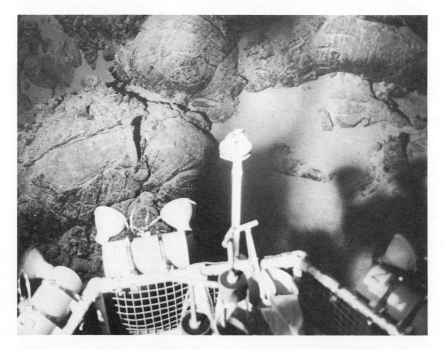

FIGURE 3-10 Bottom photograph made during Project FAMOUS. A view from *Alvin* showing bulbous pillow lavas on the sea floor in the Mid-Atlantic Ridge region. Note the sponge growing to the left of center. The devices in the immediate foreground are scientific equipment carried by the submersible. (Photograph courtesy of Woods Hole Oceanographic Institution.)

successes using submersibles have shown their considerable value for innovative research programs.

A person can function almost as a submersible with the use of scuba. Exploration of the sea bottom to depths of 90 m (about 500 fathoms or 300 ft) has been routinely performed by experienced divers. Newer techniques, using a mixture of gases and portable decompression chambers or special suits, have allowed people to work for extended periods of time at even greater depths.

Drilling Ships

Drilling ships are very useful vessels for oceanographic research, principally because of their ability to obtain samples of the sediments and rocks underlying the sea floor. They have successfully been used in the deep sea since the early 1960s during the beginning stages of Project Mohole, which was an attempt to drill completely through the earth's crust to the mantle. Although the project was ultimately abandoned, it showed that the technology for drilling in the deep sea was available. From 1968 to the early 1980s a new major program called the Deep Sea Drilling Project (DSDP) has resulted in hundreds of deep holes drilled on major geological features in the Atlantic, Pacific, and Indian Oceans. In recent years this project has developed an international flavor as countries such as West Germany, Japan, France, the Soviet Union, and the United Kingdom have contributed about $1 million annually to the project.

The Deep Sea Drilling Project uses the drilling vessel *Glomar Challenger* (Figure 3-11), which has already drilled in a water depth of 7,044 m (23,116 ft). At one site in the Atlantic Ocean a well was drilled to 1,741 m (5,709 ft) beneath the ocean floor. The principal objective of this program is to obtain long cores of the sediments underlying the sea floor. By early 1980 a total of 846 holes had been drilled at 519 different sites (to a cumulative depth of over 202 km or 126 mi). Over 38 km (24 mi) of sediment and rock have been recovered and studied by marine scientists from all over the world (Figure 3-12).

Numerous records and technical advances have been made with the *Glomar Challenger* during its history. Perhaps the most important technological advance was the development of reentry capability (Figure 3-13). Among its scientific achievements was the verification of many aspects of sea-floor spreading and plate tectonics and an understanding of the past history and evolution of the ocean basins.

The *Glomar Challenger* may be replaced in the mid-1980s by the *Glomar Explorer*, a vessel of considerable notoriety because of its use in an attempt to raise a sunken Soviet submarine (see pages 415–416). The *Glomar Explorer* will add a new dimension to the drilling program (to be renamed the Ocean Margin Drilling Program) as it will allow safe drilling into the thick sediments of the continental margin. The vessel will be equipped with a sophisticated blowout prevention system to minimize any possible pollution should it penetrate any layers containing hydrocarbons.

FIGURE 3-11 A port-side view of the drilling vessel *Glomar Challenger* Scripps Institution of Oceanography, of the University of California at San Diego, is managing institution for this program under a contract with the National Science Foundation. The drilling vessel is owned and operated by Global Marine, Inc., of Los Angeles, which holds a subcontract with Scripps to do actual drilling and coring work. The *Glomar Challenger* weighs 10,400 tons and is about 131 m or 400 ft long and the million-lb-hook-load capacity drilling derrick stands about 64 m or 194 ft above the waterline. This ship is one of a new generation of heavy drilling ships capable of conducting drilling operations in the open ocean by using dynamic positioning to maintain position over the bore hole (Figure 3-13). Forward is the automatic pipe racker that holds about 7,800 m or 24,000 ft of 12.7 cm or 5 in. drill pipe. (Photograph courtesy of the Deep Sea Drilling Project.)

FIGURE 3-12 Part of the core collection of the Deep Sea Drilling Project. Cores (cylindrical sections) are split in half longitudinally and stored in the plastic containers, one of which is being removed by the technician. (Photograph courtesy of the Deep Sea Drilling Project.)

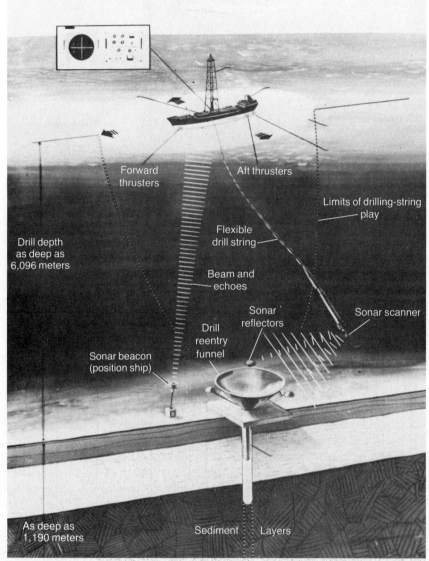

Labels within figure:

Forward thrusters

Aft thrusters

Limits of drilling-string play

Drill depth as deep as 6,096 meters

Flexible drill string

Beam and echoes

Sonar reflectors

Sonar scanner

Drill reentry funnel

Sonar beacon (position ship)

As deep as 1,190 meters

Sediment   Layers

FIGURE 3-13 Reentry and dynamic positioning system used on the drilling vessel *Glomar Challenger.* The *Glomar* uses "dynamic positioning" to hold station above a sonar sound source placed on the ocean bottom while drilling. Two tunnel thrusters forward and two thrusters aft, along with the vessel's two main propellers, are computer controlled to hold position without anchors in water depths to about 6,500 m or 20,000 ft so that drilling and coring can be accomplished. When a drill bit is worn out, it is now possible to retract the drill string, change the bit, and return to the same bore hole through a reentry funnel placed on the ocean floor. High-resolution scanning sonar is used to locate the funnel and to guide the drill string over it. The artist's concept shows a sonar beacon used for "dynamic positioning" and a sonar scanner at the end of the drill string searching for the three sonar reflectors on the reentry cone. The relative position of bit and funnel are displayed at the surface on a Positive Position Indicator Scope. The Deep Sea Drilling Project developed reentry when stopped short of scientific goals at many bore holes in the Atlantic and Pacific Oceans because the bit hit beds of chert or flintlike rocks that dulled the bit and forced early abandonment of bore holes. (Photograph courtesy of the Deep Sea Drilling Project.)

Floating and Fixed Platforms

Important contributions to the early phases of oceanography were made by *Maud* and *Fram*, two vessels that were frozen into the Arctic polar ice and allowed to drift with it for several years. These expeditions clearly showed the value of permanent or semipermanent observational stations on the ocean.

A drifting ice station was first established by the Soviet Union in 1937 in the Arctic Ocean. The United States started its first scientific camp in 1952 on Fletcher's Ice Island, more commonly known as T-3. The drift track of this ice island was observed, either by its temporary occupants or by airplane, from 1947 to 1964. One advantage of floating islands, besides permitting normal oceanographic observations in the polar region, is the large geographic coverage they afford. Many unique Arctic phenomena, such as auroras, magnetic conditions, and ice drift, can be conveniently studied from these floating oceanographic platforms.

Permanently fixed platforms that are attached to the sea floor have been used to measure numerous oceanographic parameters. These platforms can be radar and navigational towers, piers, or weather and lighthouse ships. Advantages of fixed platforms are their relatively small cost in comparison with ships and their stability for long-term, uninterrupted measurements.

A different type of oceanographic instrument is FLIP (Floating Instrument Platform). FLIP, when it is in its "flipped" position (Figure 3-14), is very stable and its up-and-down motion is only a small fraction of the waves around it. Because of its stability FLIP has been used successfully in studies of, among other things, wave movements and sound transmission in the ocean.

Another type of platform commonly used in oceanography is the anchored buoy system (Figure 3-15). Buoys may be free floating or attached to the bottom. Often one or more buoys are left at the surface as markers and to store information obtained from underwater sensors attached to the buoy line. These surface buoys can also record weather conditions, surface currents, and waves. Subsurface sensors can measure currents, temperature, or other parameters. Often floats, or glass spheres, are attached to keep the mooring line taut and reduce the horizontal and vertical motion of the system. Subsurface buoys may contain a power supply and recorders for other sensing devices. The mooring line often ends at a release

FIGURE 3-14 (opposite page) (a) FLIP, a 116-m (355-ft) long Floating Instrument Platform, was developed by the Marine Physical Laboratory of the University of California, San Diego's Scripps Institution of Oceanography. It is shown here in the horizontal position. In this position it is towed to the research site where ballast tanks are flooded to "flip" it into the vertical position for research work. (b) FLIP is photographed here in the process of "flipping" from the horizontal position to the vertical position where it will afford scientists an extremely stable platform from which to conduct scientific studies. (c) FLIP when in the vertical position as shown here displays only 17 m (about 55 ft) of its total length. The extended cranes are for lowering instruments into the water. (Photographs courtesy of Scripps Institution of Oceanography.)

(a)

(b)

(c)

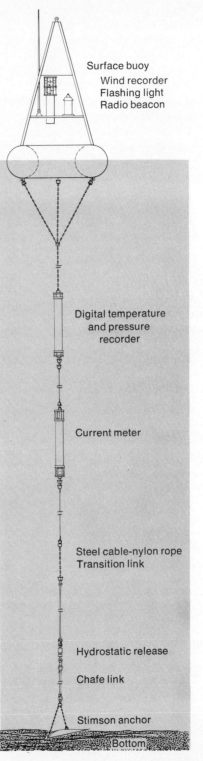

Surface buoy
Wind recorder
Flashing light
Radio beacon

Digital temperature
and pressure
recorder

Current meter

Steel cable-nylon rope
Transition link

Hydrostatic release

Chafe link

Stimson anchor

Bottom

FIGURE 3-15 A single-buoy system.

mechanism near the bottom that may be triggered at command, allowing recovery of the buoy system.

Information obtained by buoys may be collected and recorded within the system until retrieved by a surface vessel. Some more sophisticated buoy systems transmit data to an onshore station or relay them to a passing satellite.

Some Basic Instruments and Techniques

Airplanes are routinely used in meteorological and geophysical studies and can be important in some phases of oceanography. For example, they can be used to measure sea-surface temperature with an airborne infrared radiation thermometer, observe wave patterns, and track and observe schools of fish. By deploying an expendable bathythermograph (Figure 3-16), an airplane can measure the

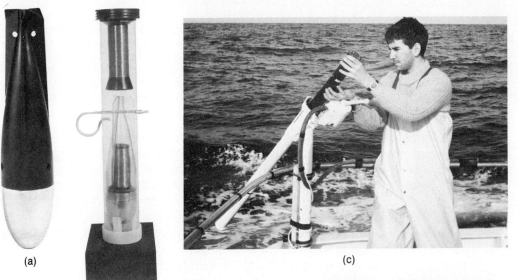

(a)

(b)

(c)

FIGURE 3-16 An expendable bathythermograph, or XBT. The key to this device is an expendable probe (a) that contains a thermistor, or temperature-measuring device, connected to a spool of fine wire; (b) a transparent view of (a). The probe is launched into the water (c) and wire is dereeled as the probe falls through the water. Changes in temperature are transmitted by the wire to a recorder (d) and since the rate of fall of the probe is known, depth can be read directly from the recorder. This device can be launched while the ship is moving or even from an airplane (in which case, a radio transmits the data to the plane). (Photographs courtesy of Sippican Corporation, Marion, Mass.)

(d)

FIGURE 3-17 Satellite photograph of Delaware Bay. The change in the patterns in and around the bay are due to sedimentation variations. [Photography courtesy of National Aeronautics and Space Administration (NASA).]

ocean's temperature to a depth of 1,800 m (about 6,000 ft). Planes can note different surface currents by water temperature variations, and they can also be used to observe the growth of waves under different meteorological conditions.

Satellites have tremendous potential for use in marine science. These devices have already been used for navigational aids, weather observations, and obtaining large-scale pictures of the earth. Some oceanographic information, such as large-scale changes in salinity when an estuary reaches the ocean, can be detected from variations in the reflectivity of seawater (Figure 3-17). Similarly, infrared techniques can be used to detect temperature changes. High-resolution photographs have been used to detect currents and locate areas of high biological productivity. Pictures of the coastal zone and nearshore regions can be used to detect the movement of sediment by currents as well as to notice some effects of pollution. With the increasing cost of fuel, satellites will become an even more cost-effective way of collecting data.

The LANDSAT series of satellites revolve around the earth and provide a systematic and repetitive coverage of the land and ocean. Such coverage can be extremely valuable in detecting changes over time. A new satellite called SEASAT was launched in 1978 especially for oceanographic studies and produced some remarkable pictures and data before it failed.

A major U.S. oceanographic project planned for the 1980s but threatened by budget cuts is the National Oceanic Satellite System, or NOSS. This project could eventually cost over $800 million. It is anticipated that by 1986 NOSS could be launched into an orbit of 300 km (186 mi) by the space shuttle and stay in space for 5 years. This system, if funded, will provide data on surface winds of the sea, sea state, water temperature, wave height, currents, chlorophyll distribution, and other aspects of the sea. The data, besides being useful for oceanography, has much practical value for ship routing, search and rescue operations, global weather forecasting, and the like.

NAVIGATION

Navigation, or knowing one's position, is of primary importance in oceanographic work; however, most navigational techniques are not very accurate.

One of the simplest methods of navigation is dead reckoning. The initial position of the vessel is determined either by star sighting or by reference to an object on land; then a course is plotted that will lead the ship to its intended objective. By knowing the speed of the vessel, one can estimate its position during any part of the voyage [Figure 3-18(a)]. The accuracy of the technique is poor because of the effect of currents, winds, and waves on the vessel. Departures from the predicted position and arrival time can, however, provide information concerning currents and surface winds.

Within visual or radar sight of land (usually less than 80 km or 50 mi) several techniques can be used. With a sextant, the horizontal angle between three well-located objects on land can be used to locate the observer's position [Figure 3-18(b)]. Radar sights taken on known objects on land provide a range (distance) and bearing (angle). These can be used, like sextant sights, to obtain a position [Figure 3-18(c)]. Positions having an accuracy of several hundred meters are possible with a good radar system, although in practical experience the error generally is larger.

If the bottom topography of an area is well known and is well-charted, it

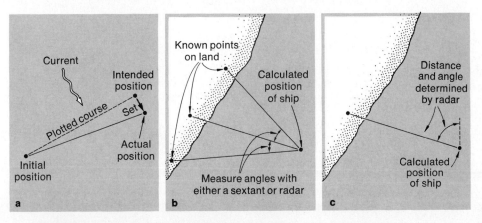

FIGURE 3-18   Some different methods of navigation: (a) dead reckoning, (b) using bearings, (c) using range and bearing.

can be used as a navigational aid. For example, a ship's position can be determined from the chart when the ship passes over a known feature of the sea floor. This technique requires the use of echo-sounding equipment (see the following section).

When the ship is out of sight of land or range of electronic devices, celestial navigation may be the main method of obtaining a position. Determining a ship's position by star sighting is a technique that has been used for many centuries. The accuracy of the method depends on the skill of the observer, and at best is usually about ±2 km (a little more than 1 mi). A potential shortcoming of celestial navigation is that the sky must be clear enough so that the stars can be seen; the sightings also must be done at dawn or dusk.

In the last few years there has been a considerable increase in the quality and use of electronic systems in oceanographic work. Several types of electronic systems allow positioning when a ship is as far as 1,600 km (about 1,000 mi) from land. The most widely used system is the Loran navigational system. Most electronic positioning systems depend on an accurate measurement of the time required for a radio signal to travel from a transmitter at a known place on land to a receiver on a ship. The arrival time is a measure of the distance the ship is from the station. Using two or more sets of land stations, a very accurate estimate (±50 m or about 165 ft) of the ship's position can be made (Figure 3-19). The main shortcoming of this and other electronic techniques is that land-based stations are not available in some parts of the world; an advantage is that they can be used in most kinds of weather.

Satellite navigation systems are used now by most oceanographic vessels. This technique requires sophisticated electronic equipment that measures the change in transmission characteristics of the satellite as it moves away from or toward the ship. These changes can be used to calculate the ship's position fairly accurately (to a kilometer or less).

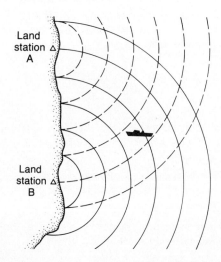

Land station A

Land station B

FIGURE 3-19 One method of electronic navigation. Two land stations transmit radio signals with a constant time delay between each signal. The ship has a receiver and can, when it has received signals from two or more stations, convert the time of arrival of the signals into a position (using appropriate tables or charts).

SOUNDING METHODS

Before the development of electronic techniques, sounding, or determining the depth of the ocean, was done by lowering a heavy line of known length and noting how much of the line had been paid out when it reached the ocean bottom. This tedious technique did not always result in correct depth values because a line, when lowered to the sea floor, does not necessarily go straight down but can be deflected either by currents or by movements of the surface vessel. Another disadvantage is that this technique gives only one depth with each lowering rather than a continuous picture of the bottom. The development of sounding techniques using electronically controlled sound impulses (called *echo sounding*) solved these problems, but introduced a few smaller difficulties.

Echo sounding is a technique whereby an outgoing signal or pulse from a ship travels through the water to the ocean bottom, is reflected off the bottom, and travels back to the ship (Figure 3-20). The time that it takes the signal to make this trip is accurately measured. Using necessary corrections and the speed of sound in water, one can calculate the depth. In other words, the water depth equals one-half the travel time (since the total travel time is actually a two-way trip—down to the bottom and back) multiplied by the speed of sound in the water.

A modern echo-sounding device produces a permanent graphic record of the returning sound (Figure 3-21). These records give the oceanographer a better idea of the character of the ocean floor. Echo-sounding records from the ocean have shown that the topography of the sea floor is at least as irregular as that of land.

There are some problems in the interpretation of the echo-sounder record. One is the exaggeration in scale. The surface vessel is traveling at a speed of about 10 or 12 kn, which represents the horizontal scale of the record. The vertical scale of the record shows the depth, commonly measured in hundreds of fathoms or meters, and is usually smaller and exaggerated relative to the horizontal scale. In other words, one scale is in hundreds of feet, the other in thousands of feet. If, however, both scales were made with the same dimensions (Figure 3-21), it would be very difficult to observe most details of the bottom features.

Another problem is the shape of the outgoing sound beam; it is a wide cone and covers a relatively large circular area when it hits the ocean bottom. The first returning echo, therefore, comes from the point closest to the ship, which, in the case of underwater canyons or mountains, may not be directly below the vessel, and the following echoes are masked or obscured by the first one. The effect makes it difficult to define small features on the sea floor accurately and to be sure that the feature is directly below the ship. There are sophisticated, and very expensive, echo-sounding devices that can solve some of these problems.

A third echo-sounding problem is determining the actual speed of sound in water. Sound velocity increases with increasing temperature, salinity, and depth

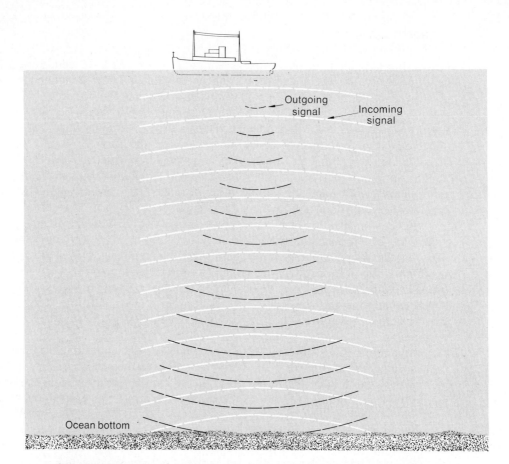

Outgoing signal

Incoming signal

Ocean bottom

FIGURE 3-20 Echo-sounding technique. Sound from the ship travels to the ocean bottom and is reflected to the ship. The time the sound takes to make the trip is used to determine the water depth.

(this is discussed in more detail in Chapter 9). These parameters have to be known and corrected for if a very accurate determination of water depth is desired.

Oceanographic vessels routinely use their echo-sounding equipment at sea. Many crossings over an area can be used to produce a model or bathymetric chart of the ocean floor. In addition, echo-sounding can provide information about the sedimentary layers beneath the ocean floor when sound energy is also reflected from layers beneath the ocean floor; the resulting record shows a cross section of the upper several meters of the ocean floor (Figure 3-22). Similar marine geophysical techniques, using considerably more sound energy, can obtain reflections from layers 2 km or more below the ocean floor (Figures 5-10, 5-11, 5-12, and 5-13).

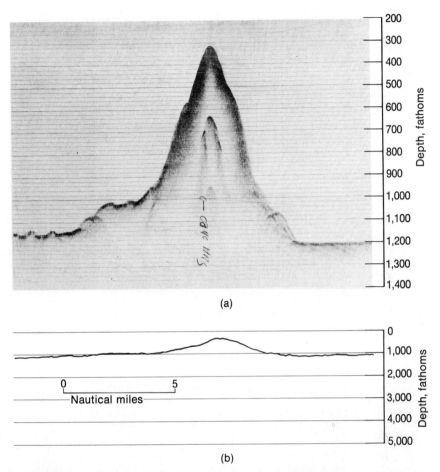

(a)

(b)

FIGURE 3-21 (a) Echo-sounding record made over a large hill, which is probably a submerged volcano. The vertical exaggeration of this record is 12 times. (Photograph courtesy of Woods Hole Oceanographic Institution.) (b) The same data shown with no vertical exaggeration.

Other Instruments

Most oceanographic instruments that are used for research vessels and lowered by cable into the sea can be considered as one of three types (Figure 3-23):

1. Those towed through the water by the movement of the surface ship.
2. Those suspended from the ship and lowered straight up and down.
3. Those attached to the ship by cable but having their own source of power to move on or near the bottom.

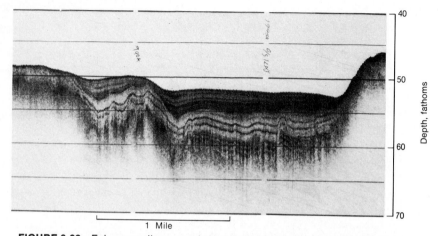

FIGURE 3-22 Echo-sounding record showing subsurface layers below the surface sediments. These layers probably are due to differences in sediment type—for example, a change from a clayish to a sandy sediment. Vertical exaggeration is about 50 times.

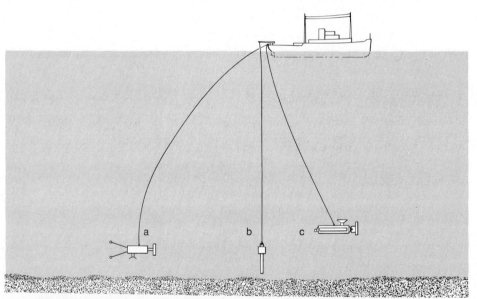

FIGURE 3-23 Different general types of oceanographic instruments: (a) towed type, (b) suspended type, (c) self-powered type. Generally, only one instrument is lowered from the ship at a time.

Instruments suspended by wire from surface ships are the most common type of oceanographic instruments (Figures 3-24 and 3-25). Towed instruments include devices like plankton nets (Figure 6-1) or special instruments used to

FIGURE 3-24 A multiple-use instrument that can be lowered from a surface vessel. The instrument package contains a nephelometer for measuring particulate matter in the water, a CTD for measuring the conductivity (equivalent of salinity) and temperature of the water and the depth of the instrument, water-sample bottles (top of figure), a bottom pinger (see text), and an acoustical signaling device. (Photograph courtesy of Dr. Lawrence Armi.)

FIGURE 3-25 Things do not always go right at sea. Hidden beneath the maze and tangle of wire is part of an instrument that became entangled with wire during an unsuccessful lowering. (Photograph courtesy of Henri Berteaux.)

make detailed studies very near the sea floor (Figure 3-26). Self-powered instruments have not been developed to any extent but have the potential of being important instruments in the near future.

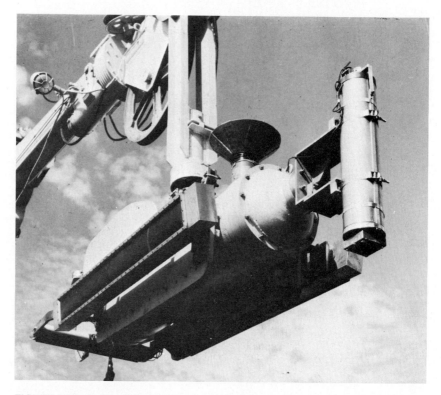

FIGURE 3-26 A deep-towed device built in cooperation with the Office of Naval Research and Scripps Institution of Oceanography. This instrument has been used to make detailed bathymetric and magnetic charts of portions of the sea floor while being towed above the bottom. (Photograph courtesy of J. D. Mudie, Marine Physical Laboratory, Scripps Institution of Oceanography.)

PINGERS

Pingers are devices used to position instruments, suspended on wires, above the sea floor. They are attached to the wire, usually near the instrument, and emit sound pulses or pings at precise intervals, usually exactly 1 second apart. The ship's echo sounder receives the outgoing (or direct) pulse and the reflected signal from the sea floor. The difference in arrival times between the direct and reflected signals indicates the height of the pinger above the bottom (Figure 3-27).

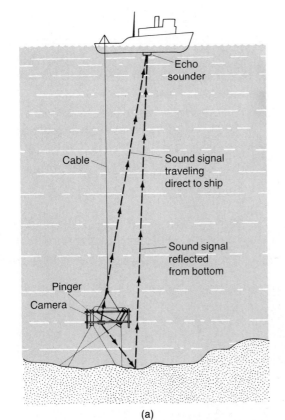

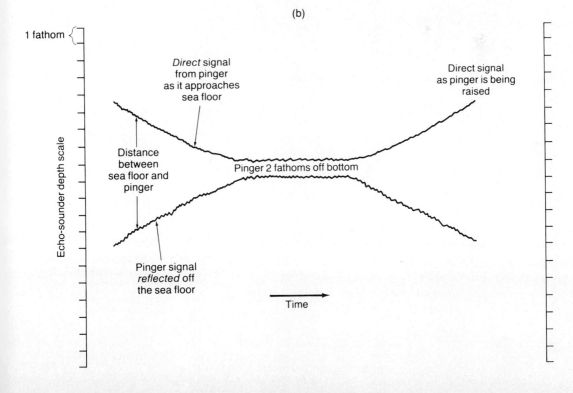

FIGURE 3-27 The use of a pinger (a) to determine relative height above the bottom. The difference in arrival time of the direct ping and the reflected ping indicates the height of the pinger above the bottom (b). If the pinger was on the bottom, both pings would arrive at the same time.

(a)

(b)

Echo sounder

Cable

Sound signal traveling direct to ship

Sound signal reflected from bottom

Pinger

Camera

1 fathom

Echo-sounder depth scale

Direct signal from pinger as it approaches sea floor

Direct signal as pinger is being raised

Distance between sea floor and pinger

Pinger 2 fathoms off bottom

Pinger signal *reflected* off the sea floor

Time

UNDERWATER CAMERAS

The old axiom that one picture is worth a thousand words is often true in ocean-ography. Underwater photography has become a very important part of ocean-ographic research. Ocean-bottom photographs can be used to study the sediments and rocks on the sea floor, examine biological activity, and observe indirect evidence of currents, such as ripple and scour marks. Camera systems (Figures 3-28 and 3-29) protected against pressure have been lowered by wire into the deepest parts of the ocean. Because most of the light in the ocean is absorbed within the top 100 to 200 m (328 to 656 ft), these camera systems must include powerful light sources. They must also have a pinger so that the camera can be accurately positioned above the sea floor. Pictures (Figure 3-30) are usually taken with the camera only a few feet above the bottom. Most camera systems take several hundred pictures with each lowering. When the instrument is near the

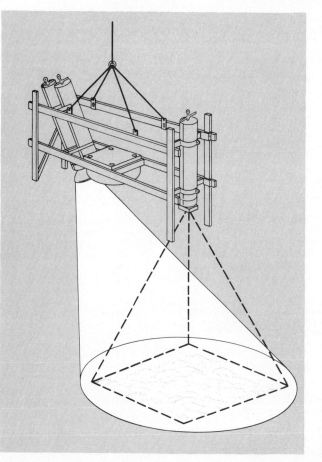

FIGURE 3-28 Diagrammatic view of how a deep-sea camera system would look while taking pictures on the ocean floor.

FIGURE 3-29 A multicamera system being lowered into the ocean. The light source for the camera is to the right, and the cameras are to the left. (Photograph courtesy of Woods Hole Oceanographic Institution.)

bottom, the winch operator, guided by the pinger returns displayed on the ship's echo sounder, tries to hold the camera a meter or so above the ocean floor. This is done to keep the distance above the bottom within the camera's optical depth of field so that the pictures will be in focus. In some instances two cameras are used to obtain stereoscopic pictures that can be used to measure the dimensions of small features on the sea floor.

SIDE-SCAN SONAR

These devices, which are similar to conventional sonar, or echo-sounding equipment (Figure 3-31), transmit sound at an angle from the ship rather than just straight down as in echo sounding. The difference in intensity of the returning sound signals from the bottom can sometimes be used to distinguish between types of sediment (such as sand from gravel) or to observe other features on the sea floor such as sunken ships (Figure 3-32, page 68). The ability of this device to detect sunken vessels could be very useful in submarine archaeological investigations (see pages 484–487) and in the search for and recovery of vessels lost at sea. It could also be used to evaluate mineral resources on the continental shelf.

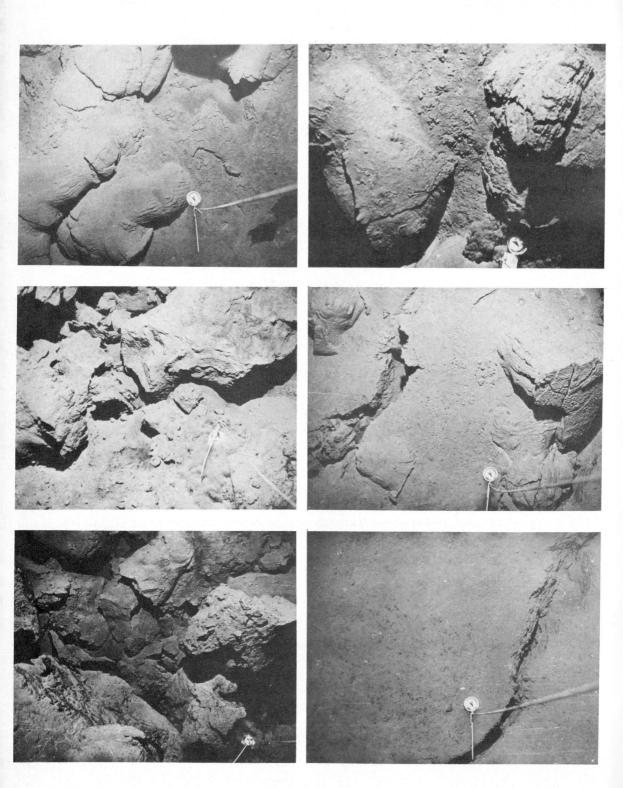

66

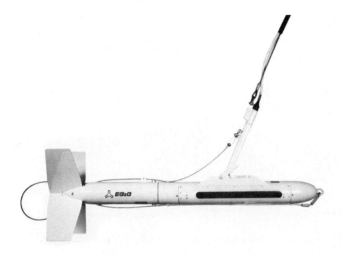

FIGURE 3-31 A towed side-scan sonar device. (Photograph courtesy of EG & G Company.)

Summary

The ships and instruments of an oceanographer are the key methods for working in and sampling the marine environment. Indeed, many oceanographic discoveries have come from a technological development. Perhaps the most important instrument is the research ship, which can range in size from the large deep-sea research vessel to the small ship for nearshore work. Recent increases in fuel costs have had an economic effect on the development of the U.S. research fleet, and several vessels have been laid up. Legal restrictions resulting from the Law of the Sea negotiations (see Chapter 15) might also influence the types of ships to be used in the future.

Research submersibles have been around for many years but their recent success in studies along spreading centers have shown how valuable they can be in research. Their use in industry (pipeline monitoring, and the like) is already well established. Drilling ships, especially the *Glomar Challenger*, have led to some major oceanographic discoveries in the 1970s and should continue their successes in coming years. Satellites, buoy systems, and fixed or floating platforms are also important oceanographic instruments. Satellites, which are becoming more adapted for marine research, may eventually end up doing some of the work now being performed from surface vessels.

Navigation at sea often is still more of an art form than a science, but the increased availability of Loran and other electronic aids, including satellites, has led to many recent improvements in this field. Echo sounding (determining the

FIGURE 3-30 (opposite page) A series of bottom photographs taken in the axial valley of the Red Sea. The object in the lower right is a compass, about 15 cm (6 in) in diameter, attached by a line to the camera. (Photographs by Robert A. Young.)

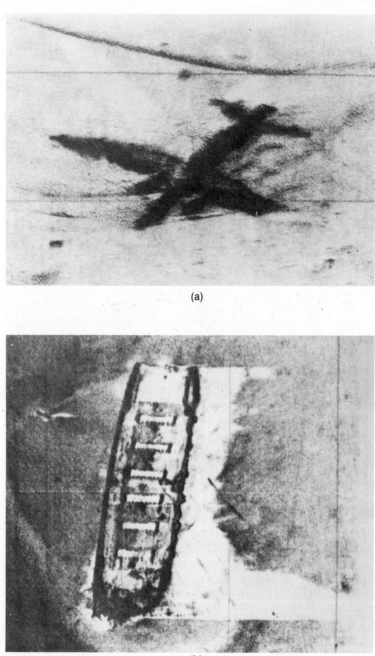

(a)

(b)

FIGURE 3-32 Pictures of side-scan sonar records that show (a) an aircraft lying on the bottom of Loch Ness and (b) an old wooden sail barge in the Great Lakes. (Photograph courtesy of Klein Associates, Inc.)

water depth) is a common procedure for almost all types of oceanographic research. In most instances it is done by electronically measuring accurately the time it takes for sound to travel from the ocean surface to the bottom and return. Some sound may penetrate the bottom and give subbottom returns that provide information on the underlying sediments and rocks.

Some other basic instruments, such as pingers, cameras, and side-scan sonars, are discussed and information about more specific tools and devices is presented in the appropriate chapter (for example, biological instruments in the chapter on biological oceanography).

Suggested Further Readings

BENDER, E., "A National Oceanic Satellite System," *Sea Technology,* no. 21 (1980), pp. 27–31.

HEIRTZLER, J. R., and A. E. MAXWELL, "The Future of Deep-Ocean Drilling," *Oceanus,* **21,** no. 3 (1978), pp. 2–12.

LOUGHRY, T., "Evolution of the Submersibles," *Surveyor,* **13,** no. 4 (1979), pp. 2–11.

LUDWIGSON, J., "Ferment in the Fleet," *Mosaic,* **11,** no. 2 (1980), pp. 2–11.

WEST, S., "DSDP: 10 Years After," *Science News,* **113,** no. 25 (1978), pp. 408–10.

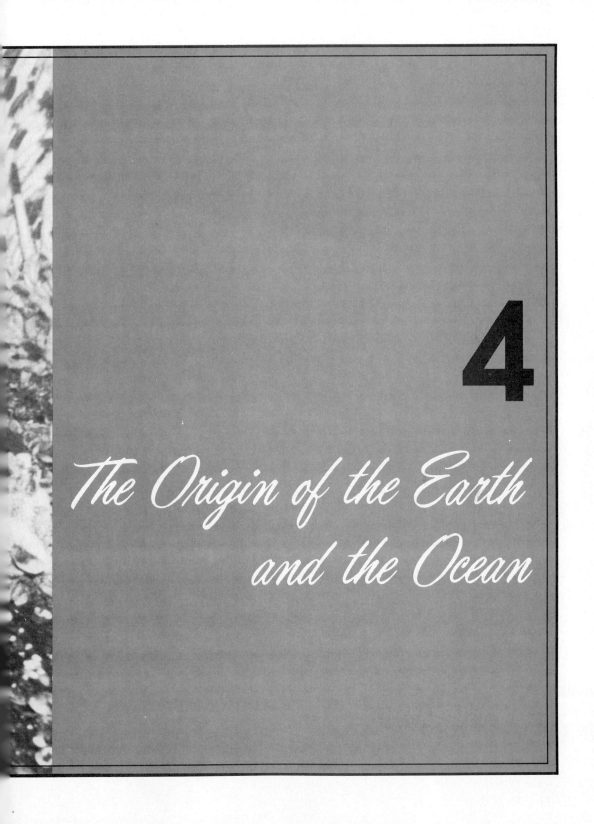

4

The Origin of the Earth and the Ocean

MUCH HAS BEEN LEARNED recently about the origin of the earth, the ocean, and our neighboring planets. This information has come from space programs such as Apollo, Pioneer, and Mariner and from new and imaginative oceanographic programs including ocean drilling and direct observation of the ocean floor by submersibles. Before considering the origin of the earth and the ocean, a brief discussion concerning the universe is useful.

Origin of the Universe

The question of the origin of the universe is of considerable interest, but unfortunately few data are available. The key questions are, Where did the matter that forms the universe come from and when was it formed?

The most commonly accepted hypothesis by astronomers and astrophysicists for the origin of the universe is the so-called big-bang hypothesis. This idea assumes that the universe began about 15 or 20 billion years ago with a gigantic explosion. The shrapnel formed by the explosion is still speeding away from the center of the blast. Some of the fragments form the Milky Way Galaxy of which one of its billion stars is our sun. A piece of evidence that the universe is still expanding is a shift toward the red end of the spectral lines observed from distant galaxies. This phenomenon, called the Doppler shift, results when one body moves toward or away from another body. If the body is moving away, it will appear to be emitting light of lower frequency (longer wavelength), or to have a red shift. The degree of this shift can be used to calculate the movement of the bodies away from the earth. Some galaxies are moving away from the earth at speeds up to 2 million miles per hour. Another important piece of evidence supporting the big-bang hypothesis was the detection of noise coming from many parts of the universe. These noises have been interpreted by most astronomers as being the reverberations from the initial explosion.

Although the big-bang hypothesis nicely accounts for the expanding universe, it does not explain the origin of the matter nor describe what the universe looked like before the explosion. These two points are the most interesting questions of all—not just to scientists but also to theologians and philosophers. The big-bang explosion has, in effect, destroyed any evidence of what existed before.

An alternate but less-favored idea, the cyclic hypothesis, does not assume a single origin of the universe but rather predicts the eventual end of the expansion of the universe, at which time the universe will start to contract, owing to gravitational forces, and will come together in a huge, high-temperature mass. An explosion will then occur, sending the matter back out to space, whereupon the entire process will be repeated. Although this hypothesis answers some parts of the previously mentioned questions, it offers no explanation of the origin of the matter that forms the universe. A recent estimate of the amount of mass in the universe indicates that the total falls short by 10 to 20 times the amount needed to eventually start contracting the universe. Some astronomers have suggested that the missing mass may be hidden within so-called black holes that are scattered throughout space. Black holes, which are invisible, are the remnants of massive stars that have been collapsed by their own gravity. This phenomenon is thought to occur after a star has burned up all its nuclear fuel and starts collapsing inward. If the star was sufficiently large, it may have had a gravitational field strong enough that nothing, not even light, could escape; thus it became an invisible black hole.

As intriguing as these ideas sound, clearly more data are needed before definitive statements can be made about the origin of the universe.

Origin of the Earth

Considerably more data are available concerning the origin of the earth. Scientists generally agree, based on studies of the abundance and distribution of radioactive elements (see Chapter 7), that the earth formed about 4.5 or 5 billion years ago. However, the exact method of the earth's formation is still debatable. Discussion has centered on two main hypotheses: the fragmentation hypothesis and the condensation hypothesis. The fragmentation hypothesis assumes that the planets of our solar system were torn from the sun when the sun collided with, or came very close to, another star and that the sun and the planets were formed at different times. This hypothesis is in disagreement with the abundance of radioactive elements of the sun and the earth, which indicates a simultaneous formation of the planets and the sun.

The condensation hypothesis suggests that all parts of our solar system were formed about the same time by the compaction of a cloud of cosmic dust and gas. Most of this material formed our sun; a smaller amount formed a diffuse cloud, or nebula, around it. Eventually the nebula flattened into a disklike shape, and in doing so it increased in density. This increase in density created an unstable condition and the nebula broke into several small clouds. These clouds were the protoplanets that developed into the present planets. The protoplanets were originally cold and gaseous, but after losing their gases they became heated by radioactivity and compression until they became molten. Eventually, 4.5 or 5 billion years ago, the earth solidified, but its core remained in a molten state. The condensation hypothesis, although it also has some complications, is presently favored over the fragmentation theory.

It was anticipated that considerable information about the origin of the earth would be obtained from the manned exploration of the moon (Figure 4-1). By the time that the Apollo program ended in 1972, over 360 kg (about 800 lb) of moon rock had been collected and returned to the earth by the 12 astronauts who walked on the moon. In addition, sophisticated instruments to measure the moon's internal structure were deployed. Even though it will take some time before the data are completely analyzed, there is no consensus about the origin of the moon. Some early ideas about the origin of the moon and the earth have been abandoned, while some new ones have had to be developed to account for unexpected data.

One idea that appears to have been discarded was that the moon originated by being torn from the earth, but some groups still cling to this possibility. Although the moon is structurally similar to the earth and appears to have formed at about the same time as the earth, it has a different overall chemical composition. The principal differences are that moon rocks have lower amounts of water, siderophiles (elements associated with iron, like nickel), and easily volatilized elements like chlorine than earth rocks. Similar oxygen isotopic composition of earth and moon rocks supports an essentially simultaneous formation of the earth and the moon from the same nebula, but if so, it is not clear why there are chemical differences. One proposed explanation is that the moon split off from

FIGURE 4-1 Apollo astronaut exploring large rock on the moon. (The *Apollo 17* photograph was provided by the National Space Science Data Center.)

the spinning earth after the earth's core (containing siderophiles) was formed—but this would not explain the difference in the volatile elements. Another possibility is that the moon did not form near the earth, but was captured into an earth orbit after its formation. This idea, however, also has many difficulties; for example, how such a capture could occur. There are some good theoretical reasons to support the belief that in the early history of the moon it was considerably closer to the earth and that it has since been slowly moving away (see the section on tidal friction in Chapter 9).

As the earth consolidated, its original atmosphere was lost. The present atmosphere developed over long periods of geologic time from gases emanating from the earth and, after plant life developed, from oxygen released by plants during photosynthesis.

Origin of the Ocean

The question of the origin of the ocean is really two problems: first, from where did the water come, and second, how did it get its unique concentration of elements? It should be noted that of the other planets in our solar system, only Mars has good evidence of water activity on its surface. Ice caps appear to exist on parts of this planet and there are topographic features that may be due to past water–flow activity (Figure 4-2). However, recent exploration by Viking landers on Mars did not find any form of life.

Concerning the water on the earth, there are three hypotheses to explain its origin:

1. From the primordial atmosphere of the earth.
2. From the decomposition of volcanic rock.
3. From the incremental addition of water throughout geologic time.

Proponents of the first hypothesis suggest that the primordial atmosphere condensed all at one time to form the ocean. If this really happened, one would expect the postulated original components of this atmosphere to be present in the ocean in higher quantities than have been observed. The so-called rare gases, such as neon and argon, are present in quantities millions to hundreds of millions of times less than expected if the first hypothesis is correct. Neon has an atomic weight of 20; water vapor, or H_2O, has a molecular weight of 18. Thus if the atmosphere was unable to maintain its neon content, it seems unreasonable to expect it to have held on to large quantities of water vapor. One possible reason that the original atmosphere was lost is that gravity was not as strong then as it is now. The present atmosphere of the earth, in fact, can hold no more than about 13,000 km³ (or about 3,100 mi³) of water, whereas the volume of water in the ocean is over 1 billion km³ (about 240 million mi³).

Advocates of the second hypothesis suggest that when the earth consoli-

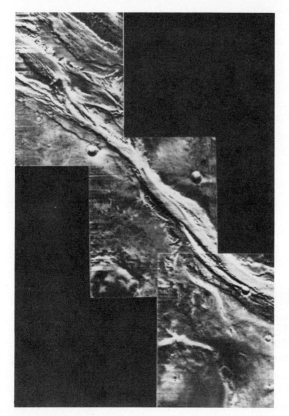

FIGURE 4-2 A channel thought to have been formed by running water in Mars' geologic past is seen in this mosaic of three pictures of the planet taken July 1, 1972, by *Mariner 9*. "Flow" of the channel was from upper left to lower right. This small segment of the channel is about 75 km (46 mi) in length and is located just north of the equator. The three pictures were taken a few minutes apart by *Mariner 9*'s narrow-angle camera. (Photograph courtesy National Aeronautics and Space Administration.)

dated, much of the original water was chemically bound into volcanic rock and subsequently has been removed by decomposition of these rocks to form the ocean. Experimental and field evidence indicate that volcanic rocks contain only about 5 percent water, and even if all the water in the volcanic rocks of the earth's crust were removed, it would amount to less than 50 percent of the water in the ocean. Thus it seems obvious that most of the water of the ocean did not come from the weathering of rocks. Nevertheless, many cations (atoms or molecules with a positive charge) in seawater, such as sodium, magnesium, calcium, and zinc, could have come from these rocks.

The third hypothesis for the origin of the water, that of incremental addition throughout geologic time, is generally the most accepted. This hypothesis proposes that ocean water was slowly, but not necessarily uniformly or continuously, added over geologic time. Probably a large amount of the water was supplied relatively early in the earth's history due to chemical and physical processes associated with the early development of the earth. Evidence for this is sediments that appear to have been deposited in a marine environment and are over 3 billion years old. The source of the water is volcanic activity, hot springs, and the heating of igneous rocks (rocks formed from the solidification of magma). In addition to water, anions (atoms or molecules with a negative charge), such as chloride and sulfide, are released by volcanic activity. Much of the volcanic activity today occurs, as it did in the past, along the mid-ocean ridges that extend through the main oceans and in the areas just landward of the ocean trenches, especially those that fringe the Pacific Ocean (Figure 1-7).

The discussions above concerning the origin of the water in the ocean are obviously a simplification of an extremely complex subject. There is solid evidence, however, that volcanic activity is the principal source of the water and the anions, while the cations come mainly from the decomposition and weathering of igneous rocks.

The relative amounts of the various elements in the ocean are determined by numerous chemical and physical processes that control and regulate the chemical composition of seawater. These processes take place mainly at the major interfaces or boundaries of the ocean:

1. The water–atmosphere interface.
2. The water–biosphere interface.
3. The water–sediment interface.

The main discussion of the reactions occurring at these interfaces will be discussed in the chapters on physical oceanography, biological oceanography, and chemical oceanography, respectively. It should be mentioned, however, that the early atmosphere and ocean probably did not contain any free oxygen. Ancient marine sediments, such as iron-rich deposits that are highly insoluble in oxygen-bearing waters, are evidence of this point. Free oxygen did not appear until the evolution of plants such as algae that were capable by the process of photosynthesis of combining water and carbon dioxide (in the presence of sunlight and adequate nutrients) to produce free oxygen and organic compounds (food for higher forms of life). Photosynthesis appears to have begun about 3 billion years ago and has profoundly changed the characteristics of the atmosphere and the ocean and made life possible.

Origin and Evolution of the Ocean and the Ocean Basins

Little is known about the early history of the ocean. Fossils found in Australia in 1980 that are about 3.5 billion years old and apparently of marine origin, indicate that some sort of marine environment existed at that time. Marine fossils about 600 million years old or younger and similar to present-day living forms suggest that the composition of seawater then may have been similar to present conditions. This similarity implies that at some time in the past the chemistry of the ocean reached a steady-state condition.

There is a common tendency to think that the earth is static, that the conditions we see today have prevailed for millions of years. In some respects, for example the general chemistry of the ocean, this view is correct. For the physical conditions of the ocean, however, one has to think in terms of a more dynamic situation. Tides raise and lower the level of the sea on a daily basis. Seasonally the surface temperature of the ocean can change by several tens of degrees. Sea level has risen and fallen hundreds of feet twice within the last 30,000 years, a phenomenon caused by major glacial advances and withdrawals on the earth's surface. The earth's magnetic field has changed numerous times within the last several million years, and the relative position of continents and oceans has been slowly changing over the last few hundred million years. Knowledge of these last three events has come from the detailed study of geologic history as it is preserved in the sediments and rocks of the ocean.

One of the most exciting scientific concepts developed during this century is that of sea-floor spreading (discussed in detail in Chapter 6). Simply said, the ocean floors are spreading apart along the mid-ocean ridges, and new sea floor is being created there by volcanic activity. Since the earth is not constantly increasing in size, some sea floor has to be destroyed or consumed; this occurs in many of the trenches of the ocean. The resulting volcanic activity of both these processes supplies many of the elements found in the oceans and eventually "recycles" the deposits on the sea floor (see Chapter 7).

Geologic History

Earth scientists have divided geologic time into four main eras, which are sub-divided into periods. This time scale and some important geologic and biological events are shown in Table 4-1.

During geologic time the geography and climate of the earth went through numerous changes. These include the major movements associated with sea-floor spreading as well as other large-scale up-and-down movements. Many areas that are now above sea level were below it in the past; in fact there is hardly any land area that was not covered by the sea at some time in its past. Ocean basins and continents were both destroyed and created. Large-scale climatic changes resulting in extensive glaciation occurred in Precambrian, Permian, and more recently in Quaternary times.

Ancient Climate Changes

Details of climatic conditions far back in the earth's history are obviously hard to obtain. Nevertheless it seems evident that major climatic changes must have followed the major movements of the continents. Over 225 million years ago there was one large supercontinent called Pangaea, extending essentially from the North Pole to the South Pole (Figure 4-3), that was surrounded by ocean. Climatic conditions and oceanic circulation were different then than at present, a point substantiated by examination of the rocks deposited during that period. These

TABLE 4-1   Geologic Time Scale and Some Major Events

Era	Period	Events	Began millions of years ago
CENOZOIC	Quaternary	Age of humans. Four major glacial advances.	2
	Tertiary	Increase in mammals. Appearance of primates. Mountain building in Europe and Asia.	65
MESOZOIC	Cretaceous	Extinction of dinosaurs. Increase in flowering plants and reptiles	140
	Jurassic	Birds. Mammals. Dominance of dinosaurs. Mountain building in western North America.	195
	Triassic	Beginning of dinosaurs and primitive mammals.	230
PALEOZOIC	Permian	Reptiles spread and develop. Evaporate deposits. Glaciation in Southern Hemisphere.	280
	Carboniferous	Abundant amphibians. Reptiles appear.	345
	Devonian	Age of fishes. First amphibians. First abundant forests on land.	395
	Silurian	First land plants. Mountain building in Europe.	435
	Ordovician	First fishes and vertebrates.	500
	Cambrian	Age of marine invertebrates.	600
PRECAMBRIAN TIME		Beginning of life, at least five times longer than all geologic time following.	

rocks show indications of large ice sheets near the South Pole that extended over much of southern Pangaea including India, Australia, South America, and Africa, while arid desert or even evaporitic conditions dominated the mid- and low-latitude regions. Following the breakup of Pangaea about 200 million years ago, the oceanic circulation slowly evolved into its present configuration. During this change there were periods (such as about 65 million years ago) when climatic conditions were less dramatic (including no ice sheets) and warmer than they are at present or were before Pangaea started separating. Eventually the east–west passage near Antarctica developed, which allowed water circulation among all oceans to develop. This eventually led to our present oceanic circulation system and modern climatic conditions.

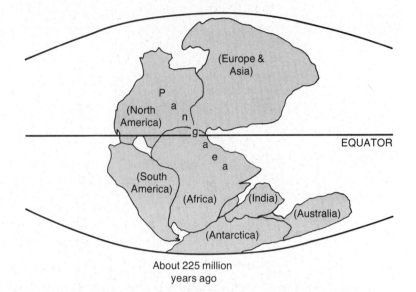

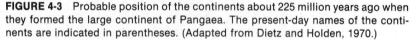

About 225 million
years ago

FIGURE 4-3 Probable position of the continents about 225 million years ago when they formed the large continent of Pangaea. The present-day names of the continents are indicated in parentheses. (Adapted from Dietz and Holden, 1970.)

Recent Climatic and Sea Level Changes

The Quaternary Period, covering the past 2 million years, was an interval of considerable climatic variation. Studies of glacial deposits, soil types, and sediments from the ocean have shown that there were several intervals of glaciation during the Quaternary. This period was marked by glacial advances with extensive areas covered by ice, snow, and glaciers followed by interglacial intervals similar to present-day conditions. The reasons for such major changes in conditions have been the source of considerable speculation; indeed, some feel we may be entering a new glacial period because of the increasing carbon dioxide (CO_2) content in the atmosphere (see discussion in Chapter 10, pages 351–353). During the most recent glaciation, which reached its maximum about 15,000 years ago, glaciers extended to New York City and large parts of northern and central United States.

Whatever the cause of an ice age, one of its principal effects is a change in sea level. The Quaternary glaciation was very significant for many features of the ocean. In North America, the glaciation period (called the Pleistocene Epoch) consisted of at least four major advances and withdrawals (interglacial stages) of glaciers over parts of the continents. The periods of glaciation had two distinct effects on the ocean: They lowered sea level by as much as 100 m or more, and they changed the temperature of the surface water by several degrees.

The amount that sea level was lowered during the most recent Wisconsin glaciation has been studied by obtaining carbon-14 (^{14}C) dates of deposits that formed at or near sea level. The deposits include peat and shells of animals that live in shallow water. Many of the deposits are now found underwater. By dating this material and noting the depth at which it was found, it is possible to reconstruct a curve of recent sea-level changes (Figure 4-4). The data show that about 15,000 years before present (B.P.), sea level was more than 100 m (328 ft) lower than today. Geologically this means that the now-submerged continental shelf was exposed to the eroding forces of waves and currents as sea level rose during the 15,000-year interval to its present position (Figure 4-5). Before the rise in sea level, rivers could cross the shelf and deposit much of their sediment directly into the deeper parts of the ocean. After 15,000 B.P., sea level rose relatively rapidly until about 7,000 B.P., when it reached a level of roughly 10 m below present sea level. Since 7,000 B.P., sea level has risen slowly and irregularly.

During the times of lowered sea level the continents of Asia and North America were connected by a "land bridge." This bridge is now covered by the shallow Bering and Chukchi Seas. But when sea level was lowered in the Pleistocene, early people and animals could have migrated from Asia to North America and South America.

Glaciation also modified conditions in the ocean. The freezing of large quantities of water increased the salinity of the ocean (most of the salt remains in the water when seawater freezes). Freezing of seawater also lowered the temperature of the surface layers of the ocean. Deep-sea geochemical and micropaleontological (the study of microscopic fossils) work has detected some of these changes. The amount of various isotopes in the water and in living organisms or in those that once lived in the water is a function of the water's temperature; therefore a study of these isotopes can reveal past oceanographic conditions. Some planktonic animals, such as Foraminifera, also undergo changes in their shell char-

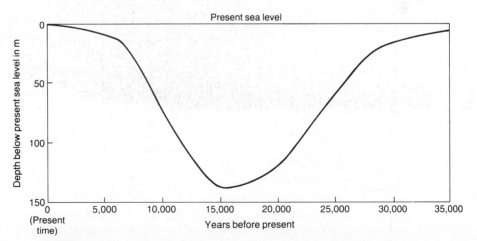

FIGURE 4-4   Recent changes in sea level. Some possible small-scale fluctuations in the last 5,000 years are not shown.

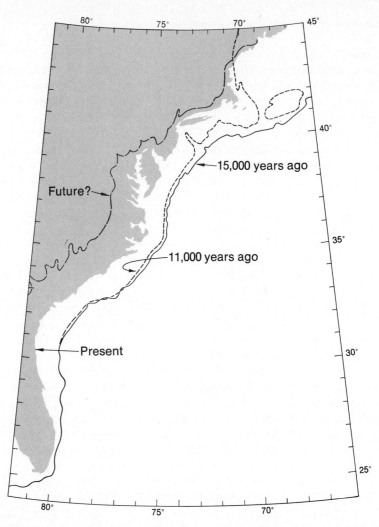

FIGURE 4-5 Shorelines of 15,000 years ago (lowest level of the sea during the most recent glacial stage) and 11,000 years ago. The present shoreline and a future one (if all ice were to melt) are also indicated. The insert shows the changing position of sea level with time as determined from radiocarbon ages of shallow-water shells and peat deposits. (Data from K. O. Emery.)

acteristics, apparently in response to temperature changes. When the shells of these organisms, obtained in deep-sea cores, are studied it is often possible to identify temperature adaptations.

These well-documented changes show that, geologically, oceanographic conditions of today are not completely representative of those of the past. The glacial influences on the chemical, biological, and physical properties of the sea are generally more subtle than the dramatic geologic changes.

It is questionable whether we are completely out of the interglacial period. Tide-gauge records from different parts of the world indicate that sea level is still slowly rising. If all the remaining ice on the earth were to melt, sea level would rise by an additional 60 m (about 180 ft). The effects of such a rise could be catastrophic since a large portion of the world's population lives at or near sea level (Figure 4-5). Alternatively, we could enter into another period of glaciation, which would result in a general decrease in temperature and a drop in sea level. These prospects are receiving serious consideration due to the awareness of the increasing content of CO_2 (because of the burning of fossil fuels) in the atmosphere. This increase, according to many scientists, may have a dramatic effect on the world's climate in coming years (see Chapter 10, pages 351–353).

Summary

Although much has recently been learned about the origin of the universe and our own solar system, many critical questions still remain. Two questions, which are perhaps unsolvable, are: Where did the matter that forms the universe come from? When was it formed? Evidence concerning the earth and its early evolution are easier to come by. There is general agreement that the earth solidified about 4.5 to 5 billion years ago. Much of the water that forms the ocean and the anions in it has come from volcanic activity; the cations in the ocean have come from decomposition and weathering of rocks of volcanic origin. The early atmosphere and ocean did not have any free oxygen until the evolution of plants and the photosynthesis process produced it.

The relative amounts of the various elements are related to chemical and physical processes that occur mainly at the three major interfaces of the ocean (water–atmosphere interface, water–biosphere interface, and water–sediment interface). Little is known about details of the early history of the ocean but the similarity of fossil forms to living organisms suggests that conditions similar to present-day conditions may have been reached about 600 million years ago. Changes in the ocean's configuration, such as those due to sea-floor spreading, have had other affects, such as on climate and oceanic circulation.

One of the more dramatic changes in the recent history of the ocean were the glacial advances and withdrawals during the Pleistocene Glacial Epoch. Sea level has recently risen by over 100 m in the last 15,000 years and in the process inundated the present continental shelves of the world. If all the remaining ice were to melt (which might be a possibility due to the increasing CO_2 content in the atmosphere) sea level could rise an additional 60 m.

Suggested Further Readings

Barghoorn, E. L., "The Oldest Fossils," *Scientific American*, **224,** no. 5 (1971), pp. 30–53.

Hammond, A. L., "Paleoceanography: Sea Floor Clues to Earlier Environments," *Science*, **191,** no. 4223 (1976), pp. 168–70, 208.

MacIntyre, F., "Why the Sea is Salt," *Scientific American*, **223,** no. 5 (1970), pp. 104–15.

5

*Marine Geology
and Geophysics*

THE SCIENCES OF geology and geophysics are principally concerned with the characteristics of the earth and occasionally of other planets. Geologists tend to be more interested in the earth's history and processes, whereas geophysicists more often focus on its structure. Differences between the two fields actually are often obscure or minor. The marine geologist (or geologic oceanographer) and marine geophysicist are both primarily interested in that portion of the earth now covered by water. This can include beaches, marshes, and tidal areas that are only sometimes covered by water as well as the continental margin and the deeper portions of the ocean. The coastal region is an extremely important portion of the ocean and is treated separately in Chapter 11.

Marine geology and geophysics have recently gone through a dramatic evolution due to the development and acceptance of the concepts of sea-floor spreading and plate tectonics. These concepts have given earth scientists a startling insight into how the ocean basins have formed and evolved. As a result, many of the geologic ideas developed over the past two centuries have had to be changed. These concepts are specifically treated in Chapter 6, although some references are made to them in this chapter.

Among the major objectives of marine geology and geophysics are to describe and determine the origin of the topography of the ocean floor, to ascertain the thickness and composition of the sediments and underlying rocks in the ocean, and to explain the origin of the structure below the ocean floor. Marine geologists and geophysicists have also been actively involved in the exploration of the continental shelf and slope looking for areas having hydrocarbon potential, as well as for other mineral deposits (see Chapter 12). It should be noted that most features found on land were originally formed or at least spent some time underwater; thus, by studying the ocean we can better understand our own habitat.

Techniques for dredging up the animals and rocks of the sea floor were developed about 1750. By 1773, mud was obtained from a depth of 1,248 m (4,095 ft) in the Arctic region. One of the first charts of the sea floor was made by Maury in 1854 (Figure 2-3). Mud that had collected in the weight at the end of sounding devices used in obtaining Maury's data was studied by Bailey and Pourtales, and by 1870 they had collected more than 9,000 samples of the ocean bottom. At about the same time Charles Darwin was making studies of coral reefs that in later years were to lead to an interesting controversy in marine geology (see pages 25–26). The Challenger Expedition (1872–1876) is usually thought to have marked the beginning of marine geology as a formal science (see pages 26–28).

After the Challenger Expedition, numerous studies of the ocean floor were made but the next major advance in marine geology did not occur until the German Meteor Expedition of 1925. This expedition was one of the first to use electronic echo-sounding equipment rather than to determine depths by the tedious and time-consuming method of lowering a line with heavy weights to the bottom. Working in the South Atlantic, the *Meteor* made over 70,000 depth soundings. These soundings clearly showed that the ocean bottom was not a single featureless plain but consisted of mountains, valleys, and flat areas. In later years it was found that the ocean floor has a topography every bit as diverse as topography on land.

During World War II many studies of the sea floor and beaches were made for military purposes. After the war, large quantities of explosives were available for seismic refraction studies at sea. These techniques were used on land in the 1920s, and by 1938 Maurice Ewing and his co-workers had extended them to shallow parts of the ocean to make studies of the continental shelf. In later years improved technology made seismic refraction possible in the deep sea and allowed its previously unknown structure to be studied. Other techniques and technology were developed for measuring the earth's magnetic and gravity fields.

Photography of the sea floor became possible in the early 1940s and was, at one time, used in attempts to detect enemy submarines in shallow water during World War II. Sophisticated cameras (Figures 3-28, 3-29, and 3-30) can now take thousands of color pictures of the deepest part of the ocean.

During more recent years the fields of marine geology and geophysics have benefited from several major technological developments. Perhaps the most note-worthy of these is the deep-ocean drilling vessel *Glomar Challenger* (Figures 3-11, 3-12, and 3-13), which can drill in water depths down to 7,000 m (about 23,000 ft). By 1980 over 840 holes had been drilled in the Atlantic, Pacific, Indian, and Antarctic Oceans as well as in several smaller seas such as the Gulf of Mexico, Red Sea, and Mediterranean. In the drilling of these holes an extensive collection of cores was obtained (the total length of the cores exceeds 38 km or 23 mi) and has been studied by scientists from all over the world (Figure 5-1). This program

FIGURE 5-1 Scientists aboard *Glomar Challenger* examining layers of salt obtained from drilling beneath the Red Sea.

may enter a new phase in the early 1980s called the Ocean Margin Drilling Program and probably will use the Hughes's *Glomar Explorer* (Figure 12-14). This vessel is considerably larger than the *Glomar Challenger* (Figure 5-2) and can drill deeper into continental margins.

Submersibles (Figures 3-6 to 3-10) have recently become another exciting piece of technology for marine geologists and geophysicists. They have been used, among other purposes, in the exploration of ocean ridges and have been responsible for some remarkable discoveries (see pages 174–176 in Chapter 6). Improved seismic systems have also been used to obtain new information about the earth's structure.

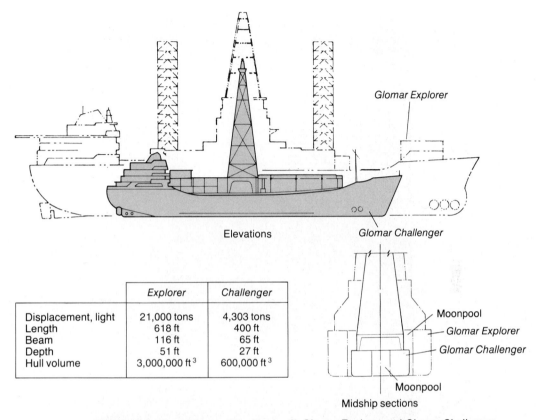

Elevations

	Explorer	Challenger
Displacement, light	21,000 tons	4,303 tons
Length	618 ft	400 ft
Beam	116 ft	65 ft
Depth	51 ft	27 ft
Hull volume	3,000,000 ft^3	600,000 ft^3

Midship sections

FIGURE 5-2 Comparison of the Hughes's *Glomar Explorer* and *Glomar Challenger*. (From Ocean Margin Drilling—A Technical Memorandum, 1980, Office of Technology Assessment, Washington, D.C.)

Tools of the Marine Geologist and Geophysicist

One objective of a marine geologist is to obtain a sample of the sediment or rock on or below the sea floor. To obtain samples (other than by drilling), generally one of three types of instruments is used: (1) snappers or grab-type samplers, (2) coring devices, or (3) dredges.

Snappers or grab samplers (Figure 5-3) obtain only a surface sample, which could be disturbed by the instrument during the sampling process. These devices are generally used only in shallow water. If a rock or some other object gets wedged in the jaws of the sampler, it will not close properly and the sediment sample may be lost.

Coring devices are used to obtain a long vertical section (or core) of the sediment. This is achieved by forcing a long pipe, usually with an inner plastic

FIGURE 5-3 An orange-peel sampler in the open and lowering position (right) and the closed position (after sampling) with its canvas cover removed (left). The jaws of the sampler close after the device strikes the bottom and penetrates the sediment. Once the jaws close the sampler is lifted to the surface; the canvas cover prevents the loss and washing of the sediment on the trip to the surface. (Photograph courtesy of Woods Hole Oceanographic Institution.)

liner, vertically into the sediment. The simplest type of coring device is the gravity corer, which is a pipe with a heavy weight at one end that penetrates the bottom by its own weight (Figure 5-4). This instrument will usually obtain only short 2- or 3-m (about 6 to 10 ft) cores.

To obtain longer cores a piston corer (Figures 5-5 and 5-6) is used. This device has a piston inside the core tube that reduces friction and utilizes hydrostatic pressure during the coring operation and therefore can take up to 20-m-long (about 62 ft) cores. The piston corer is usually triggered by a gravity corer. These sampling devices can be positioned above the bottom with the use of a pinger (Figure 3-27).

The weight portion of a piston corer can be used to hold other instruments, such as cameras. Photographs taken by these cameras during the coring operation can be used to measure bottom currents and to evaluate the coring operation (Figure 5-7).

An ingenious coring device is the free-fall corer. This instrument is not attached to the ship's wire but is dropped free over the side of the ship. When the device hits the bottom, buoyant glass spheres are released that pull the plastic

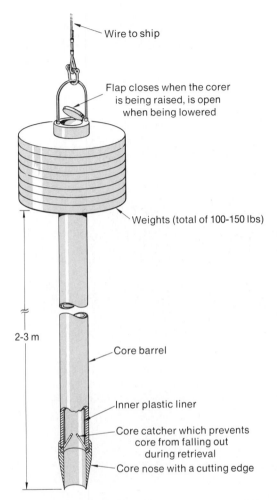

Wire to ship

Flap closes when the corer
is being raised, is open
when being lowered

Weights (total of 100-150 lbs)

2-3 m

Core barrel

Inner plastic liner

Core catcher which prevents
core from falling out
during retrieval

Core nose with a cutting edge

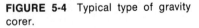

FIGURE 5-4 Typical type of gravity corer.

liner containing the sediment core out of the core barrel, and both float to the surface where they are retrieved (Figure 5-8). The main advantages of this device are that many samples can be taken within a short period of time (by just throwing them over the side) and the cores can be accurately positioned with respect to certain features on the sea floor. For example, if one wanted to take a core with a piston or gravity corer in a canyon on the sea floor, first the ship would have to be stopped and then the corer would have to be lowered from the ship to the bottom. This operation, even in waters of a few hundred fathoms' depth, could take as long as a half hour during which time the vessel would probably drift away from the canyon because of surface currents and winds. The corer itself, while being lowered, could be affected by subsurface currents and deflected from its target. The free-fall corer, on the other hand, would immediately be dropped from the moving ship when over the selected sample area. The device falls rapidly

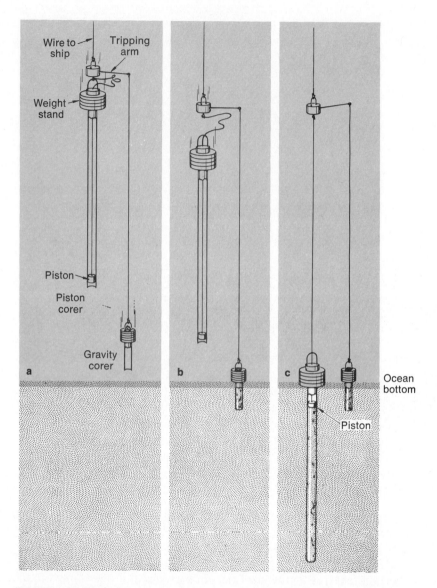

FIGURE 5-5 Operation of a piston corer that is tripped with a gravity corer. (a) Lowering position. (b) When the gravity corer hits the bottom, the piston corer is released and free-falls to the bottom. At moment of impact the line to the piston tightens and the core barrel moves past it; this reduces friction within the core barrel. (c) Completion of the coring operation prior to retrieval.

enough to avoid the effect of subsurface currents. This free-falling technique has been used for other instruments, such as those that measure water characteristics (Figure 9-5).

FIGURE 5-6 Retrieval of a piston corer. A line is attached to the barrel of the corer and is lifting it aboard the ship. The weight stand is already secured, and the gravity corer is already on board. (Photograph courtesy of Woods Hole Oceanographic Institution.)

The third type of geologic sampler is the dredge, which often is a large-diameter pipe partially closed at one end or a large metal frame with a chain "bag" at its end (Figure 5-9). Dredges are rock-sampling devices dragged at slow speed along the ocean floor. If the open end of the dredge encounters a rock, the pulling power of the ship on the rock sometimes breaks it off, so part of the rock is caught in the dredge (a biologic dredge is shown in Figure 8-5).

Other important instruments for the marine geologist are bottom cameras, side-scan sonar, and echo-sounders. These are discussed in some detail in Chapter 3.

Marine geophysicists are interested primarily in the internal structure of the earth. Their instruments are designed to measure the earth's gravity and magnetic field, to examine its subsurface layers and structure beneath the ocean, and to

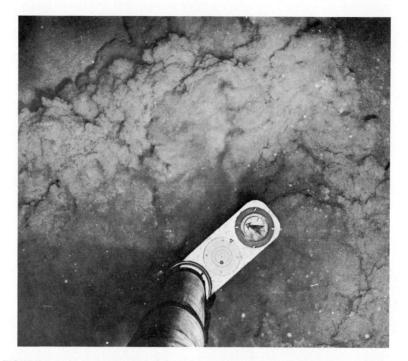

FIGURE 5-7 Bottom photograph taken by a camera mounted in the weight portion of a piston corer. The large sediment cloud has been produced during penetration of the corer into the bottom. Note the compass and inclinometer mounted on the core barrel to record the attitude of the piston corer in the bottom, which at this moment has penetrated about 8 m (about 26 ft) into the bottom sediment. Movement of the sediment cloud, as observed in subsequent pictures, provides information about bottom currents. (Photograph courtesy of F. McCoy.)

determine the amount of heat being lost from the earth to the ocean and atmosphere. In some instances modification of the same instruments used on land can be used at sea. Usually, however, the ocean environment necessitates a new instrument system.

An example is the measurement of gravity. The force of gravity, which varies from place to place, is measured by noting either the period of a pendulum or the pull of gravity against a delicately calibrated spring. On land these measurements can be made easily and quickly. Because of waves at sea, however, a ship experiences up-and-down accelerations that produce errors in the gravity measurements thousands of times greater than the anticipated variations in gravity. F. A. Vening Meinesz partially solved this problem by devising a pendulum system that could be used on a submarine, below the waves. Gravity-measuring instruments, or gravimeters, have been successfully used at sea when they are mounted in devices that keep them effectively motionless regardless of the movements of the ship.

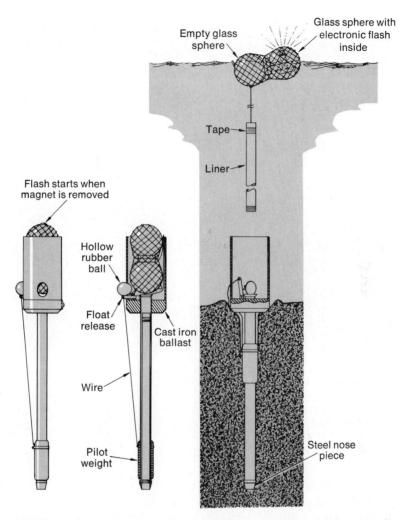

FIGURE 5-8 Free-fall corer built by Benthos Corporation. The corer is freely dropped from the vessel, without being attached to a wire. When the corer penetrates the bottom, the pilot weight is pushed up, releasing the hollow glass spheres that due to their buoyancy pull out the plastic liner containing the core. Both then float to the surface where they can be retrieved by the surface vessel.

Gravity differences are usually interpreted as being due to some change in the underlying geological structure. Because of the many possible causes of variations, the gravity data alone cannot provide a unique interpretation of the subsurface structure and must be used in combination with other geophysical data, such as magnetic and seismic refraction data.

Measurement of the earth's magnetic field at sea is a relatively simple operation. A magnetometer can be towed through the ocean by ship (Figure 5-11)

FIGURE 5-9 A chain bag dredge being prepared for lowering. (Photograph courtesy of Woods Hole Oceanographic Institution.)

or over the ocean by airplane. Magnetic measurements can reveal information about the composition of the rocks forming the upper parts of the earth's crust. Igneous rock bodies, such as volcanoes, can be identified by their distinctive magnetic properties. The magnetic properties depend mainly on how much magnetite (a magnetic mineral) the rock contains, the thickness of the rock body, and its depth below the surface. Magnetization is also a function of the direction and strength of the earth's magnetic field when the rock cooled. A rock will have a magnetization similar in direction and proportional in magnitude to the earth's magnetic field at the time the rock solidified or was deposited. This fact is very important because the earth's magnetic field changed often in geologic times (that is, the magnetic poles reversed). Thus the rocks on the sea floor have magnetic patterns that are characteristic of the condition at the time they were deposited.

This point is especially relevant in the sea-floor spreading concept (see Chapter 6).

The amount of heat coming through the crust from the interior of the earth can be measured by heat probes attached to piston coring devices or by separate probes. Heat flow is the product of the vertical temperature gradient in the sed-

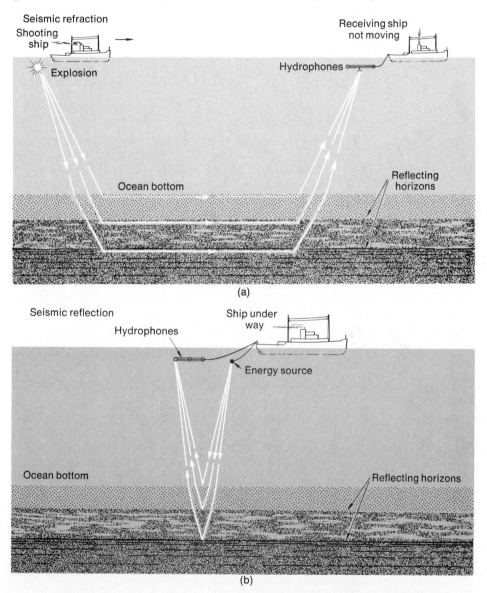

(a)

(b)

FIGURE 5-10   (a) Procedure for seismic refraction. (b) Procedure for seismic reflection.

iment multiplied by the thermal conductivity of the sediment; usually the heat flow is very small—about 1 microcalorie (a millionth of a calorie) per square centimeter per second. A calorie is the amount of heat necessary to raise the temperature of 1 g (0.035 ounces) of water at atmospheric pressure 1°C. Areas where the heat flow is several times higher than the average value may be areas of crustal instability. For example, the mid-ocean ridges usually have higher than average heat flow, which reflects the hot volcanic material rising to the surface in these areas.

Studies of the layering and structure of the ocean bottom are done mainly by seismic reflection and refraction techniques. Both these techniques are based on essentially the same principle (Figure 5-10). Sound waves generated by an explosion will travel to the ocean bottom and through it to subsurface layers; some energy will be reflected back to the surface ship (reflection). Some energy will also travel along the various subsurface layers of the crust and could be received by a second ship (refraction). The receiving devices are called *hydrophones*. A plot of the distance between the ships and the time of the first arrival of refracted sound energy can be used to determine the depth and velocity of sound in the different layers. By knowing the sound velocity of a specific layer, it is possible to speculate about its composition. In the reflection technique only one ship is necessary.

A more recent technique, called *continuous seismic profiling*, utilizes the reflection technique and as many as several hundred hydrophones towed in series (in one cable) behind the ship. The ship can travel at speeds of 8 kn (about 15 km per hour) or more, collecting data as it proceeds. The sound source, generally

FIGURE 5-11 Surface air bubble resulting from discharge of air gun towed about 10 m (about 33 ft) below the surface. The line to the left is attached to a magnetometer that is being towed about 100 m (about 330 ft) behind the ship.

FIGURE 5-12 Marine technician adjusting recording device that is printing the continuous seismic profile.

either an electrical discharge or the release of air under high pressure (air gun), usually is fired every 10 or 12 seconds (Figure 5-11). The returning signals, after being amplified and filtered, are printed by the ship's echo sounder (Figures 5-12 and 5-13).

A recent effort of marine geophysicists has been to study the motion of small-scale natural seismic waves as they pass through the earth's ocean crust by placing sensitive listening packages, called *ocean-bottom hydrophones*, on the sea floor (Figure 5-14). Such studies also produce information on the earth's structure and help place some limits on the physical properties of the crust. The hydrophones can be placed in various patterns on the sea floor to detect regional variations in crust composition and structure.

Crustal Structure

After World War II the fields of seismic refraction and seismic reflection advanced very rapidly. One of the first questions studied was the amount of sediment on the ocean floor. Prior to the use of seismic techniques at sea, there had been considerable speculation on this question. Most estimates were that about 2 or 3 km (1.2 to 1.8 mi) of sediment should be found.

It was known that the earth can be divided into three basic layers: crust, mantle, and core [Figure 5-15(a)]. The general characteristics of these layers are summarized in Table 5-1. For oceanographers, however, it was the top layer of the crust that was of most interest. Seismic refraction and reflection studies made in the early 1950s showed that the earth's oceanic crust could be divided into three major layers on the basis of the velocity of sound within these layers [Figure

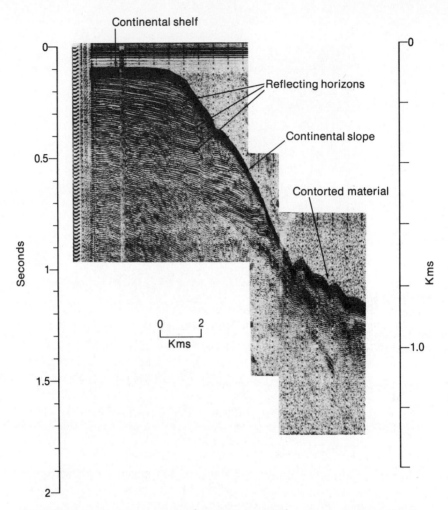

FIGURE 5-13 A continuous seismic reflection profile made across the outer continental shelf and slope off the west coast of Mexico. The time scale is the time the sound takes to travel through the water to a reflecting horizon [see Figure 5-10(b)] and back to the ship. Note the abrupt termination of the continental shelf and the contorted reflections on the slope, which may be material that slumped down from above.

5-15(b)]. In many instances smaller units or layers could be detected within the major layers. The first layer, which we know by sampling to be unconsolidated sediments, has a sound velocity of about 2 to 3 km per second (1.2 to 1.8 mi per second). This sedimentary layer is commonly about 300 m (984 ft) thick in the Pacific and 600 m (1,968 ft) thick in the Atlantic. In all oceans, sediment thickness decreases toward the mid-ocean ridges for reasons that will be shown in Chapter 6.

The overall generally thin layer of marine sediments surprised most scientists, who had expected a considerably thicker accumulation. The original estimate

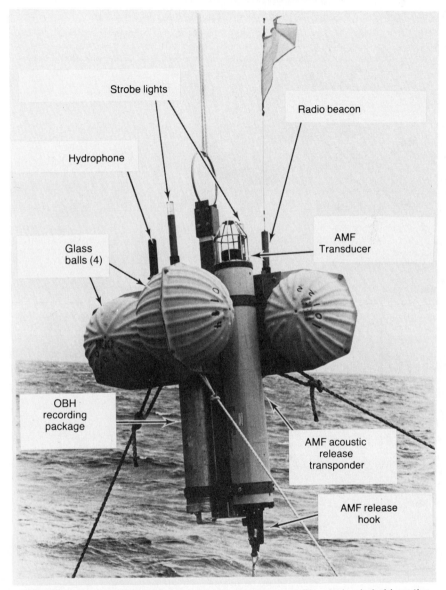

FIGURE 5-14 An ocean-bottom hydrophone package. The device is held on the bottom by an anchor attached to the release hook. (Photograph courtesy of G. M. Purdy, Woods Hole Oceanographic Institution.)

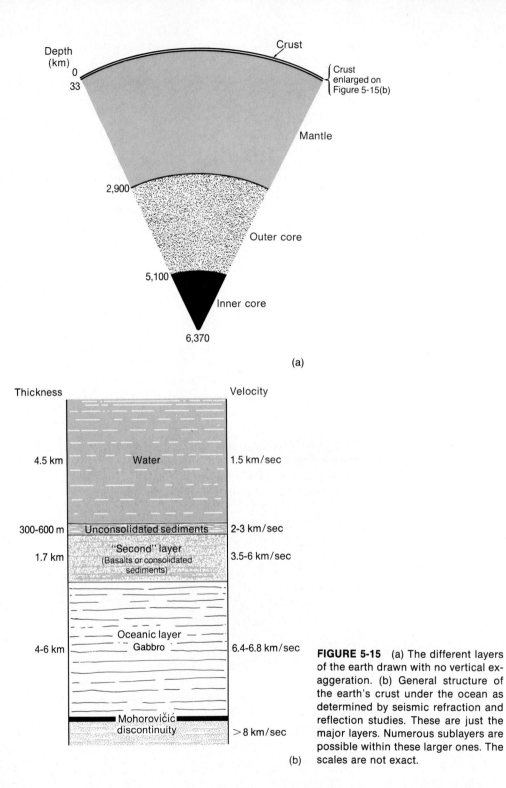

Depth (km)

0
33

Crust

Crust enlarged on Figure 5-15(b)

Mantle

2,900

Outer core

5,100

Inner core

6,370

(a)

Thickness

Velocity

4.5 km

Water

1.5 km/sec

300–600 m

Unconsolidated sediments

2–3 km/sec

1.7 km

"Second" layer
(Basalts or consolidated sediments)

3.5–6 km/sec

4–6 km

Oceanic layer
Gabbro

6.4–6.8 km/sec

Mohorovičić discontinuity

>8 km/sec

(b)

FIGURE 5-15 (a) The different layers of the earth drawn with no vertical exaggeration. (b) General structure of the earth's crust under the ocean as determined by seismic refraction and reflection studies. These are just the major layers. Numerous sublayers are possible within these larger ones. The scales are not exact.

TABLE 5-1 General Characteristics of the Main Layers of the Earth

Layer	Average density (g/cm^2)	General composition	Average thickness (km)	Percentage of total mass of earth
Crust	2.8 (continents) 3.0 (oceans)	Magnesium and aluminum silicates	35 (21.7 mi) (continents) 11 (6.8 mi) (ocean, including the water)	0.4
Mantle	4.5	Iron and magnesium silicates	2,900 (1,802 mi)	68.1
Core: outer inner	11.8 17.0	Liquid iron and nickel Solid iron and nickel	2,200 (1,367 mi) 1,270 (789 mi)	31.5

of 2 to 3 km (1.2 to 1.8 mi) was based on the sedimentation rate in the ocean extrapolated throughout geologic time and on estimates of the quantity of material removed from the continents. Using these estimates it would have taken no more than 400 million years for the sediments now found in the ocean to have accumulated. This amount of time is less than 10 percent of the geologic time during which the ocean is thought to have existed. Assuming that the assumptions above are correct (and there is good evidence that they are), then there are only two possible escapes from this dilemma: Either the sediments have been consolidated and occur as a deeper layer within the earth's crust or some sediments have been removed from the ocean.

E. L. Hamilton suggested that as sediments accumulated on the sea floor, the weight of the overlying sediments would cause a compaction and consolidation of the deeper sediments, which would result in their having a higher sound velocity. Thus in some areas the so-called "second layer" beneath the sediments may consist of consolidated sediments. Even if this idea were applicable everywhere, it would not account for all the missing sediments.

When *Glomar Challenger* started drilling in the different oceans of the world, it was found that usually this so-called second layer was often basaltic rock rather than consolidated sediment. Consolidated sediment was also found, however, but generally it constituted only a portion of the second layer. Thus marine geologists were left with the alternative that large amounts of sediment had been removed from the oceans. Initially, this was not a comforting idea, but it was later found to fit very well into the developing concept of sea-floor spreading.

The second layer has a sound velocity of about 3.5 to 6.0 km per second (2.2 to 3.7 mi) with an average value of about 5.0 km per second (about 3.1 mi per second). The thickness of this layer is about 1.7 km (about 1.05 mi). These numbers can vary due to differences in topography and measurement techniques.

The third major layer of the ocean crust is a relatively uniform layer, with a sound velocity between 6.4 and 6.7 km per second (3.9 to 4.1 mi per second) and a thickness between 4 and 6 km (2.5 to 3.7 mi). Its uniformity suggests that it is a major feature of the oceanic crust. This layer is called the *oceanic layer* and is assumed to be composed of gabbroic rocks. This layer occurs near the surface, at places like the Mid-Atlantic Ridge, where gabbroic rocks are dominant.

These three layers constitute the oceanic crust of the earth. There is a fundamental difference between the crust under the land and under the ocean, both in thickness and composition (Figure 5-16). The continental crust averages about 35 km (21.7 mi) in thickness and the ocean crust averages about 11 km (6.8 mi) including the overlying water. The main rock type of the continents is granite, while that of the oceanic crust is gabbro. Granites and gabbros are both rocks that have formed from the cooling of magma. Granites tend to be relatively lighter in color and have a lower density as well as a lower sound velocity than gabbros. Gabbros also have a relatively high content of magnesium and a low content of silica compared with granites.

At the base of the third layer, under both the oceans and the continents, is the Mohorovičić (Moho) discontinuity where the speed of sound increases to about 8.0 km per second (about 4.9 mi per sec). This discontinuity marks the base of the crust and the beginning of the earth's mantle. There is considerable controversy over whether this layer represents a chemical change or a phase change (like water changing to ice) due to increased pressure and temperature.

The Moho generally occurs at a depth of about 6.5 km (about 4 mi) below most of the floor of the ocean. There is, however, some variation. In most of the marginal seas the Moho occurs at a depth intermediate to that of the ocean and the continents. The Moho also occurs at intermediate depths along the margins of the continents.

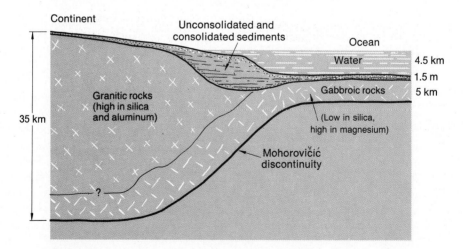

FIGURE 5-16   Crustal structure under the land and the ocean basin as determined by geophysical techniques. Note that the abrupt change between continent and oceanic structure occurs at the continental slope area.

The mantle has a density (mass per unit volume) of about 4.5 g per square centimeter, whereas the rocks of the ocean and continent average 3.0 g per square centimeter and 2.8 g per square centimeter, respectively. Thus the rocks of the ocean and continents can be thought of as floating on the denser underlying mantle. The ocean basins float lower than the continents because they are composed of denser material. This fact explains the basic difference in elevation between the oceans and the continents.

The Continental Margin

The ocean floor can be divided into two main parts, based on either its depth (see Figure 1-7) or its crustal structure (see Figure 5-16). These two main parts are the continental margin and the ocean basin (Figure 5-17). The continental margin includes the coastal region (including beaches), the continental shelf, the continental slope, and the continental rise or borderland—in other words, that portion of the ocean immediately adjacent to the continents. On an areal basis, the continental margins make up only about 21 percent of the total ocean (Table 5-2), but for people they are the most valuable part of the ocean.

THE COASTAL REGION

The coastal region is that part of the continent immediately adjacent to, and obviously influenced by, the ocean. It includes the coast and shoreline, beaches, estuaries, lagoons, marshes, and deltas. A detailed discussion of this portion of the ocean is deferred until Chapter 11. Suffice it to say here that although the coastal zone occupies only a small portion of the continental margin, it is easily one of the most important parts of the marine environment. For example, close to 70 percent of the U.S. population lives within 320 km or 200 mi of the ocean or the Great Lakes, and many of the large cities of the United States and the

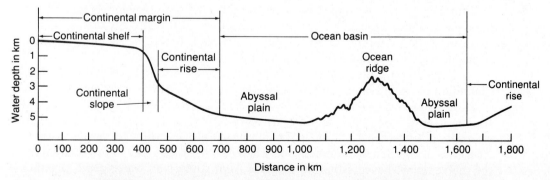

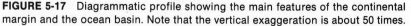
FIGURE 5-17 Diagrammatic profile showing the main features of the continental margin and the ocean basin. Note that the vertical exaggeration is about 50 times.

TABLE 5-2 Percentage of Continental Margin in the Main Oceans and Adjacent Seas

Oceans and adjacent seas	CONTINENTAL MARGIN[a]		OCEAN BASIN			PERCENTAGE OF TOTAL OCEAN
	Continental shelves and slopes	Continental rise and partially filled sedimentary basins	Abyssal plains	Oceanic ridges	Other areas	
Pacific and adjacent seas	13.1	2.7	43.0	35.9	6.3	50.1
Atlantic and adjacent seas	17.1	8.0	39.3	32.3	2.7	26.0
Indian and adjacent seas	9.1	5.7	49.2	30.2	5.8	20.5
Arctic and adjacent seas	68.2	20.8	0	4.2	6.8	3.4
Percent of total ocean in each group	15.3	5.3	41.8	32.7	4.9	100.0

Source: Data from Menard and Smith, 1966.
[a] The continental margin has an area of about 74.5 million km² (28.8 million mi²).

world are near the ocean—many on estuaries. Estuaries, besides being good areas for ports, are also often sites of especially high biological productivity and can serve as the breeding grounds for many marine species. Beaches are valuable both as recreational areas and as easily minable areas of sand and gravel. Deltas in many instances have large accumulations of oil and gas.

CONTINENTAL SHELVES

Continental shelves are the shallow part of the sea floor immediately adjacent to and surrounding the land. They are relatively smooth platforms that terminate seaward with an abrupt change in slope—the shelf break, or shelf edge, which leads to the continental slope (Figure 5-18).

Continental shelves cover a large area—almost one-sixth of the world. The definition or limit of the continental shelves in the past was important for many reasons: Some so-called rights of the sea, such as free passage of ships, mineral rights, fishing rights, and military jurisdiction, were determined by the position of the continental shelf. These rights are in the process of being renegotiated in the Law of the Sea Conference (see Chapter 15). In the past, many definitions of shelves were based either on distance from land or on water depth. Neither of these criteria is completely acceptable since some shelves are exceptionally wide or narrow while others are unusually shallow or deep.

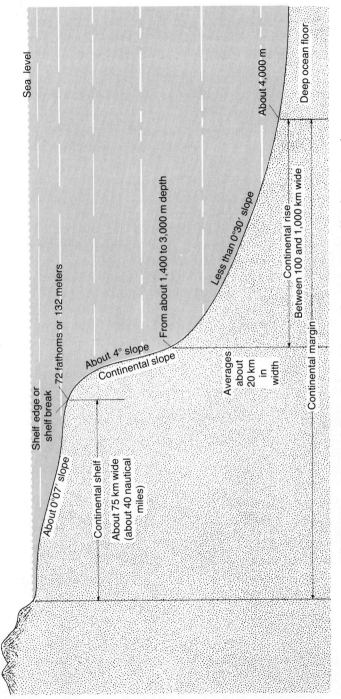

Sea level

Shelf edge or
shelf break

72 fathoms or 132 meters

About 0°07' slope

About 4° slope

Continental slope

From about 1,400 to 3,000 m depth

Less than 0°30' slope

About 4,000 m

Deep ocean floor

Continental rise
Between 100 and 1,000 km wide

Continental margin

Averages
about
20 km
in
width

Continental shelf
About 75 km wide
(about 40 nautical
miles)

FIGURE 5-18 General characteristics of the continental margin. Not drawn to scale.

The basic topography of the continental shelf is fairly well known in most areas and some generalizations can be made. Shelf topography is generally irregular, consisting of many small hills, valleys, and depressions. Shepard in 1973 noted the following statistics of continental shelves:

1. The average width of the shelf is 75 km (40.5 n mi).
2. The average slope of the shelf is 0°07' (a slope undetectable by the naked eye).
3. The average depth of the flattest portion of the shelf is about 60 m (197 ft).
4. The average depth where the greatest change of slope occurs is 130 m (426 ft).
5. Hills with relief of 20 m (65 ft) or more were observed on 60 percent of the profiles he examined; depressions of 20 m or more were present on 35 percent of the profiles.

Although the continental shelves of the world are quite variable, two types seem to prevail: (1) glaciated shelves and (2) unglaciated shelves. One should remember that most shelves were recently above sea level because of the lower stand of sea level in the Pleistocene. When sea level started rising, about 15,000 years ago, the shoreline migrated from what is now the submerged outer edge of most continental shelves to its present position. Some shelf areas themselves have been geologically unstable in that they have moved up or down independent of sea-level changes. For example, parts of the California coast where beaches formed a few thousand years ago are now found several meters above present sea level, indicating that this area has been rising more quickly than sea level.

Both glaciated and unglaciated shelves have been affected by the rise of sea level, but the glaciated ones show direct effects. An example is the northeastern coast of the United States. In the northern part of this area glaciers covered and eroded the shelf (Figure 5-19). The erosion is shown by the numerous banks, basins, and valleys on the shelf. The irregular topography of the shelf off Nova Scotia and in the Gulf of Maine (Figure 5-20) is an indication that the glaciers may have reached to Georges Bank. Georges Bank, however, is very shallow and exposed to the smoothing effects of waves and tides; therefore, little topographic evidence of the glaciation remains. The channel between Georges Bank and the Nova Scotian shelf was cut by a glacier. That glaciers covered this area is also indicated by the mixture of coarse- and fine-grained sediments found in the Gulf of Maine (Figure 5-21). Sediments having such a wide range of grain size are typical of glacial deposits.

South of Cape Cod, particularly in the Long Island area, there is an abrupt change in the character of the shelf. The deep basin, channels, and banks found to the north are absent and the topography becomes relatively smooth. This is an unglaciated shelf, which is generally smoother than the glaciated variety, although some local relief may appear because of strong currents, submarine canyons, or folding and faulting.

Off coasts where there are very strong currents, the continental shelf can be narrow or almost absent. An example is the east coast of southern Florida [Figure 5-20(b)] where the Gulf Stream, flowing sometimes with speeds up to 6 kn (about 11 km per hour), comes close to the mainland. The current is strong

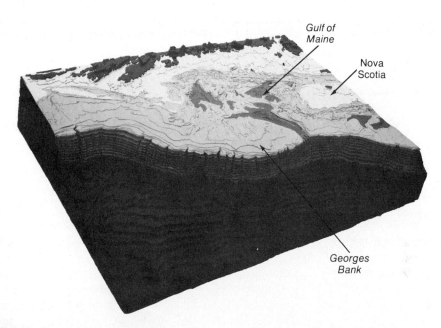

Gulf of
Maine

Nova
Scotia

Georges
Bank

FIGURE 5-19 Topographic model of the northern part of the east coast of the United States from Nova Scotia to south of New York. Note the irregular topography of the Gulf of Maine and the relatively smooth topography of the shelf off Long Island. Color changes occur at 200 m (656 ft), 2,000 m (6,562 ft), and 4,000 m (13,124 ft). Vertical exaggeration is 20 times.

enough to prevent deposition of most sediment and apparently has not permitted normal development of a shelf in this area. The Gulf Stream is so strong here that even at depths of several hundred fathoms on the Blake Plateau [Figure 5-20(b)], it has scoured and removed much of the bottom sediment.

Another variety of a continental shelf found near large rivers is deltas. For example, the Mississippi River has extended across most of its continental shelf and is now supplying sediment to the deeper parts of the Gulf of Mexico.

Most of the shelves off west coast of the United States are unglaciated and relatively narrow, and often covered with only a thin veneer of sediments overlying rocks. Some glaciation is evident off the coast of Washington and British Columbia.

The sediment types of the eastern U.S. continental shelf are often dominated by sand- or silt-sized fluviatile (deposited by rivers) sediment, except for some biogenic (of biological origin) sediments on the Blake Plateau (Figure 5-22), and some authigenic sediments (minerals that formed in the place where they are found, usually the result of some chemical or biological reaction) on the continental slope. Some sediments are called residual, which implies that they have resulted from weathering of the underlying rock. Most of the sediments of the shelf are relict, meaning that they do not represent the present-day environment,

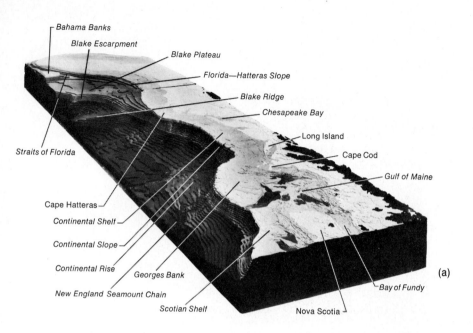

Bahama Banks
Blake Escarpment
Blake Plateau
Florida—Hatteras Slope
Blake Ridge
Chesapeake Bay
Long Island
Cape Cod
Gulf of Maine
Straits of Florida
Cape Hatteras
Continental Shelf
Continental Slope
Continental Rise
Georges Bank
New England Seamount Chain
Scotian Shelf
Bay of Fundy
Nova Scotia

(a)

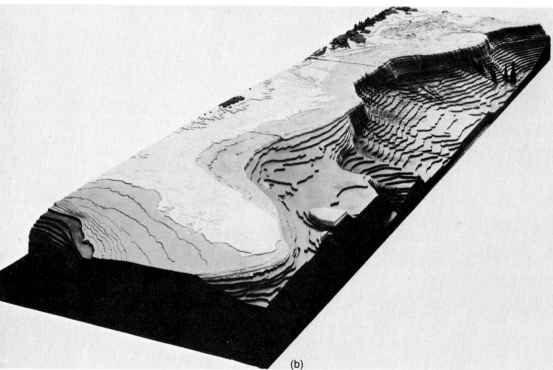

(b)

FIGURE 5-20 Topographic model of the east coast of the United States (a) looking south, (b) looking north. Color changes occur at 200 m (656 ft) (about the edge of the continental shelf), 2,000 m (6,562 ft) and 4,000 m (13,124 ft). Vertical exaggeration is 20 times.

110

FIGURE 5-21 Coarse-grained sediments photographed in the Gulf of Maine. Such sediments are typical of glacial deposits. The compass in the upper left hand corner of the figure is about 10 cm (about 4 in.) in width.

FIGURE 5-22 The texture, mode of deposition, and age of the surface sediments on the continental shelf and slope of the east coast of the United States. (From Emery, 1966.)

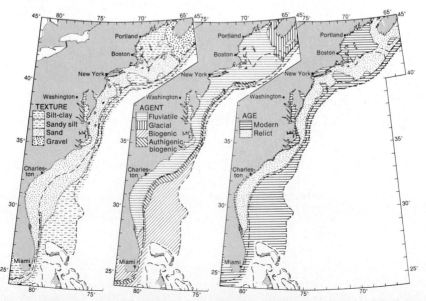

but are left from a previous one—in this instance because of the recent rise of sea level. An example would be a beach deposit found out on the continental shelf. In many areas the sediments of the continental shelf have been reworked by waves and currents during the rising sea level.

Before development of elaborate geophysical equipment, especially continuous seismic profiling (see Figure 5-13), the usual hypothesis for the origin of continental shelves was that they were created by a combination of wave-cutting and wave-building processes. As the waves reached the coastline, they would erode it, forming a terrace, and the sediment removed would be carried seaward to form the outer part of the shelf and the slope (Figure 5-23).

This hypothesis was found to be lacking, particularly in light of data revealed by continuous seismic profiling (CSP). The CSP records showed that the structure of most continental shelves varies considerably and that a simple hypothesis was not adequate. Those CSP records from the east coast of the United States (Figure 5-24) show many different structures to the shelf. Profile 34 off Nova Scotia shows a thin sequence of sedimentary layers on the continental shelf, thickening toward the deeper continental slope. Profile 55 off Georges Bank and profile 130 off Cape Hatteras show an abrupt truncation of the sedimentary layers at the edge of the continental slope. Some truncation, although of a different nature, is shown in profile 118 off Delaware. Profile 63 off Long Island is intermediate in that the reflecting layers are parallel to the slope and shelf. Profiles TW 9 and 81 off South Carolina and Florida show an outbuilding of the continental shelf and slope away from the land. Profile 71 off southern Florida is complex, showing terraces covered by sediments. A buried coral reef may be present about 20 km from the shelf break. The CSP records made along the east coast of the United States show that development of the continental shelf has been due mainly to upbuilding by sedimentation but that the process is complex and differs from area to area. Some ideas of shelf development are shown in Figure 5-25. From the data summarized above one must conclude that numerous methods of continental shelf formation are possible. Recent changes of sea level due to Pleistocene glaciation and

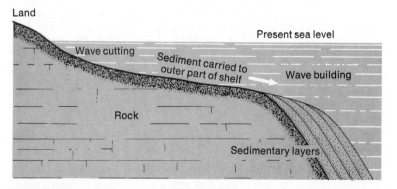

FIGURE 5-23 The old wave-cutting and wave-building hypothesis for the origin of continental shelves.

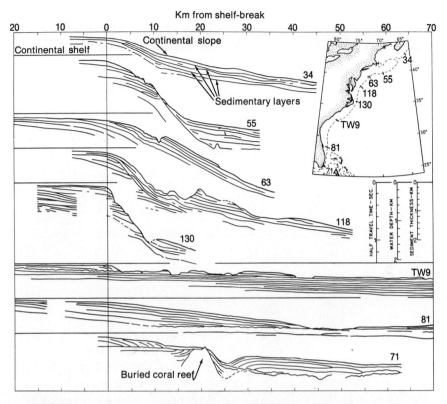

FIGURE 5-24 Continuous seismic profiles made off various parts of the continental shelf and slope of the eastern United States. The lines indicate sedimentary layers. (After Emery, 1967.)

variations in type and rate of sediment supply obviously play important, if not dominant, roles in most of these methods.

CONTINENTAL SLOPES

Continental slopes occur between the two major topographical features of our planet (see Figure 1-4): the land including the continental shelf and the ocean basin. The continental slope extends from the outer part of the continental shelf usually down to the deep-sea floor; in some instances the junction with the deep-sea floor may be hard to define, either because of a gradual decrease in slope or because of the presence of a deep-sea trench bordering the continental shelf. The region where the slope flattens out, near the deep-sea floor, is called the continental rise (see Figure 5-18). In some regions, such as off California and on the Blake Plateau off southeastern Florida, there is an intermediate area between the deep sea and the continental shelf called a *continental borderland* (see Figure 5-20).

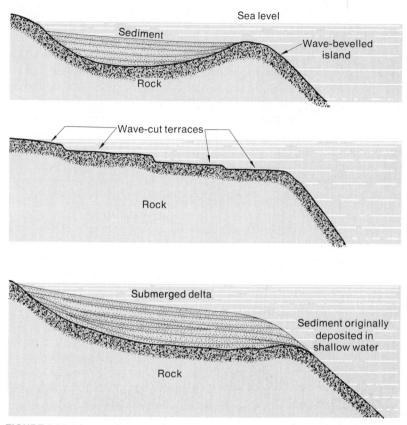

FIGURE 5-25 Some different methods of shelf formation. (Adapted from Shepard, 1973.)

The average inclination of the continental slope is about 4° (a slope of about 70 m in 1 km or 70:1,000), although in some areas the inclination can be 20° or more. The inclination of the continental slope off large rivers and deltas, by contrast, can be as gentle as 1°. Shepard noted that slopes off coasts of fault origin average about 5.6°; slopes off mountainous coasts, about 4.6°; slopes off stable coasts, about 3°; and slopes off major deltas, about 1.3°. The Pacific coast, which follows a major earthquake zone, tends to have somewhat steeper slopes than the Atlantic or Indian Oceans.

The sediments of the continental slope are usually finer grained than the continental shelf; they are generally mud (Figure 5-22). Much of the mud was carried to the slope during times of lowered sea level when erosion by rivers, wind, and other processes could have carried material directly to the continental slope rather than depositing it on the shelves that were exposed at that time. In some areas, especially those off fault or mountainous coasts, rock outcrops are common.

The structure of some continental slopes has been examined by continuous seismic reflection techniques. Results (see Figure 5-24) show considerable variations, but most slopes have some slumping of sediments. Slumping undoubtedly explains the rock outcrops common to many continental slopes. Some seismic sections show large sequences of sediments, especially near the lower part of the slope where the continental rise begins.

Many hypotheses have been proposed for the origin of the continental slopes; they include the ideas proposed for the continental shelf (Figure 5-25) and various folding and faulting hypotheses. In many areas, especially where the slopes lead to deep-sea trenches, continental slopes are clearly related to sea-floor spreading (see Chapter 6).

CONTINENTAL RISES

Many continental slopes end in a gently inclining, broad topographic feature called the *continental rise*. The continental rise usually has an inclination of less than 0.5° (or less than 1 part per 100), whereas the average inclination of the continental slope is about 7 parts per 100. The width of the rise is usually between 100 and 1,000 km (about 62 to 620 mi), whereas the continental slope averages about 20 km (about 12 mi) in width (Figure 5-18). The relief of most rises is very small, generally being interrupted only by submarine canyons or seamounts.

Continuous seismic profiles from continental rise areas generally show a wedgelike sequence of sediments that thickens toward the continental slope. These sediments can sometimes be as thick as 10 km (about 6 mi) and frequently have evidence of material that has slumped down from the continental slope.

As will be seen in the next chapter, the location of continental rises is strongly controlled by sea-floor spreading. Where the ocean floor is moving under the continents, rises generally do not form (Figures 6-5 and 6-8), but where such motion is absent, thick continental rises may occur. For example, continental rises are found along 85 percent of the Atlantic and Indian Oceans but along less than 30 percent of the Pacific Ocean (Figure 5-26). Areas where these thick sediment sequences are found are also areas of possible future petroleum potential (see pages 403–407).

The thick sequence of sediments in continental rises has been compared with the rocks found in many mountain ranges, suggesting that the continental rise may, after considerable time, be uplifted above sea level and eventually form a mountain range of sedimentary rocks. The long linear trend of most continental rises and their parallel position to the edge of the continent are similar to the position and form of many mountain ranges, such as the Appalachian and Sierra Nevada before they were uplifted. This suggestion, consistent with sea-floor spreading ideas, indicates that continents may grow at their borders by the accretion of the sediments on the continental rise to the continent.

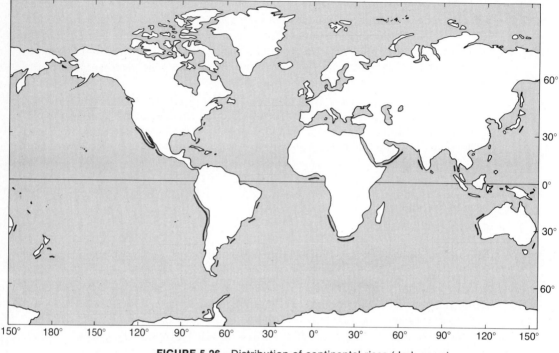

FIGURE 5-26 Distribution of continental rises (dark areas).

SUBMARINE CANYONS

Many areas of the continental shelf and slope are cut by submarine canyons. These features may even cross the continental rise and extend into the deep-ocean basins. Although canyons or valleys of glacial origin (see Northeast Channel in Figures 5-19 and 5-20) or obvious fault origin have been observed, the term submarine canyon is generally applied only to those canyons having winding, rock-walled, V-shaped profiles, and often tributary canyons that extend down the continental slope. The dimensions of some submarine canyons are very impressive: Monterey Canyon off southern California has a greater relief than the Grand Canyon; Hudson Canyon off the Hudson River extends seaward 240 km or about 150 mi and then continues for another 240 km as a leveed channel across the continental rise.

Among the best known canyons are those off the coast of southern California, especially two that are within a mile of Scripps Institution of Oceanography—La Jolla and Scripps Canyons (Figure 5-27). Scripps Canyon, which is a tributary of La Jolla Canyon, lies about 200 m (about 656 ft) offshore of a land canyon. Sediment often accumulates in the head or nearshore parts of the submarine canyon and during certain times of the year moves out of the canyon into deep water because of slumping or other types of movement. This sediment movement

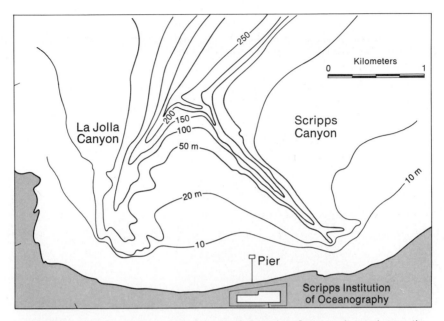

FIGURE 5-27 The Scripps and La Jolla Submarine Canyons located near the Scripps Institution of Oceanography. (After Shepard et al., 1964.)

has enlarged and scoured Scripps Canyon. Such movements may result in turbidity currents (discussed in following section). La Jolla and Scripps Canyons continue as steep-walled features out to a depth of about 300 m (about 984 ft), about 1.6 km or 1 mi from land, where they join. A little farther seaward the relief of the steep, precipitous walls is less, and the canyon becomes more of a small channel cut into the underlying sediments. The channel continues seaward until it ends in a thick sedimentary deposit, or fan. Large quantities of stratified coarse sand similar to that found near the head or beginning of the canyon constitute the fan, suggesting that a turbidity current or similar mechanism has carried the sediment a distance of about 35 km (or 21.7 mi).

Submarine canyons found along the northeast coast of the United States (Figure 5-20) generally do not have tributaries and usually extend straight down the continental slope (Figure 5-28) rather than cross it at an angle as do most of the California canyons. Atlantic coast canyons also tend to have a more V-shaped profile and contain rock outcrops.

Questions about the origin of submarine canyons are not completely or easily resolved. Numerous indications of erosion by strong currents, presumably turbidity currents, have been observed by divers and from submersibles (Figure 5-29). The large accumulation of terrestrial (of land origin) sediment at the seaward end of most canyons also argues for the existence of turbidity currents. Some scientists have noted the similarity of submarine canyons to those canyons on land that were cut by streams and have suggested that perhaps submarine canyons were cut during times of lowered sea level, when rivers would have crossed and cut into most continental shelves and parts of some continental slopes. Proponents of this hypothesis have extreme difficulty reconciling the fact that many canyons exist below the depths of even the most extreme suggested limits of lowered sea level. On the other hand, it is hard to imagine how turbidity currents could cut a winding canyon or how the currents could be strong enough to erode the hard

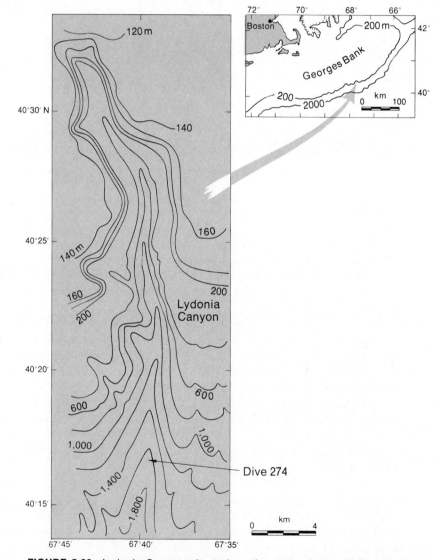

FIGURE 5-28 Lydonia Canyon, situated on the seaward side of Georges Bank. Note the generally straight trend of the canyon. The arrow points to the location of a dive the author made with *Alvin* into the axis of the canyon (Figure 5-30).

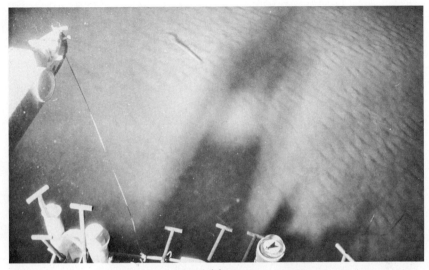

(a)

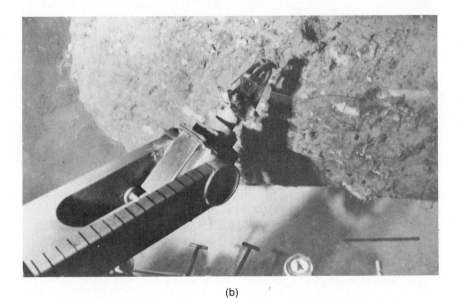

(b)

FIGURE 5-29 Pictures taken in Corsair Canyon, another canyon on the seaward side of Georges Bank. This canyon is northeast of Lydonia Canyon (Figure 5-28). The pictures were taken from the submersible *Alvin*; note the mechanical arm and instruments of the submersible in the figures. (a) Shows ripple marks on the ocean-bottom sediments in the axis of the canyon at a depth of 1,604 m (5,262 ft). (b) Shows a rock outcrop that has been undercut. The ripple marks and the undercutting are both probably due to strong currents.

FIGURE 5-30 Diagrammatic representation of the structure in the axis of Lydonia Canyon (Figure 5-28) at a depth of about 1,600 m (about 5,250 ft). This interpretation is based on a dive made into the area with *Alvin.* It is most reasonable to assume that this structure formed by slumping.

rock into which some canyons are cut. Perhaps both hypotheses are partially correct: The upper parts of some canyons could have been cut during times of lowered sea level and are subsequently maintained, enlarged, and deepened by turbidity currents.

Other hypotheses of submarine canyon formation include submarine springs and slumping (Figure 5-30). The increased use of submersibles for research in canyons should provide definitive answers concerning the origin of submarine canyons.

TURBIDITY CURRENTS

Sediments when suspended in water may produce a dense mixture that can flow down a slope on the ocean bottom. When such a flow results, it is called a *turbidity current*. If the velocity and turbulence of the current are sufficient to prevent settling of the sediment particles, the current can flow for long distances. Eventually it will stop and its sediment load will be deposited; the deposit is called a *turbidite*.

Turbidity currents, their origin, and their effects are controversial topics in marine geology, especially since few direct observations have been made. However, considerable indirect evidence, such as the loss of instruments or movement of large bodies of sediment, indicates that turbidity currents do exist.

One of the first indications of the possible effects of turbidity currents was noted after the Grand Banks earthquake in 1929 when B. C. Heezen and M. Ewing called attention to a regular sequence in the breaking of submarine telephone cables south of the Grand Banks area: The closer the cables were to the earthquake epicenter, the sooner they broke. The cable breaks may have been caused by a turbidity current moving down the slope; this current, Heezen and Ewing suggested, had an average velocity of 50 kn in the earlier phases of the flow and an average of about 25 kn for the duration of the complete flow. Others have questioned these velocities and suggested that 15 kn (about 27.8 km per hour) may be a more accurate average velocity. Both of these values are based on the timing of the cable breaks and the possible route of the turbidity current. Another scientist, K. Terzaghi, proposed that rather than turbidity currents a progressive spontaneous liquefaction of the sediments could have broken the cables. Subsequent observations in other areas have also indicated that earthquakes can produce movements of large amounts of sediment that possibly result in turbidity currents.

The best evidence for turbidity currents is provided by numerous deep-sea cores collected from the abyssal plains region of the ocean. These cores often contain layers of coarse-grained, sand-sized sediment interlayered between normal deep-sea, fine-grained clay deposits. These sand layers, because of their grain size and mineral composition and their numerous shells of shallow-water organisms, are probably derived from a nearshore source. The only known mechanism that could transport and deposit them as layers is a turbidity current. The sand layers generally have finer-sized material toward the top of the layer and coarser-sized material toward the bottom of the layer, which is what one would expect if the sediment was settling out of a current.

Many areas of the ocean, protected by trenches or distant from land, are not affected by turbidity current deposits (Figure 5-31). The topography of these areas is generally hilly and rough. Deep-sea regions where turbidity current deposits have smoothed the topography by their blanketing effect are called abyssal plains.

The origin of turbidity currents is also controversial. The Grand Banks evidence suggests that turbidity currents can be initiated by earthquakes that cause sediment to slump and flow. F. P. Shepard, R. F. Dill, and other workers have noted that coarse-grained sediment will accumulate in the nearshore parts of submarine canyons and then, usually after a large storm, be transported seaward by some form of mass movement. The deposits resulting from such movements are distinguished by their coarse-grained texture overlying the normal fine-grained deep-water sediments. Similar types of mass movements may also occur on the continental slope, although the resulting sediment deposit would not be

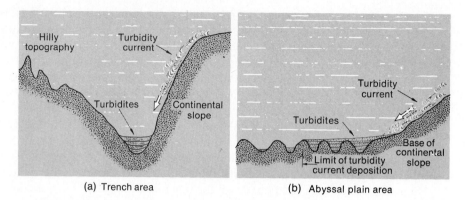

(a) Trench area   (b) Abyssal plain area

FIGURE 5-31 Turbidity current deposition in the ocean. (a) In a trench area the turbidity current will be intercepted and will not reach the more seaward areas. (b) In an abyssal plain area.

called a turbidite since the sediment may never obtain a fluid state; rather, it would travel as a solid mass—that is, a slump deposit.

MARGINAL SEAS

Marginal seas are a common feature of the ocean (Figure 1-6), but do not easily fall into a geologic classification of the ocean: continental margin or ocean basin. These seas, of which the Mediterranean, Red, Black, and Caribbean are examples, often have a crustal structure intermediate in thickness between those of the continental margin and the ocean basin although their water depths are usually more similar to ocean basins.

Some marginal seas are associated with volcanic islands, such as the Aleutian Sea and the Sea of Okhotsk (along the east coast of Russia), while others are almost surrounded by land, such as the Mediterranean, Red, and Black Seas and the Gulf of Mexico. Most lie near present or past structurally active areas.

Marginal seas generally have thick accumulations of sediment because of three factors: their age (that is, they have been around a long time to receive sediments); their closeness to continents, which are a source of sediment; and the many large rivers flowing into them.

Many marginal seas have a restricted connection with the ocean and thus, in turn, have a restricted exchange of water with the open ocean; this favors the accumulation of organic material (because of the lack of oxygen replenishment by bottom waters that could oxidize the organic material; see page 315). This accumulation combined with common coarse-grained sediments found in marginal seas results in an environment favorable for the accumulation of oil and gas. Many of these seas may hold large unfound quantities of hydrocarbons; significant concentrations have already been found in the Persian Gulf and in the Gulf of Mexico.

Several marginal seas are the direct result of sea-floor spreading. The best examples of this are the Gulf of California and the Red Sea. The Red Sea appears to have resulted from the moving apart of the African continent and the Arabian Peninsula (Figure 6-13). This separation may have begun about 30 million years ago and is still continuing. Presently these two continental areas are moving apart at a rate of about 2 cm per year (a little less than an inch per year). A similar movement is occurring in the Gulf of California where Baja California is moving west with respect to the Mexican coast.

The Ocean Basin

This section first describes the major features of each of the oceans and then the characteristics of deep-sea sediments. The various hypotheses proposed to account for the features of the deep sea are briefly treated here but in more detail in Chapter 6.

THE ATLANTIC OCEAN

The Atlantic Ocean is the second largest ocean, and probably the most explored one. (A bathymetric chart of the North Atlantic was made over a century ago; see Figure 2-3.) The most obvious feature of the Atlantic is its shape. It is an elongated, sinuous basin that extends more than 11,000 km (over 6,800 mi) in the north–south direction. Early studies showed that its bathymetry is dominated by the Mid-Atlantic Ridge, a continuous ridge that extends down the central part of the ocean. The symmetrical shape of the ocean on either side of this ridge (Figure 5-32) and the outline of the adjacent continents, which would fit together like pieces of a jigsaw puzzle if they were pushed together (Figure 6-2), have long fascinated scientists and support the hypotheses of continental drift and, more recently, sea-floor spreading (see Chapter 6).

The bathymetry of the Mid-Atlantic Ridge shows that it is part of a continuous major oceanic feature that can be traced more than 55,000 km (over 34,000 mi) through the Atlantic, Indian, Arctic, and Pacific Oceans (Figures 1-7 and 5-33). In the Atlantic Ocean the ridge is especially wide and occupies the middle third of the ocean; it is highly fractured, consisting of numerous mountains and hills. The most rugged topography occurs in the central part of the ridge, where a large central rift or crack, which may be continuous over most of the ridge system, appears on most profiles. The central rift can be as much as 2,000 m (6,562 ft) below the peaks of the adjacent surrounding peaks. These peaks or mountains often rise to within a thousand or so meters of sea level (the average depth of the ocean is 3,279 m). In some instances the mountains protrude through the surface of the ocean forming islands, such as the Azores and Iceland.

Studies of the central rift in the Mid-Atlantic Ridge, using submersibles, showed considerable evidence of volcanic activity and recent faulting (Figure 5-

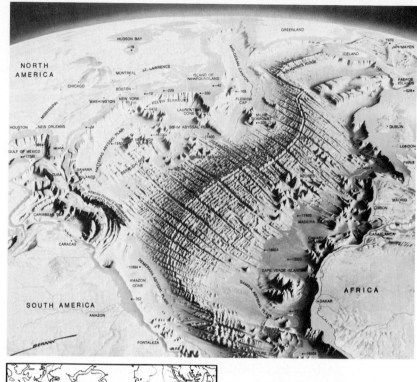

(a)

(b)

FIGURE 5-32 (a) Physiographic diagram of the North Atlantic. [Photograph courtesy of Aluminum Company of America (ALCOA).] (b) General topographic features of the Atlantic Ocean. Note how the Mid-Atlantic Ridge dominates the topography of this ocean.

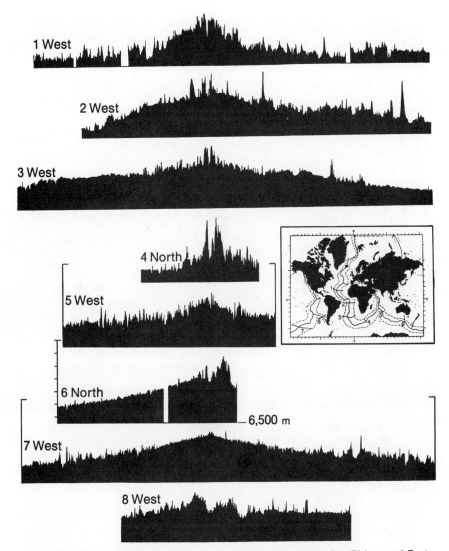

FIGURE 5-33 Profiles across the Mid-Atlantic Ridge, Mid-Indian Ridge, and East Pacific Rise. Base line is 6,500 m (21,326 ft) for all profiles. (From Heezen and Ewing, 1963.)

34). This type of activity is expected in the central rift region because of its sea-floor spreading origin. Further evidence of sea-floor spreading is seismic activity in the ridge area and the observation that most of the earthquakes are centered under the central rift area. Measurements of the rate of heat being discharged from the Mid-Atlantic Ridge show relatively high values on the crest of the ridge that may likewise be explained by the intrusion of volcanic rocks along the crest of the ridge.

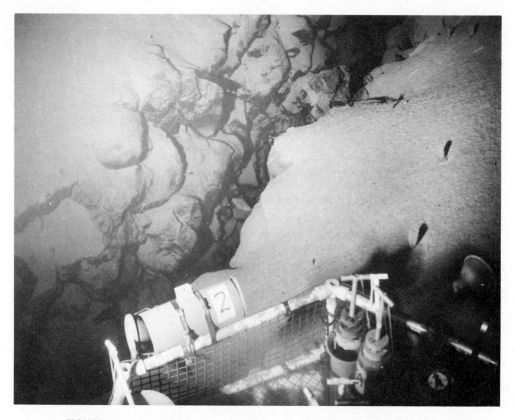

FIGURE 5-34 A large crack or fissure on the sea floor as observed from *Alvin* during Project FAMOUS. Some of these cracks extend for 50 to 100 m (164 to 328 ft) and are as deep as 100 m (328 ft) or more. They result from tensional forces associated with sea-floor spreading. (Photograph courtesy of Woods Hole Oceanographic Institution.)

Igneous rocks dredged from the ridge are basaltic or gabbroic; granites are the typical igneous rocks on the continents. Sediments on the ridge are generally young in age and not too abundant (usually less than 100 m or 328 ft thick).

In many places the ridge is offset in an east–west direction by transform faults (Figure 6-7). These faults are horizontally oriented faults that cause a relative displacement of the axis of the ridge to the east or to the west (Figure 5-32).

On either side of the ridge are large, flat basin areas, called abyssal plains (Figures 5-17 and 5-32) that are common to many parts of the deep sea. These plains, however, are not necessarily connected but may be separated by ridges or other topographic features. The flatness of abyssal plains (slopes of 1 : 1000 or less are typical) has been ascribed to the effects of turbidity currents. These sediment-laden currents, originating from the shallower continental shelf and slope, cover and smooth out most previously existing topographic irregularities

TABLE 5-3 Dimensions of Some
Oceanic Trenches

Trench	Depth (m)	Length (km)
Marianas	11,022	2,550
Tonga	10,882	1,400
Kuril–Kamchatka	10,542	2,200
Philippine	10,497	1,400
Puerto Rico	8,385	1,550
Peru–Chile	8,055	5,900
Java	7,450	4,500
Middle America	6,662	2,800

as they deposit their sediments. Continuous seismic profiles taken across abyssal plains clearly indicate that a deeper irregular topography has been buried by sediments.

Two deep-sea trenches are found in the Atlantic: the Puerto Rico Trench in the Caribbean region and the Sandwich Trench off the southern part of South America. The largest is the Puerto Rico Trench, which is about 1,500 km long (about 932 mi) with a maximum depth of about 8,300 m (27,232 ft). Impressive as that may seem, the trenches of the Atlantic are small when compared to their more numerous counterparts in the Pacific Ocean (Table 5-3).

A difference between the Atlantic and Pacific Oceans is the relative absence of seamounts and coral atolls in the Atlantic. Seamounts are isolated submarine hills with steep sides often standing more than 1,000 m (3,281 ft) above the surrounding sea floor. Seamounts can be very large; one of the largest is the Great Meteor Seamount in the northeastern Atlantic. It has a diameter of 110 km (about 68 mi) at its base and reaches an elevation of over 4,000 m (over 13,100 ft) above the sea floor. The area of its summit is about the size of Rhode Island.

Rocks dredged from seamounts are usually basaltic, suggesting a volcanic origin. Seamounts also have typical volcano shapes. When seamounts have flattened tops, they are called guyots. The flattening apparently resulted when the seamount was near sea level and was subjected to wave erosion. If the seamount is located in the equatorial area and the right oceanographic conditions prevail, a coral atoll may develop on it (see page 238).

THE PACIFIC OCEAN

The Pacific, the largest of the world's oceans, has a nearly circular shape. The topography and structure of the Pacific are different from those of the other oceans. Its most striking topographic feature is the almost continuous series of trenches along its outer edge (Figure 5-35). On the continent side of these trenches are usually found folded mountain ranges or a ridge (or ridges) formed by a series of islands (Figure 6-8). Trenches are seismically active and many earthquakes

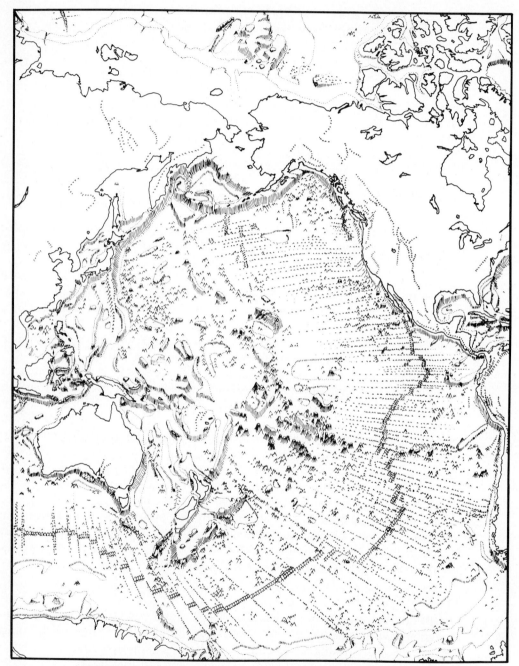

FIGURE 5-35 General topographic features of the Pacific Ocean. Note the trenches along the western and northern parts of the ocean and along the coast of South America, and that the East Pacific Rise trends into the Gulf of California.

occur in or near the trench axis. The circular area around the Pacific has been called *the ring of fire* because of the many earthquakes that occur there and the many volcanoes associated with those earthquakes. There is a gap in the ring along the west coast of the United States and Canada where no trench is found, and fortunately the earthquake incidence there is also low. Volcanic activity, however, still occurs there as evidenced by Mount St. Helens (Figure 5-44).

The deepest trench and area of the ocean is the Marianas Trench (Table 5-3), where the bathyscaph *Trieste* made its famous descent in 1960 to the depth of 10,910 m (35,795 ft) near the Challenger Deep.

Trenches tend to be V-shaped with narrow and sometimes flat floors. The flatness of the floor may be due to turbidity or slump deposits from shallower areas. The width of the trench floor usually does not exceed a few kilometers. Trenches near land usually have wider floors because of a higher inflow of sediments.

It has been previously mentioned that trenches are in seismically active zones (compare Figures 5-32, 5-35, and 5-37 with 6-10). The focus is the point within the earth where the earthquake disturbance has occurred. A focus is generally near the surface in the vicinity of the trench but it will occur at a greater depth in the continental direction. Some geophysicists believe that these focuses indicate a fault zone that may extend as much as 700 km (about 435 mi) below the earth's surface, thus dipping below the continents.

Seismic refraction studies show another interesting feature of trenches: The Mohorovičić discontinuity (the interface between the earth's crust and the underlying mantle) considerably deepens under trenches (Figure 6-8). Thus trenches in many areas seem to mark the boundary between oceanic crust and the continental crust. These are very important facts that are critical in the sea-floor spreading concept. Trenches, as we shall see in the next chapter, are zones of subduction, which means that the new sea floor that was produced at the oceanic ridges eventually is forced under the continent area in the vicinity of the trench. This subduction accounts for the geophysical characteristics of the trench.

Another major feature of the Pacific is the East Pacific Rise, a continuation of the oceanic ridge system observed in the Atlantic Ocean (the Mid-Atlantic Ridge). It extends northeasterly as a broad rise from off New Zealand to the Gulf of California. This ridge differs from the Mid-Atlantic Ridge in some respects; for example, it extends across the ocean basin rather than running down the center of the basin. Its topography is relatively smooth and shows no indication of a distinct central rift valley (Figure 5-33). Its crest rises 2,000 to 3,000 m (about 6,562 to 9,843 ft) above the general level of the floor of the Pacific. Like the Mid-Atlantic Ridge, however, its crest is offset in many places by transform faults. The East Pacific Rise is also seismically active and has higher than normal heat flow.

The East Pacific Rise turns in toward land near the Gulf of California (Figure 5-35) and extends into the Gulf and runs along the western part of the United States, returning to the sea near Cape Mendocino, California.

Seamounts and guyots are especially abundant in the Pacific Ocean. H. W. Menard has listed more than 1,400 and probably many more are still to be found. Seamounts are defined as individual features, although in most instances they are concentrated in certain areas of the ocean and are called provinces or chains. This clumping together and linear trend of seamounts suggests that they are related to major patterns of stress within the oceanic crust.

The volcanic origin of seamounts is well accepted now, but some questions about the sequence of events still remain. Most seamounts show evidence of some submergence; that is, they once stood higher than they do now, maybe even above sea level. In some areas a whole group of seamounts may have been submerged. Other seamounts have remained near sea level long enough for waves to have eroded and flattened their tops to form guyots. In some instances a coral reef was established on the seamount and then grew as the seamount subsided, forming an atoll (see discussion of coral atolls in Chapter 8, page 238). Drilling on Eniwetok atoll has revealed about 1,200 m (4,000 ft) of coralline material over the basaltic rocks of the seamount. The coralline material originated in shallow water and clearly shows that the seamount was submerging during deposition. During part of its history the atoll was elevated above sea level, as indicated by weathered zones of coral material obtained during the drilling.

Abyssal plains are found in some areas of the Pacific but they are not as common as in the Atlantic for two reasons. First, abyssal plains are formed by turbidity current deposits covering the previous topography. The presence of an almost continuous belt of deep-sea trenches around the Pacific Ocean prevents many turbidity currents from reaching the more seaward but shallower ocean basins. In other words, the current would have to run uphill to get out of the deep trench axis. Second, a much smaller amount of sediment is carried by rivers to the Pacific than to the Atlantic (most large rivers flow into the Atlantic Ocean; see Table 5-4). It therefore follows that the incidence of turbidity currents (containing land-derived sediments) should be considerably lower in the Pacific than in the Atlantic. Turbidity currents can, however, also occur far from land if marine sediments fail and slump. The resulting deposits will also tend to cover the existing topography and can locally produce abyssal plains. This mechanism is of lesser importance, however, than the effects of land-derived turbidity currents.

TABLE 5-4 Oceanic Areas and Complementary Land Areas Draining into Them

Ocean	Area (thousands of km^2)	Land area drained (thousands of km^2)	Percentage drained
Atlantic	98,000	67,000	68.5
Indian	65,500	17,000	26.0
Arctic	32,000	14,000	44.0
Pacific	165,000	18,000	11.0

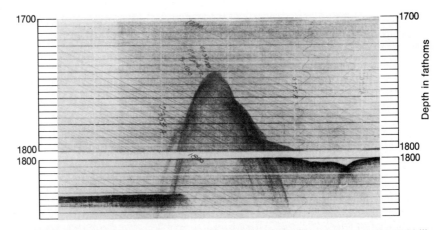

FIGURE 5-36 Echo-sounding record showing the contact between an abyssal hill area (to the right) and an abyssal plain area. Depth is in fathoms; the width of the hill is about 2 mi (about 3.6 km). This is a photograph of the actual echo-sounding record, and the markings on the record are shipboard notations. (Photograph courtesy of Woods Hole Oceanographic Institution.)

Because most of the Pacific has not been covered by thick sequences of turbidity current material, its topography has not been smoothed and much of the original topography is exposed. A prominent feature is abyssal hills, which are hills from 30 to 1,000 m (about 98 to 3,281 ft) above the sea floor and up to several miles wide (Figure 5-36). In some areas abyssal hills occur at the seaward ends of abyssal plains, where they have not been completely covered by turbidites (Figure 5-31). The recent use of sophisticated instrument packages has allowed a few detailed studies to be made of abyssal hills. In these instances they were found to be of volcanic origin with a thin sedimentary cover. The sediments were normal pelagic material (see next section).

THE INDIAN OCEAN

Somewhat less is known about the Indian Ocean than the Atlantic or Pacific. The continental margins (shelf, slope, and rise) of the Indian Ocean are similar to those of the other oceans (Figure 5-37). Submarine canyons, although fewer in number than in the other oceans, do occur. The lesser number of canyons may reflect only the smaller amount of surveying and studies made in the Indian Ocean. Some very large submarine canyons are found off the Indus and Ganges rivers. They are cut into extensive areas of thick sediment that have been deposited by these rivers.

The Java Trench, with a maximum depth of about 7,400 m (about 24,280 ft), extends northwest from western Australia. Shallower than the Atlantic trenches, it is the major trench in the Indian Ocean (Table 5-3).

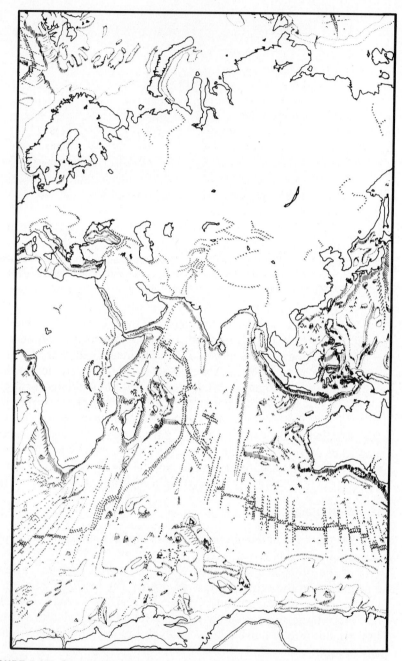

FIGURE 5-37 General topographic features of the Indian Ocean. Note the large number of abyssal plains (black area on figure) and that the mid-ocean ridge splits, with one limb going into the Red Sea.

The ocean basin floor in the Indian Ocean is dominated by abyssal plains having gradients that range from 1 : 1,000 to 1 : 7,000 and a relief usually of only a few meters. In some areas small channels, possibly of turbidity current origin, cut across the abyssal plains.

The major feature of the Indian Ocean is its mid-oceanic ridge that appears to split in the center of the ocean, forming an inverted Y. One of the split limbs goes south of Australia into the Pacific where it intersects with the East Pacific Rise. The other split limb continues south of Africa and joins with the Mid-Atlantic Ridge. The third part of the inverted Y turns and continues to the Gulf of Aden and the Red Sea. In the Indian Ocean the ridge has characteristics similar to those of the Mid-Atlantic Ridge. It is rugged, seismically active, and may have a continuous median rift (Figure 5-33). The Mid-Indian Ridge is also situated essentially in the central part of the ocean.

Linear, aseismic ridges are characteristic of the Indian Ocean (Figure 5-37). They differ from the mid-oceanic ridge in their lack of seismicity and in their general shape, being somewhat higher and more blocklike. Fracture zones are evident in the Indian Ocean and, in some instances, contain small trenches.

THE ARCTIC OCEAN

The Arctic Ocean is the smallest of the major oceans. Most of this ocean is covered by ice, and so it has not been fully explored. Some observations have been made by nuclear submarines traveling under the sea ice. The Arctic Ocean is roughly circular in shape and is divided by three submarine ridges (Figure 5-38). The ridges, the Alpha, the Lomonosov (named after a Russian scientist), and a possible extension of the Mid-Atlantic Ridge, divide the ocean into a series of basins or deeps.

Between one-third and one-half of the Arctic floor consists of continental shelves, some of which extend to a depth of 300 m or about 984 ft (for example, off the coast of Greenland). This depth may be due to depression from the weight of the ice sheet that covers Greenland.

The basins or deeps of the Arctic Ocean are separated from the Pacific and Atlantic Oceans by shallow submarine ridges. The basin depths and general topography are similar to those of the other oceans. The extension of the Mid-Atlantic Ridge into the Arctic Ocean is not completely confirmed but is highly likely. The area has not been sufficiently surveyed to reveal where this ridge goes when it leaves the Arctic region. Perhaps it extends across the Siberian Sea into Russia.

Marine Sediments

Marine sediments can be divided into two major groups: terrigenous sediments and pelagic sediments (Table 5-5). The distinction is based in part on the origin

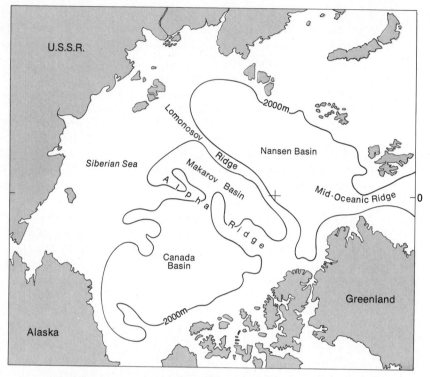

FIGURE 5-38 General topographic features of the Arctic Ocean.

TABLE 5-5 Classification of Marine Sediments

Terrigenous sediments	Pelagic sediments
Terrigenous muds	Biogenous deposits
Slump deposits	Inorganic deposits
Turbidites	Authigenic deposits
Glacial deposits	Volcanic deposits

and in part on the method of deposition of the sediment. This classification does not include all possible types; among those missing are sediments enriched in heavy metals and having potential economic values. These are discussed in Chapter 12.

TERRIGENOUS SEDIMENTS

Terrigenous sediments are those obviously derived from land and, as expected, they tend to be found mainly near or on the continental margin close to land. These sediments can have small quantities of biogenous (of biological origin) material. The relatively high rate of deposition of the terrigenous sediments often dilutes the more slowly deposited biogenous material. In some instances terrigenous sediments may be moved and redeposited in a catastrophic manner, such as by submarine slumping or by turbidity currents. In the case of the latter, they can be carried into the deep sea far from land and their place of origin.

Terrigenous Muds

Terrigenous muds are generally variable in color because of differences in source or the depositional environment. For example, black muds are common in areas where there is a large supply of organic matter but insufficient oxygen to oxidize it; red or brown muds are typical of oxygenated areas.

Terrigenous muds may contain quartz, feldspar, and micaceous minerals common to land rocks. The transportation of the minerals to the ocean is mainly by ocean currents and, to a lesser extent, by wind. The rate of deposition is variable depending on nearness to the source. The deposition rates are, however, several times higher than those of pelagic sediments. Near large rivers a rate of 100 cm per 1,000 years (about 39 in. per 1,000 years) or more is common.

Slump Deposits

Slump deposits are those sediments that by some mechanism have moved or slumped down from a topographically higher area. They may be difficult to distinguish by sampling techniques but sometimes can be recognized from seismic reflection profiles, especially those made across the continental rise, which frequently show large blocks of deformed sediment that have slumped down from the higher continental slope. Under some conditions, slumped material may form a turbidity current, and then sediment moves more as individual particles rather than as a large mass.

Turbidites

The coarse-grained sand and silt layers found interbedded with the typical fine-grained muds of the deep sea are generally attributed to turbidity currents. These deposits are called turbidites and are common to many areas of the deep sea. Turbidite beds range in thickness from a few centimeters to several meters, generally showing a variation in grain size with coarser material at the bottom of the layer and finer material above.

Turbidites commonly contain material such as wood fragments, shells of organisms that live in shallow water, and other materials that indicate a shallow-water origin for the sediment. In many instances, especially off large submarine canyons, large parts of the deep sea are blanketed with these sediments forming flat abyssal plains.

Much debate has occurred concerning the erosive and carrying powers of turbidity currents and their role in the formation of submarine canyons (see page 121). The evidence that turbidity currents exist is very compelling; the main questions are how fast they move and whether they cause significant erosion.

Glacial Deposits

Glacially derived sediments are common to many parts of the continental shelf (Figure 5-21). Glacial sediments are also found in the deep sea where they can be recognized by their high sand, silt, and occasional gravel content. The sediments were initially carried by glaciers. As the glaciers grew, parts were broken off and ended up in the ocean. Once in the ocean, the ice could be moved around by currents, and when the pieces eventually melted, any sediment or debris that they carried fell to the sea floor.

PELAGIC SEDIMENTS

Pelagic sediments are generally found in and predominate the deep part of the ocean far from land. Some types of pelagic sediment have settled down through the water in the absence of strong currents; other types have directly formed on the sea floor. The deposits may contain the skeletal material of plants (mainly phytoplankton) and animals, be fine-grained clays, or be a mixture of the two. The clays can be land-derived and carried to their deposition site by wind (Figure 5-39), water, or currents, or all of these. Other sources of fine-grained clays are volcanoes and fragments from meteorites.

Biogenous Deposits

Biogenous deposits, sometimes called *oozes*, by definition contain more than 30 percent skeletal material (fragments or shells of different plants or animals). Biogenous deposits may also contain material of inorganic origin, such as clay particles. Oozes are named after the organism that is most prevalent; common types of oozes are foraminiferal, coccolith, pteropod, radiolarian, and diatomaceous. (Figures 5-40 and 5-41; also Figures 8-25 and 8-26.) Diatoms and coccoliths are plants (phytoplankton) and the others are animals (zooplankton). Radiolaria and diatoms have shells made of siliceous material. The others have calcareous shells.

Most of these organisms live in the surface waters of the ocean and when they die, their shells settle to the ocean floor forming a sedimentary deposit. The distribution of these sediments on the ocean bottom (Figure 5-42) generally reflects the conditions of the surface waters, indicating that the settling rate of the shells to the deep-sea floor is sufficiently fast to prevent the shells from being moved far in a horizontal direction from below the waters where the organisms lived. Calcareous oozes are usually found at shallower depths than siliceous oozes because the calcareous shells are more soluble (that is, they will dissolve). This solubility increases with depth and decreasing temperature; thus calcareous shells are rarely found below a depth of 5,000 m because most have been dissolved. The effect is more obvious on the small and delicate coccoliths than on robust Foraminifera.

Biogenous sediments are common under regions of high biological produc-

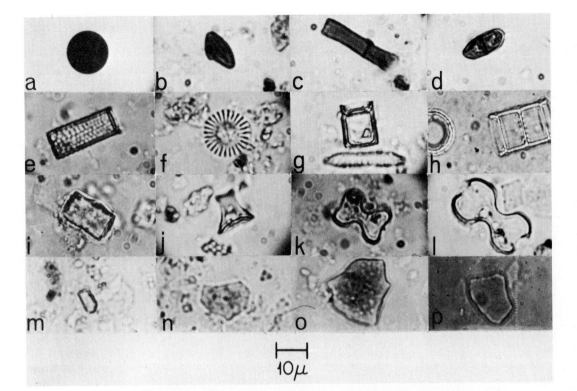

FIGURE 5-39 Some typical land-derived material that has been carried by wind and found in the surface water and air over the North Atlantic. (a) Is an opaque spherule, probably of industrial origin; (b) through (d) are fungus spores; (e) through (h) are freshwater diatoms; (i) through (l) are opal particles precipitated by plants; and (m) through (p) are mineral grains. The scale line is 10μ (one-hundredth of a millimeter). (Photograph courtesy of D. W. Folger.)

tivity, such as upwelling areas, or where the other sediment types are absent or are deposited at a relatively slow rate. The sedimentation rate of biogenous sediments is about 1 to 5 cm per 1,000 years (about 0.4 to 2 in. per year), which is about 10 times higher than the sedimentation rate of the inorganic clays.

Biogenous deposits are valuable to oceanographers because they can be used to tell something about the environment in which the organisms lived. The organisms best suited for these studies are pelagic Foraminifera, which live mainly in the subsurface water layers of the ocean. Certain species are typical of low latitude and warm water, while other species are typical of middle or high latitude and cold water. Cores of sediments containing foraminiferal ooze, in most instances, show a change in the dominant species at depth as a result of major changes in the past of the surface-water conditions (Figure 5-43). These changes indicated by the different species of Foraminifera can be dated using carbon-14, magnetic characteristics, or other procedures. Some species of Foraminifera show

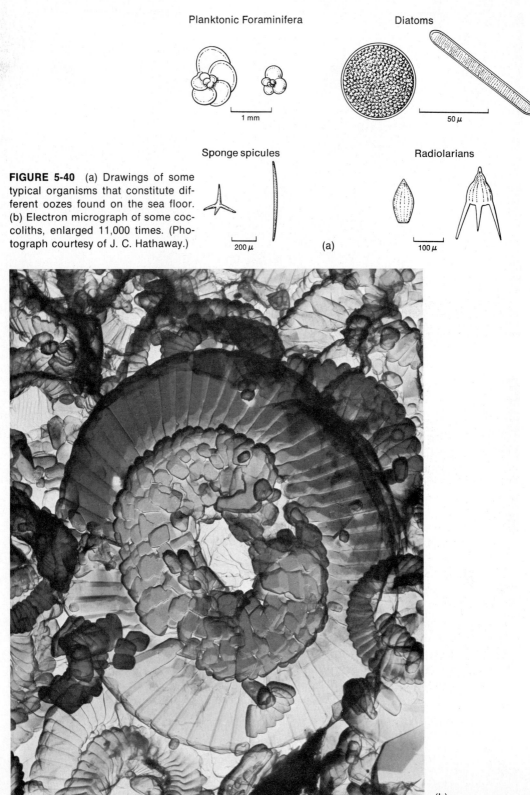

Planktonic Foraminifera

Diatoms

1 mm

50 μ

FIGURE 5-40 (a) Drawings of some typical organisms that constitute different oozes found on the sea floor. (b) Electron micrograph of some coccoliths, enlarged 11,000 times. (Photograph courtesy of J. C. Hathaway.)

Sponge spicules

200 μ

Radiolarians

100 μ

(a)

(b)

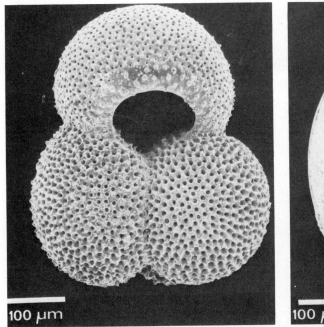

(a)

(b)

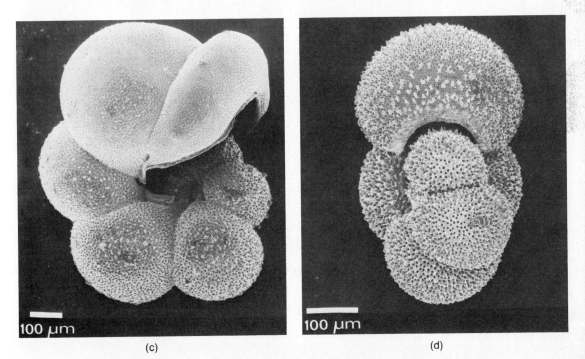

(c)

(d)

FIGURE 5-41 Some examples of Foraminifera often found in marine sediments (a) *Globigerinoides ruber*, (b) *Globorotalia truncatulinoides*, (c) *Hastigerina pelagica*, (d) *Globigerinella aequilateralis*. (Photographs courtesy of Dr. Werner Deuser.)

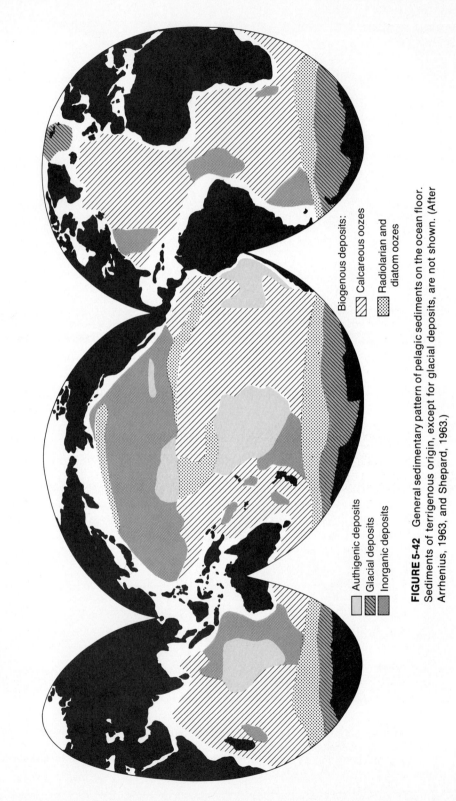

FIGURE 5-42 General sedimentary pattern of pelagic sediments on the ocean floor. Sediments of terrigenous origin, except for glacial deposits, are not shown. (After Arrhenius, 1963, and Shepard, 1963.)

Biogenous deposits:

▨ Calcareous oozes

▨ Radiolarian and diatom oozes

Authigenic deposits

Glacial deposits

Inorganic deposits

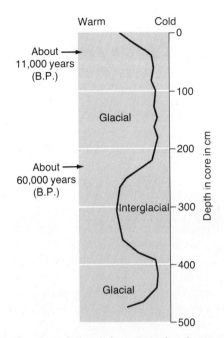

Warm   Cold

About → 11,000 years (B.P.)

Glacial

About → 60,000 years (B.P.)

Interglacial

Glacial

Depth in core in cm

0
100
200
300
400
500

FIGURE 5-43 Generalized temperature variations as observed in a deep-sea core. The ancient temperatures are estimated from the relative abundance of pelagic Foraminifera. Certain species indicate a warm temperature of the surface waters; other species, cold temperature. Major changes occurred about 11,000 and 60,000 years before the present.

an evolutionary change that can also be used as a time marker and that allows correlation from one core to another. In addition, the coiling of the shells of some Foraminifera species has changed direction, apparently in response to temperature changes in the waters where they lived. Another technique for studying the ancient environment is measurement of the oxygen isotopic composition of the shells; the oxygen-18/oxygen-16 ($^{18}O/^{16}O$) ratio can be used to determine the temperature of the water in which the organisms lived (see pages 215–216). Results of these techniques are not absolute; in some instances they are even in conflict with each other. One reason for discrepancy is that Foraminifera may live in a different part of the water column (and therefore at a different temperature) during different stages of their life histories.

The combination of all these techniques has made possible an acceptable description of recent changes in the ocean. The general picture is that the last glacial period ended about 15,000 years ago and that the surface waters of the ocean have been warming since that time. Prior to the last glacial period, there was a warm period (called an interglacial period) that ended about 60,000 years ago. Between 11,000 and 60,000 years ago, the oceans were relatively cold. The warm period that ended 60,000 years ago extended back to 100,000 years ago and was preceded by another glacial period.

Another type of biogenous deposit can result from coral reefs that have been exposed to the destructive powers of waves or other erosive forces. If reworking by waves is especially intense, a white mud composed of fine-grained debris may result. The deposits are generally localized around the reef itself, perhaps even in the inner lagoon that is typical of most atolls.

Inorganic Deposits

Inorganic deposits are the very fine-grained muds found in most of the deep ocean basins far from land. By definition they contain less than 30 percent biogenous material. They generally have a brown color due to having been oxidized. Very little is known about the origin and source of these clays. It has been suggested that some may be derived from the Gobi Desert of Mongolia and carried to the northern Pacific by wind and ocean currents. Other possible sources are meteoric dust or volcanic ash. In some instances the clays are altered to a mineral called phillipsite and should be considered in the next group, the authigenic deposits.

The sedimentation rate of inorganic deposits is very low, usually 1 to 2 mm (about 0.04 to 0.08 in.) per 1,000 years, or 1 m (3.3 ft) per 1,000,000 years. Such a low sedimentation rate limits the occurrence of such deposits. Obviously they will be diluted and negligible in areas where other types of sediments are forming or accumulating more rapidly. Only in the deep areas of the ocean far from any land-derived sediments and where solution has removed many of the calcareous shells are inorganic deposits significant. Even under such restrictive conditions there are large portions of the ocean, notably in the Pacific, that have inorganic deposits (Figure 5-42).

Authigenic Deposits

Authigenic deposits are those that have formed in the place where they are found, usually before the sediment is buried. These deposits are commonly precipitated directly from the ocean water. The mineral phillipsite found in large areas of the South Pacific may be in this group. There is, however, some question whether it precipitated directly from seawater or if it is an alteration product from another clay mineral. Some sediments undergo chemical and physical alteration after their accumulation on the sea floor. This alteration is a different process than the formation of authigenic deposits.

Other interesting authigenic deposits include manganese nodules (really not a sediment but a rock) (Figures 7-5, 7-17, 12-11, 12-12, 12-13, 12-14, and 12-15). These deposits have been found over large areas of the ocean and economically may be the most important deposit of the deep sea (see Chapter 12). They commonly occur as rounded nodules the size of a baseball and can also be found as slabs or as coatings on rocks. The nodules frequently form around small objects on the sea floor, such as a shark's tooth or the ear bone of a fish. Since the deposits form by precipitation from seawater, their deposition essentially stops when the nodule or rock is covered with sediment. The distribution of manganese nodules is usually restricted to areas having an otherwise low sedimentation rate or to areas where strong currents prevent the deposition of other sediments. An example of the latter is the Blake Plateau off Florida, where the swiftly flowing Gulf Stream prevents sediment deposition. The origin of manganese nodules is discussed in Chapter 7 (pages 205–207) and their economic importance in Chapter 12 (pages 410–418).

Another authigenic deposit of possible economic significance is phosphorite. Large deposits of this mineral are not common to the deep sea but in general are restricted to depths of 500 m (about 1,540 ft) or less (Figure 12-3). This mineral apparently forms in basins or areas that contain virtually no oxygen—an anaerobic environment. This absence of oxygen can result when large amounts of organic matter, produced in the surface waters, sink and are oxidized (consuming oxygen) in the deeper waters.

Volcanic Deposits

Sediments of volcanic origin, such as volcanic dust, are common in some areas of the deep sea. One can argue that perhaps they are not of true pelagic origin because the volcano may be on land, but many of these volcanoes are submarine. In either case, the particles settle away from the influence of land and a pelagic classification seems justified.

In some areas, volcanic dust settled at such a high rate that distinct ash layers covered large portions of the ocean. Some layers may be related to a historical volcanic eruption, thus serving as a convenient time horizon with which to relate other deposits.

Recent volcanic explosions, such as the one of Krakatoa in 1883 and Mount St. Helens in 1980 (Figure 5-44), formed large dust clouds that, in the case of Krakatoa, eventually covered much of the world. The heavier material settled and accumulated near the eruption area. Pumice, which can float, stayed on the ocean surface until it became water saturated and sank. The ash layers near the eruption site generally are composed of shards of volcanic material and smaller fragments of volcanic rock. Farther away, finer-grained volcanic minerals, often highly altered, are found.

ROCKS ON THE SEA FLOOR

Rocks exposed on the sea floor are volcanic rather than the granitic rocks typically found on land. Basalt is composed mainly of the minerals feldspar and pyroxene, and is found on many oceanic islands, on the mid-ocean ridges, and along some fracture zones. Other rocks found on the sea floor include the previously mentioned manganese nodules and cemented fragments of calcareous sediments.

AREAS OF NONDEPOSITION

Apparently there are many places on the ocean floor that can be considered as areas of nondeposition, or areas where no sedimentation is presently occurring. Shallow-water localities where strong currents on the bottom prevent the accumulation of sediments are an example. There are regions in the Atlantic and Pacific Oceans where recent sediments are absent and the surface sediments are several million years old. One must assume, under these conditions, that essen-

FIGURE 5-44 A view of one of the eruptions of Mount St. Helens, taken on April 1, 1980. (Picture courtesy of Dr. P. D. Snavely, Jr., U.S. Geological Survey.)

tially no deposition has occurred for many millions of years or that any younger deposits have been removed by erosion.

Stratigraphy

As we have seen, sediments can vary over horizontal distances, but they can also change in character in the vertical sense. In other words, there can be variations that form layers or sequences with depth below the bottom. A study of these sequences is called stratigraphy; its objectives can include ascertaining past conditions of the ocean such as worldwide climatic events or correlating sediments in one area with those in another, and relating events in the ocean to events on land. One of the ways these aims are accomplished is by use of dating methods such as radioactivity (see Chapter 7, pages 216–218) or magnetic reversals (Figure

6-6). Stratigraphic studies are more effective when long cores from the sea floor are obtained. Commonly, piston cores (Figure 5-6) or sediment sections obtained from drilling, such as by the *Glomar Challenger*, are used.

The magnetic reversal procedure is based on the facts that the earth's magnetic field changes in polarity over time, and that these changes are recorded in the magnetic minerals included within the rocks or sediments formed at that time. These sediments or rocks can be dated by the potassium–argon method and give a time pattern for the periods of magnetic reversal. For example, about 700,000 years ago there was a major change or reversal in the earth's magnetic field and it essentially stayed reversed until about 1.7 million years ago, with only one minor interval of change. By taking a long sediment sample from the sea floor and measuring the magnetic orientation of its minerals, one may identify these reversals, and by relating this pattern to that of the known dated sequence, a good estimate of the age can be made of the sediment. A modification of this technique, as we shall see in Chapter 6, was most valuable in the development of the sea-floor spreading concept.

The magnetic technique is an extremely good one since the position of the poles affects all sediments being deposited in all parts of the ocean. Reversals, when they occurred, took place rather quickly—perhaps over a hundred years or so. Therefore, not only can the magnetic technique be used as a correlation device between cores far apart, but it also can be used as an absolute date for that particular horizon (Figure 5-45).

The combination of the techniques of radioactive dating, magnetic stratigraphy dating, and oxygen isotope measurement (see pages 215–216), and the types of fossils obtained in the sediments can lead to a relatively good approximation of the physical characteristics of the ocean at times in the past. Techniques like this, besides helping to identify the past glaciations, have been used to determine major variations in ocean circulation and world climate.

A key aspect in stratigraphy is that many organisms that live in the ocean have evolved into different species over periods of time. Therefore, individual species have time ranges that can sometimes be used both for dating and for correlation. Unfortunately most species have limited geographic ranges, which means that one species cannot be used for correlation in all parts of the ocean. In other words, there are species that tend to live in the equatorial, temperate, or polar regions, but very few live in all three.

One problem in developing the stratigraphic record of the sea floor is that sediment does not accumulate everywhere at a constant rate, and in some places there may be erosion or a considerable difference in the sediment type and amount that reaches and remains on the sea floor. Therefore, dating and correlating sediments of different types but of similar ages are important aspects of working out the history of the ocean. A gap in the sedimentary record is called an *unconformity*. It can be due to periods of nondeposition or erosion when bottom currents may remove sediments already on the sea floor. It should be noted that in developing the sedimentary history of a region one does not directly observe

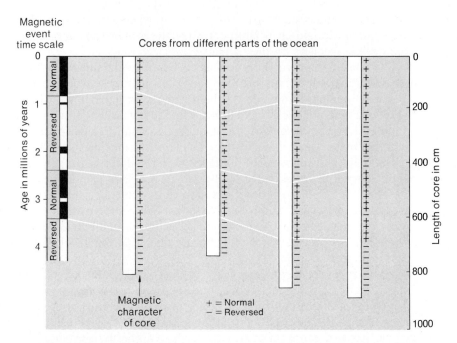

FIGURE 5-45 Use of the effects of the reversal of the earth's magnetic field to correlate deep-sea cores taken in different parts of the ocean. The magnetic time scale was established by dating the reversals on land using radioactive isotopes; by measuring the magnetic character of the cores they can be corrected to the time scale and to each other.

the layers but rather looks at small samples the width of a coring tube or drilling pipe. This often can make the unraveling of the past harder than it might be on land where the rocks can often be seen. Another complicating factor in stratigraphic studies is that organisms on the sea floor often rework the bottom sediments by their normal life activities. In doing so, they can smooth out or mix up sediment characteristics such as a thin ash layer that might have existed. The presence of key or unique layers can be very helpful in working out the stratigraphy in a region. For example, if a major volcanic event occurred that deposited ash over a large portion of the ocean basin, that layer can be correlated whenever it is found.

Summary

The fields of marine geology and geophysics are principally concerned with the history, processes, and structure of the sea floor. The two fields are usually closely related. Among the key objectives are to describe and understand the origin of the physiography of the sea floor, to determine the type and thickness of the sediments in rocks in the ocean, and to determine and explain the origin of the deeper structure. Both disciplines have been dramatically changed by the concept of sea-floor spreading and how it can explain the evolution of the ocean basins.

The sea floor can be divided into two principal components based either on depth or structure. These are the continental margin, immediately adjacent to land, and the ocean basin. The continental margin, which is strongly influenced by land, includes the coastal region, the continental shelf, slope, and rise and makes up about 21 percent of the total ocean. It is one of the more economically valuable parts of the ocean. The ocean basin comprises the deeper portions of the ocean and includes the abyssal plains, abyssal hill regions, trenches, and oceanic ridges.

Continental shelves and slopes are extremely important areas of study for marine geology and geophysics because they generally mark the boundary between oceanic and continental structure and also are the site of many of the mineral and most of the fishing activities in the ocean. Basically there are two types of shelves, those that are glaciated and those that are unglaciated. Most parts of the continental shelf were recently below sea level because of the lowering of sea level as a result of glacial periods. With the rise of sea level many are just in the process of adjusting to the present environmental conditions. Seismic records from continental shelves and slopes show that there is not a unique method of formation but more often various types of processes are responsible for these features.

Marginal seas, small areas isolated from the main oceans, are a relatively common feature of the ocean although they do not fall into any easy geologic classification. In structure they are generally intermediate between that of the ocean basin and continental margin. They often are situated in strategic locations and since they generally have thick sequences of sediment are favorable regions for the accumulation of hydrocarbons. Some marginal seas, such as the Gulf of Mexico and the Persian Gulf, are presently active sites for exploration of marine resources.

Probably the most dominant feature of the ocean basin is the continuous oceanic ridge system that winds through the Atlantic, Pacific, Indian, and Arctic Oceans. This ridge, which dominates the physiography of the areas, is a result of the process of sea-floor spreading, is earthquake prone, and can be traced for over 55,000 km (over 34,000 mi).

Sediments in the ocean can be divided into two major categories: terrigenous and pelagic. The terrigenous sediments are those that are clearly derived from land and often are the dominant sediment on the continental margin. They can include terrigenous muds, slump deposits, turbidites (from turbidity currents), and glacial deposits and are often variable in color and characteristics. Pelagic sediments are more typical of the deep ocean and areas far from land. These are sediments that have settled down through the water column usually in the absence of strong currents or have formed directly on the sea floor. Biological components often may dominate. There are several basic sources for the sediments on the sea floor. They may initially come from the erosion of continental material that can, in early stages, come to rest on beaches or shelves or be carried into the deep sea or, if fine-grained material, be directly carried out into the pelagic environment. Some material from the continents may also be carried to the ocean by winds, but rivers are probably the main source of the material's getting into the ocean. Some of the deposits resulting from erosion on land enter the ocean in the dissolved matter. There it can be precipitated as sediments or used in the biological processes whereby animals and plants form shells that eventually accumulate on the sea floor. Another source involves volcanic activity either on land or in the ocean. Finally, a small amount of material can end up in the ocean from extraterrestrial sources, such as parts of meteorites that enter the atmosphere.

Sediments can vary in the horizontal sense and also with depth. A study of their changes with depth by looking at the layers in sequences below the bottom is called stratigraphy and is of value in interpreting changes on an oceanwide basis. Various techniques can be used to correlate sediments, including radioactive dating and use of the facts that magnetic reversals have occurred during the history of the earth and that these reversals have been recorded in the sediments and rocks on the sea floor. A combination of many of these techniques can help in understanding some of the past conditions over the recent history of the earth.

Suggested Further Readings

BULLARD, EDWARD, "The Origin of the Oceans," *Scientific American,* **221,** no. 3 (1969), pp. 66–75.

Continents Adrift: Readings from Scientific American. Introduction, J. Tuzo Wilson, San Francisco: W. H. Freeman & Company, Publishers, 1972.

DIETZ, R. S., "Geosynclines, Mountains, and Continent-Building," *Scientific American,* **22,** no. 3 (1972), pp. 30–45.

EMERY, K. O., "The Continental Shelves," *Scientific American,* **221,** no. 3 (1969), pp. 107–22.

ERICSON, D. B., and G. WOLLIN, *The Ever-Changing Sea.* New York: Knopf, 1967.

GASS, I. G., *et al., Understanding the Earth: A Reader in the Earth Sciences,* Cambridge, Mass.: The M.I.T. Press, 1971.

HEEZEN, B. C. and I. D. MACGREGOR, "The Evolution of the Pacific," *Scientific American,* **229,** no. 5 (1973), pp. 102–15.

JORDAN, T. H., "The Deep Structure of the Continents," *Scientific American,* **240,** no. 1 (1979), pp. 92–107.

MCKENZIE, D. P., and J. G. SCLATER, "The Evolution of the Indian Ocean," *Scientific American,* **228,** no. 5 (1973), pp. 62–74.

MENARD, H. W., "The Deep-Ocean Floor," *Scientific American,* **221,** no. 3 (1969), pp. 127–42.

REVELLE, R., "The Ocean," *Scientific American,* **221,** no. 3 (1969), pp. 127–42.

RONA, P. A., "Plate Tectonics and Mineral Resources," *Scientific American,* **229,** no. 1 (1973), pp. 2–11.

SHEPARD, F. P., *Submarine Geology* (3rd ed.), New York: Harper & Row, Pub., 1973.

SHEPARD, F. P., *The Earth Beneath the Sea.* Baltimore: Johns Hopkins Press, 1959.

TAPSCOTT, C., "The History of the Atlantic," *Scientific American,* **240,** no. 6 (1979), pp. 156–75.

WETHERILL, G. W., and C. L. DRAKE, "The Earth and Planetary Sciences," *Science,* **209** (1978), pp. 96–104.

6

*Sea-Floor Spreading
and Plate Tectonics*

AMONG THE MORE EXCITING scientific discoveries of recent years has been the concept of sea-floor spreading or, as it is more often called, plate tectonics. The concept is simple to understand and seems to explain most of the major geological and geophysical features of the earth's crust and, equally important, how they have evolved over geologic time. Important to this discussion are the four basic geophysical measurements: seismics, magnetics, gravity, and heat flow described on pages 93–99. These techniques have given marine geologists and geophysicists a very good, but indirect, view of the structure of the earth (Figures 5-15, 5-16 and Table 5-1).

Continental Drift—Early Aspects

Keeping the geophysical techniques and the data obtained in mind, we can proceed to the development of the concept of sea-floor spreading and plate tectonics. The history and evolution of the idea is very fascinating (see, for example, the article by Frankel, 1979). Some early geologists, noting the long linear trends to many mountain ranges and the thick folded sequence of sediments that they contained, suggested that the earth was contracting. In the process of contracting its outer skin, the earth's crust became wrinkled—that is, compressed and in this manner folded some sediments into mountain ranges. Other geologists noted the presence of rifts, both on the ocean floor (especially along ridges like the Mid-Atlantic Ridge) and on land (east Africa and the Middle East, for example) and argued that only an expanding earth, causing tension, could form these features. More important than who was right or wrong was, How could *both* types of features be present?

An especially important point was that scientists for over a century had noted the similar shapes to the edges of the continents on either side of the Atlantic, especially western Africa and eastern South America (Figures 5-32 and 6-2), and suggested that they were once connected and had split apart some time in the past, in the process forming the Atlantic, Indian, and Southern Oceans and reducing the size of the Pacific. This idea, generally called continental drift, was probably first advanced by von Humboldt in the early 1800s, and later modifications were proposed by Suess in 1888, Taylor in 1910, and Wegener in 1912. These hypotheses argue that the continents essentially float on the deeper and heavier subcrustal or mantle material, and that they are moved around by deep-seated convection currents operating within the earth's mantle (Figure 6-1). The convection currents are driven by the uneven heating of the earth's mantle, with most of the heat coming from decay of radioactive elements. The mantle is hotter near the core, whereas it tends to be cooler near the thin crust; thus circulation cells develop, as in a pan of water heated at the bottom. These slow movements may drag and move parts of the lighter crust (continents) and concentrate them over the downward-moving part of the convection cells. Oceanic ridges, which appear to be tensional (pulled-apart) features, form at the upward-moving part of the convection cell. The relatively high heat flow observed on ocean ridges supports a convection cell aspect.

Although the idea that continents could move around on the earth's surface had been proposed over a century ago, it took a very long time to become popular. Part of the reason is that there were few places then recognized where evidence of such a major movement was evident. It was only recently that such evidence became obvious, in large part because of the dramatic change in the way earth scientists and oceanographers viewed the continents and the oceans.

In the following years some geologic data favored continental drift and some did not. Some scientists argued that the forces needed to move continents around were immense and that in many places the evidence for drifting was poor or nonexistent. For example, if continents were moving around like a ship, new sea floor without any sediment cover should be exposed in its wake. Another objection focused on the completely different structure underneath the oceans and conti-

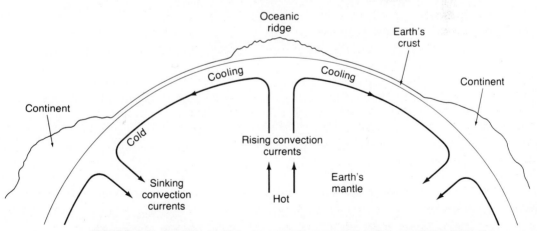

FIGURE 6-1 Possible model of convection current cells operating within the earth's mantle. Note that the continents are over the downward, or sinking, part of the convection cell and that the oceanic ridges are over the upward-moving part of the cell.

nents (Figure 5-16), which suggested some degree of permanency to both areas. Actually, this is a critical question: Why does one portion (the continents) of the earth have a thick crust, while such a thick crust is absent from the remainder (the ocean basins) of the earth? One answer proposed was that the earth is expanding and that at one time the whole earth was covered with this thick continental crust. The crust then split and expanded along zones of weakness that are now occupied by the ocean basins. The problem with this hypothesis is that the ocean basins should then contain much more sediment than they do.

Another objection to continental drift was that it involved large-scale movements (involving thousands of kilometers) of parts of the oceans that seemed unreasonable since no evidence of a similar type of massive displacement was observed on land. This objection was later partially overcome by studies off the

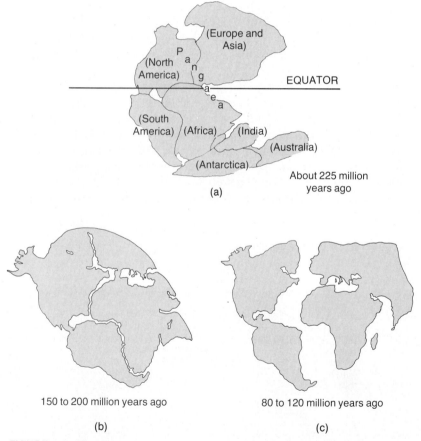

FIGURE 6-2 The changing positions of the continents according to the concept of sea-floor spreading. (a) About 225 million years ago (adapted from Dietz and Holden, 1970). (b) Period from 150 to 200 million years ago. (c) Period from 80 to 120 million years ago.

California coast and along the San Andreas Fault, which showed that such large-scale displacements are indeed possible.

Proponents of the continental drift hypothesis often differed with each other on details, but they usually felt that the continents were joined in one supercontinent (called *Pangaea*—named after the Greek earth goddess). This supercontinent started to split apart about 225 million years ago. When reconstruction (Figure 6-2) of the position of the continents at that time is made, many large-scale features on the earth's surface seem to fit together and extend across several continents. There are also many data to suggest that the magnetic poles have migrated (actually moved—not the same as reversing) in the geologic past. Such migration would help explain the fact that glaciers in the past covered what are now tropical areas. In other words, part of the original supercontinent may once have been situated at one of the poles. Some opposing scientists note, however, that almost any arrangement of the continents still leaves a considerable portion of apparently past-glaciated land near what was then the equator.

The main difficulty with the continental drift hypothesis is this question: How, if the continents are light masses floating on a deeper and denser layer, can a continental area become an oceanic area and vice versa? The central portions of most continents are relatively old (rocks as old as 3.5 billion years have been found), being composed of a conglomeration of sediments and rocks that have been folded, deformed, melted, and remelted. These rocks pose a very complex mixture of types and processes that makes understanding them very difficult. The oceans, on the other hand, have relatively young rocks and sediments (rarely older than 200 million years) that have had comparatively little deformation. Erosion and depositional processes, which are often very dramatic and variable on land, are usually relatively constant and continuous in the ocean. For these reasons, then, it is not surprising that it was the ocean that gave us the important evidence concerning the origin and evolution of major features on the earth.

Sea-Floor Spreading—Plate Tectonics

A new hypothesis, called *sea-floor spreading*, was first proposed by Harry Hess and Robert Dietz in separate articles published in the early 1960s. The hypothesis incorporated several older ideas and was able to overcome most of the previous objections to continental drift. The idea has considerable supporting data and can be tested (something that was difficult to do with other hypotheses). A basic part of the concept is that the sea floor *itself* is moving (the moving force being convection currents). The hypothesis that the sea floor is moving is what distinguishes this idea from others and gives it its name—sea-floor spreading. The plate tectonics part comes from a later refinement, which shows that the surface of the earth is actually composed of a series of rigid plates that are all moving relative to each other. In the sea-floor spreading concept, movement starts at the mid-ocean ridges and continues out away from the ridge. The resulting space at the

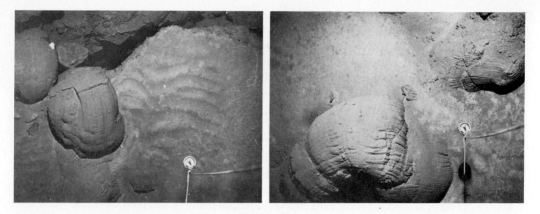

FIGURE 6-3 Two photographs from the bottom of the Red Sea. Note the rifting in the figure to the left and the material that looks as if it was squeezed out of a tube of toothpaste—which actually is volcanic rock. Compass in both figures is about 10 cm (about 4 in.) in diameter.

ridge is filled with new sea floor formed by volcanic material (containing magnetic minerals) coming from deep within the earth (Figures 5-34 and 6-3). The process is similar to the movement of a conveyor belt away from the mid-ocean ridge toward the continent and then eventually either moving under it or carrying the continent along with it (Figure 6-4). The oceanic ridge is therefore an area of crustal upwelling, or divergence, where new volcanic material is being added from the deeper parts of the earth's crust and mantle. The high heat flow and seismicity of the oceanic ridges are consequences of this process. The outer edges of some continents, especially where there are trenches, are areas of downwelling or subduction.

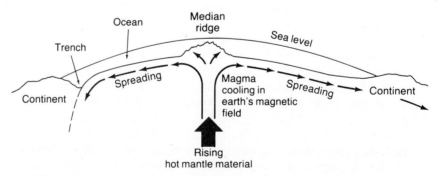

FIGURE 6-4 A diagram showing the basic parts of the sea-floor spreading hypothesis. As the magnetic minerals cool, they are oriented along the earth's magnetic field. The resulting magnetic patterns (due to reversals of the magnetic field) can be used to estimate the rate of spreading (Figures 6-5 and 6-6). On the right-hand side of the figure the continent is being carried along with the spreading, whereas on the left-hand side the oceanic material is being thrust (subducted) under the continent, forming a trench.

The new hypothesis of sea-floor spreading was especially supported by broad-scale magnetic observations made over the sea floor. These observations showed that long, linear magnetic patterns tended to parallel the mid-ocean ridges. These patterns, when followed perpendicularly out from the ridge, showed reversals in polarity on both sides of the ridge (Figure 6-5) and a mirror image pattern. The patterns were interpreted as showing that the underlying magnetic rock was either solidified in the earth's present magnetic field, in which case it would have a positive value, or solidified at a time when the field was reversed (north and south poles interchanged), in which case the rock would have a negative value. In a way, the formation of these patterns is similar to a tape recording: the changes in the earth's magnetic field are the signal being recorded on a tape coming out from the spreading area.

Thus according to the sea-floor spreading concept, slow convection currents from within the earth's mantle are gradually bringing volcanic (and magnetic) molten rock or lava to the ridge areas. As this material cools, it starts to solidify, at which time the magnetic minerals align themselves with the earth's field. Material cooling and solidifying at the present time will be normally oriented in the present magnetic field and will have a positive pattern. Rocks solidified at times in the geologic past when the poles were reversed would have a negative or reversed pattern (Figure 6-6). Since the rising basaltic material is slowly being carried away from the ridge in a conveyor-belt fashion, measurements of the magnetic pattern show an alternating pattern of normal and reversed values. The patterns observed also generally show a high degree of symmetry around the ridge area; that is, the pattern on one side is the reverse image of that on the other side (Figure 6-5). This is to be expected since the material is moving out and away from either side of the ridge (Figure 6-4).

These magnetic patterns around the mid-ocean ridges can be correlated with magnetic reversals observed in rocks on land that have been dated by radioactive techniques (Figure 6-6). If a correlation is possible, an age can be ascribed to the various parts of the sea-floor pattern and thus an estimate can be made of the rate of spreading. This procedure has been done in many areas, and a spreading rate of about 1 to 5 cm (about 0.4 to 2 in.) per year (actually an average rate since motion is probably not continuous) is typical. A spreading rate of 1 cm (about 0.4 in.) per year is equivalent to the formation of 10 km (about 6 mi) of new sea floor in a million years. Such a rate of motion is compatible with the idea that the continents on either side of the Atlantic were joined about 200 million years ago and since that time the continents have been slowly moving apart, forming the Atlantic Ocean in the process (Figure 6-2).

The rising volcanic rock at the oceanic ridge is hot and causes an upward expansion of the ridge area. However, as the rocks move away from the ridge axis they gradually cool and contract, causing the sea floor to sink. This phenomenon explains why the ridge areas are relatively high compared to the adjacent regions.

There are numerous places along the oceanic ridge where its trend is offset

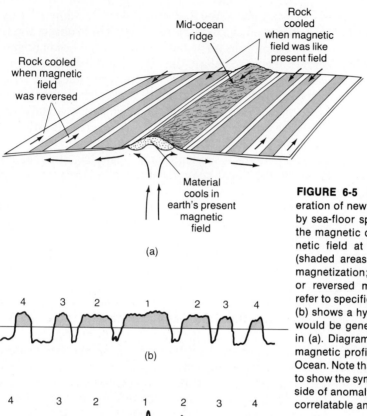

Rock cooled
when magnetic
field
was reversed

Mid-ocean
ridge

Rock
cooled
when magnetic
field was like
present field

Material
cools in
earth's present
magnetic
field

(a)

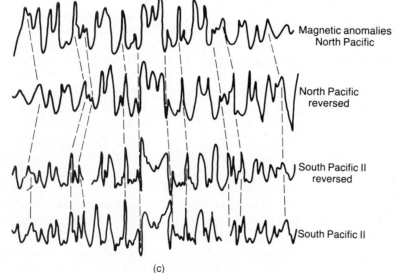

4 3 2 1 2 3 4

(b)

4 3 2 1 2 3 4

Magnetic anomalies
North Pacific

North Pacific
reversed

South Pacific II
reversed

South Pacific II

(c)

FIGURE 6-5 Diagram (a) shows the generation of new sea floor at the oceanic ridge by sea-floor spreading and how it assumes the magnetic character of the present magnetic field at the time the rock solidifies (shaded areas indicate present or positive magnetization; clear areas indicate negative or reversed magnetization). The numbers refer to specific positive anomalies. Diagram (b) shows a hypothetical magnetic field that would be generated by the scenario shown in (a). Diagram (c) shows a series of actual magnetic profiles obtained from the Pacific Ocean. Note that each profile is also reversed to show the symmetry of the pattern on either side of anomaly 1. The dashed lines indicate correlatable anomalies.

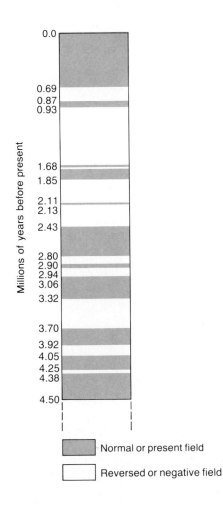

Millions of years before present

0.0
0.69
0.87
0.93
1.68
1.85
2.11
2.13
2.43
2.80
2.90
2.94
3.06
3.32
3.70
3.92
4.05
4.25
4.38
4.50

█ Normal or present field

☐ Reversed or negative field

FIGURE 6-6 The geomagnetic time scale. This scale is determined by measuring the age of reversals and positive anomaly patterns on land, which can then be extrapolated to the patterns observed from ocean rocks (Figure 6-5). The shaded areas indicate times when the poles were in their normal position; the clear areas, when the poles were reversed.

either to the right or left—a result of variations in the spreading rate along the ridge. The offset creates a fracture zone with a fault (called a *transform fault*) situated between the active areas of the ridge (Figure 6-7). Fracture zones, forming long escarpments, can extend over 1,000 km (over 621 mi) and reach the continents (Figures 5-32, 5-35, and 5-37).

The character of the edge of a continent is often influenced by its position relative to areas of spreading and subduction. If the continental edge sits at the boundary of a subducting oceanic plate, a trench and sometimes an island arc or coastal mountain range will form (Figure 6-8). Examples are the trench–island–arc systems along the west coast of the Pacific and the Peru–Chile Trench along the west coast of South America. The subducting plate forms the trench and in the process has much of its sediment scraped off onto the landward side of the trench. This material, and the subducting plate, may be reheated, perhaps even become molten, and can eventually become part of the continental plate forming a coastal mountain range.

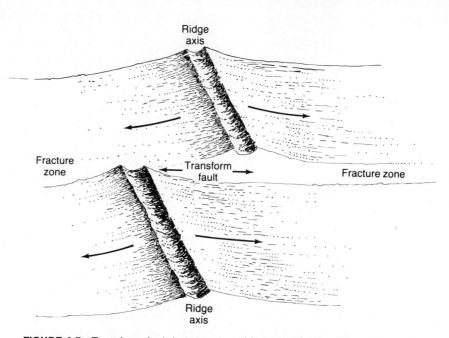

FIGURE 6-7 Transform fault between two ridge axes due to different spreading rates along the different axes. Fracture zones extend beyond the active parts of the fault.

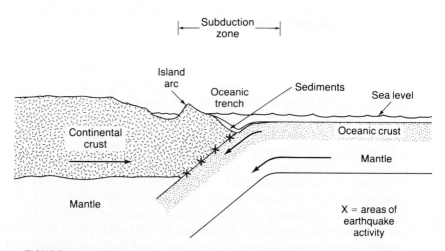

FIGURE 6-8 Collision of oceanic and continental plate resulting in a subduction zone. The denser oceanic crust is depressed under the lighter continental area. Increased earthquake activity results from the contact between the plates (Figure 6-10).

An intellectual advantage of the hypothesis of sea-floor spreading is that marine scientists are no longer faced with the embarrassing problem of making a continental area become an oceanic one; instead, new sea floor is constantly being produced at the mid-ocean ridges and old sea floor is disappearing at areas of subduction (Figure 6-4). Continents can grow at their edges by the accumulation of material (as with the formation of coastal mountain ranges), while the ocean basins can also increase by spreading along the oceanic ridges.

As the "conveyor belt" carries much of the missing sediment down below the continents or accretes it onto the continents, the sediment remaining in the ocean is just what has been deposited within the last few hundred million years. While the process of sea-floor spreading is occurring, there is a slow, continuous rain of shells of microscopic organisms and clay particles from the surface waters to the depths. These materials accumulate at the rate of a few centimeters per 1,000 years. Where the sea floor is new, the amount of accumulated sediments will be relatively thin, and thicker toward older areas (they have been around longer to receive sediments). In this manner the sediments will thicken in a direction away from the spreading region or the oceanic ridge. If the sediments that first accumulated over the new sea floor when it was initially formed can be dated (by isotopic techniques or by knowing the ages of the fossil shells), then a good approximation can be made for the age of that sea floor or, in other words, the time of its formation. This age should be similar to that indicated by the magnetic method of dating (Figures 6-5 and 6-6). Since the youngest part of the ocean basin is the ridge area, this is also the area where one would expect the thinnest amount of sediments. Studies of sediment thickness do indeed show that relatively small quantities of sediment are found in the ridge area and that this thickness increases as one proceeds away from the ridge.

The basic mechanism whereby "old" sea floor is consumed or subducted seems well documented. When regions of sea floor collide with continents, the lighter continents (Table 5-1) override the heavier ocean floor (Figure 6-8). This process results in the formation of a trench. The sediments of the sea floor may be "scraped" off and accreted onto the continental plate. If the sea floor and continents are both moving together, as in the case of the eastern and western sides of the Atlantic Ocean, there is no subduction and thick sedimentary sequences (sediments mainly from land) probably develop at the base of the continental slope—that is, continental rises (see right side of Figure 6-4). If at a later time subduction occurs, the sediments that have formed the rise will be carried into the trench and possibly accreted onto the continent. Subduction, although not as spectacular a process as spreading, is nevertheless an extremely important part of the sea-floor spreading concept. If subduction did not occur the earth would have to be continually expanding to accommodate the new sea floor being formed along the ridge axis. However, it appears that the amount of new crust formed by spreading is essentially balanced by the material consumed by subduction or by another process that occurs when plates of similar composition (both oceanic or both continental) collide.

When plates of similar composition collide, the result can be quite different than subduction. In this case, since neither easily overrides the other, unusually intense folding and faulting of the sediments and rocks result, often leading to the formation of a major mountain range. The Himalayan Mountains are believed to have formed in this manner from a past collision of India with Asia. The folding of the sediments effectively is the same as shortening the crust since a given length of sediments will be folded into an area of smaller length.

A different type of motion, called *strike slip*, can occur when plates move parallel by each other, creating a shear or fault zone. In this instance neither plate is overriding nor being thrust under another. The three basic types of motion, spreading, subduction, and strike slip are shown in Figure 6-9.

More recent thoughts concerning sea-floor spreading and the origin and evolution of the ocean basins are incorporated into an even more encompassing hypothesis called plate tectonics. This concept covers the broader structural

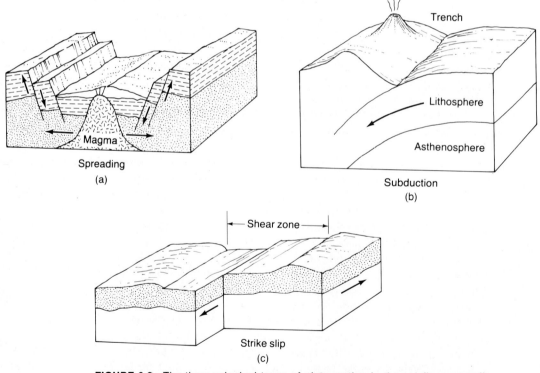

FIGURE 6-9 The three principal types of plate motion in the sea-floor spreading hypothesis: (a) spreading, or extension, (b) subduction, and (c) strike slip, or translation. Arrows indicate direction of relative motion.

features of the earth and is principally based on earthquake studies and sea-floor spreading. In plate tectonics the entire earth's crust is considered to be composed of a series of large plates, up to 185 km or 100 mi thick (called the *lithosphere*), that move essentially as rigid blocks on and over the earth's surface. The lithosphere sits upon the less rigid part of the earth called the *asthenosphere*. Boundaries of individual plates are usually seismically active areas.

An examination of the world's seismic activity clearly shows that most earthquakes are restricted to narrow but continuous zones that enclose large, relatively stable zones (Figure 6-10). In areas of spreading and strike-slip motion, the seismic activity tends to be relatively shallow and moderate. In subduction zones, deeper earthquakes of 100 km (62 mi) or so may also occur. This is consistent with the downplunging aspect of the plate (Figure 6-8). The seismic activity is due either to spreading and the addition of new crustal material along the mid-ocean ridges, the sliding motion along plates (strike slip), or because of subduction in trench regions. The plates are thus defined by the seismic activity (Figures 6-10 and 6-11).

The earthquake activity and the type of motion indicated usually confirm the relative motion of the plates as predicted by the sea-floor spreading concept. Compressional types of movement are indicated by earthquakes in areas where plates collide, whereas tensional motion is found where plates move apart. This relationship between seismic activity and plate boundaries is valuable in earthquake prediction, since most earthquakes occur along plate boundaries, whereas the interior portions of plates are among the most seismically quiet areas of the earth. These interior regions, especially in oceanic areas, have been suggested as possible sites for nuclear waste disposal (see pages 479–480). One plate boundary well known to those along the west coast of the United States is the San Andreas Fault, a zone of active faulting and earthquakes.

Some geophysicists have suggested that there are several relatively permanent areas of volcanic activity in the earth's crust. These areas, or "hot spots," have an almost continuously rising plume of magma that comes to the surface and results in volcanoes and eventually islands. If the overlying plate moves over one of these hot spots (which is essentially stationary with respect to the overlying plate), a series or chain of volcanoes or islands may form. In the Pacific Ocean there are several such submerged or sometimes elevated (islands) volcanoes, such as the Hawaiian Chain, the Emperor Seamount Chain, and the Line Islands, that seem to have been formed in this manner. An interesting aspect is that as these northward trending chains were being formed, the direction of movement of the Pacific Plate over the hot spot changed, resulting in a more northwestern trend to the island chains.

One basic result of the sea-floor spreading and plate tectonics hypotheses is that the ocean basins are not fixed in their position or size but rather are in the process of opening in one area and possibly closing in another. It appears that the Atlantic Ocean is still opening, whereas the Pacific Ocean is closing. The

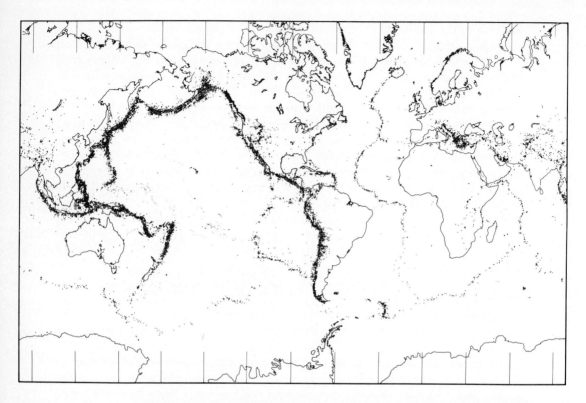

FIGURE 6-10 Worldwide distribution of earthquake epicenters. Each dot (there are about 30,000 of them) represents a point on the surface of the earth that is directly above an earthquake for the period 1961 to 1967 as compiled by the U.S. Coast and Geodetic Survey. (Adapted from Barazangi and Dorman, 1969.)

collision of ocean plates against continental plates, besides forming the trenches typical of the Pacific Ocean, may also play a major role in the formation of the mountain ranges along many of these coasts. These ranges represent ancient marine sediment accumulations that have been folded and thrust up above sea level. A similar situation would result if the Atlantic Ocean were to stop spreading and a subduction zone started forming along the continental rise of the east coast of the Atlantic. The sediments there could be thrust and folded up forming a long mountain range that would parallel the east coast of the United States, similar to the old Appalachian Mountains. Indeed, perhaps that is how the Appalachians formed—that is, during a prior period of opening and closing of the Atlantic Ocean.

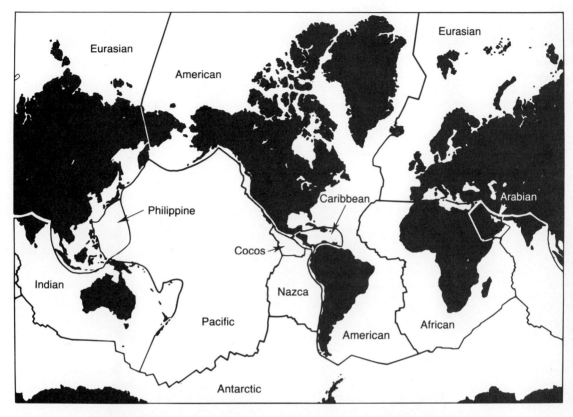

FIGURE 6-11 The major plates of the ocean. Numerous smaller divisions can be made and in some classifications as many as 26 plates are possible. Note how the boundaries tend to follow the seismically active areas (Figure 6-10).

A Tour Around One Plate

Perhaps the best way to visualize the concept of plate tectonics is to look at a single plate. A good one to consider is the Arabian Plate—a relatively small plate situated between the larger African and Eurasian plates (Figures 6-11 and 6-12). Sea-floor spreading is clearly shown in the Red Sea region, which is often used as a type example of what an ocean should look like in its early stages of evolution. The opening of the Red Sea has resulted from the movement of the Arabian Peninsula away from Africa (Figure 6-13). The Red Sea has a deep central rift (Figure 6-14) that is the result of the most recent movements. Bottom photographs

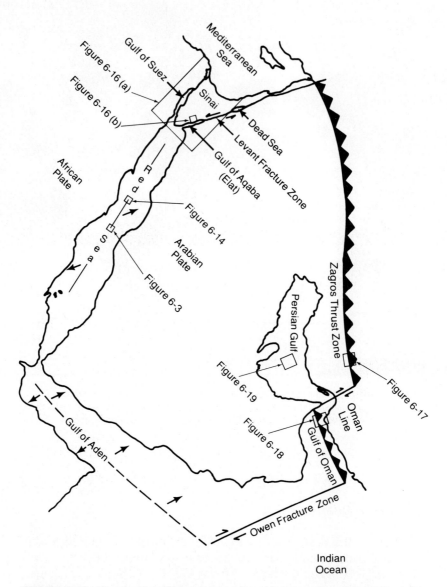

FIGURE 6-12 The Arabian Plate and location of some other figures from this chapter. Hatched areas in the Gulf of Oman and along the Zagros thrust zone are regions of subduction. Spreading is indicated by full arrows, as in the Gulf of Aden and the Red Sea. Strike-slip motion is shown by half arrow as along the Levant fracture zone and the Owen fracture zone.

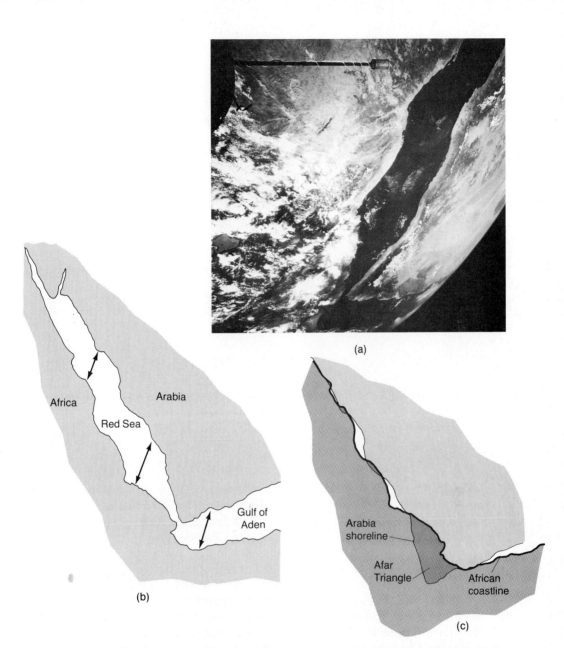

(a)

Africa

Red Sea

Arabia

Gulf of
Aden

(b)

Arabia
shoreline

Afar
Triangle

African
coastline

(c)

FIGURE 6-13 (a) *Gemini 11* satellite photograph of the Red Sea and the Gulf of Aden, taken from an altitude of about 620 km or 390 mi. The Red Sea and the Gulf of Aden have formed by sea-floor spreading in the directions indicated in the interpretation diagram (b). (Photograph courtesy of NASA.) (b) Line drawing of the present-day configuration of Red Sea and Gulf of Aden. Arrows indicate direction of present sea-floor spreading. (c) Fit of Arabian and African shorelines. The area of overlap, the Afar Triangle, has many characteristics of the ocean floor and may just represent an uplifted part of the ocean. If the interpretations are correct, the fit of the two shorelines is almost perfect.

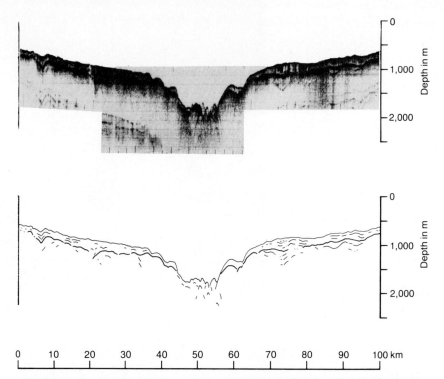

FIGURE 6-14 Seismic reflection profile (top) and interpretation (bottom) from the Red Sea. The central rift area is believed to have resulted from recent spreading that occurred over the last 3 million years.

taken in the rift show clear evidence of recent volcanic action (Figure 6-3) and in some regions there are discharges of hot water similar to that of the Galapagos region of the Pacific (described later). In the case of the Red Sea, these discharges are forming a valuable mineral deposit (see Chapter 12).

Further evidence of the spreading of the Red Sea is seen in the similarities in the geology of the land regions on either side of the sea and in the similar shape to the coastline (Figure 6-13). It appears that the Red Sea probably formed from a doming up of the entire region, perhaps similar to what is now happening in the African rift valley, followed by subsidence and rifting, then eventually sea-floor spreading (Figure 6-15). Spreading has also occurred in the Gulf of Aden at the southern end of the Red Sea. Apparently as the Arabian Plate is moving toward the northeast, the Red Sea and the Gulf of Aden are slowly opening at rates of 1 to 2 cm per year. The movement at the northern end of the Red Sea in the Gulf of Aqaba (or Elat) is of the strike-slip variety (Figure 6-9) and the region is marked by very rugged relief [Figure 6-16(b)]. This type of motion continues, forming the Dead Sea and extending into Syria. A similar type of motion is occurring along the Owen fracture zone. As the Arabian Plate is moving toward the northeast, its eastern margin is meeting a relatively immovable object—the Eurasian Plate.

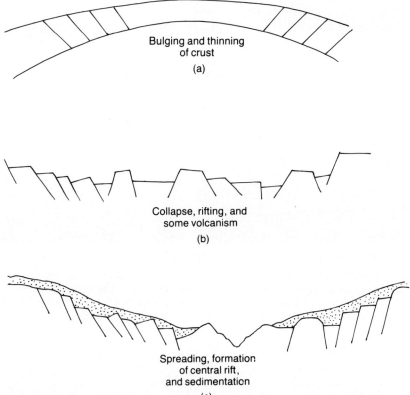

Bulging and thinning
of crust

(a)

Collapse, rifting, and
some volcanism

(b)

Spreading, formation
of central rift,
and sedimentation

(c)

FIGURE 6-15 Simple model for the origin of the Red Sea or, in general, a new ocean. Compare phase (c) with Figure 6-14.

The result is a continent-to-continent collision. The effect of this is best seen in the Zagros Mountains of Iran, which have been formed by this collision. In this case, since both parts are continents one is not thrust under the other but rather they are folded to form a mountain range (Figure 6-17). In the Gulf of Oman, however, the collision is between two oceanic parts, but the folding has not yet reached the mountain-forming stage (Figure 6-18). The Persian Gulf is far from the present subduction activity, and the sedimentary layers are relatively under-formed (Figure 6-19).

Some Implications of Plate Tectonics

Sea-floor spreading and plate tectonic ideas have been valuable in locating areas of hydrocarbon potential. For example, thick sedimentary sequences necessary for petroleum generation would rarely be found in subduction zones but are

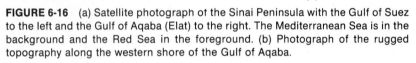

(a)   (b)

FIGURE 6-16 (a) Satellite photograph of the Sinai Peninsula with the Gulf of Suez to the left and the Gulf of Aqaba (Elat) to the right. The Mediterranean Sea is in the background and the Red Sea in the foreground. (b) Photograph of the rugged topography along the western shore of the Gulf of Aqaba.

FIGURE 6-17 An aerial view of the Zagros Mountains of Iran.

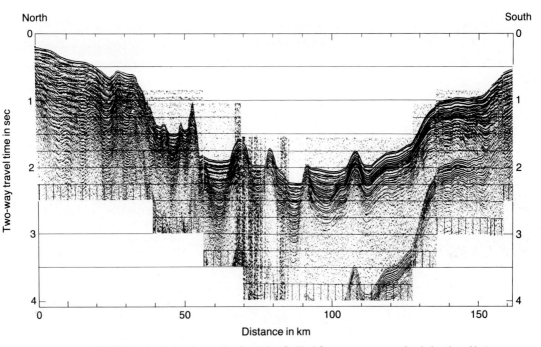

North ... South

Two-way travel time in sec

Distance in km

FIGURE 6-18 Seismic profile from the Gulf of Oman—an area of subduction. Note the large amount of faulting, especially when compared with essentially horizontal layers typical of the Persian Gulf (Figure 6-19). Travel time scale on the left is equivalent to depth in kilometers.

common in continental margins in marine areas where subduction does not occur. These points are discussed further in Chapter 12.

These concepts can also be useful in finding other types of mineral deposits. The similarities of rocks on either side of the South Atlantic was one of the initial reasons for believing that the continents were joined. This point can also be applied in searching for resources: For example, mineral or oil deposits found in coastal or nearshore regions on one side of the Atlantic, and formed before the spreading, may also be found in similar deposits on the other side. In other words, the spreading may have cut or separated the resources. In addition, the actual process of spreading or subduction can form mineral deposits. An example of the former is the Red Sea brine deposits (see Chapter 12). Some recent discoveries in the Galapagos region of the Pacific may also indicate a mineral deposit in an early stage of development.

Sea-floor spreading can even affect life styles. For example, California is undergoing a series of movements similar to those occurring along the Gulf of Aqaba. The lower portion of California is splitting or moving away from the rest of California. The motion is taking place along the San Andreas Fault and the Imperial Valley, in the process causing many of the California earthquakes, and

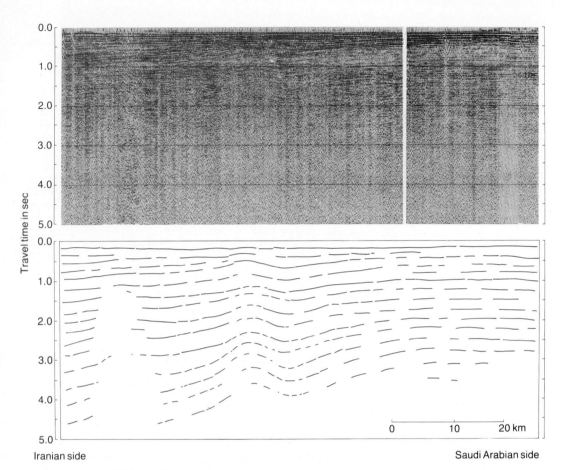

FIGURE 6-19 Seismic profile and interpretation (lower figure) across the Persian Gulf. Note the general lack of folding, and flatness to the layering. Travel time scale on left is equivalent to depth in kilometers.

leading to much concern about the possibility of a major destructive quake in the near future.

The implications of sea-floor spreading are far reaching and may have affected ocean circulation, climate, sea level, and the like. The presence of new ridges, the opening of new basins, or the establishment or closing of passageways could have changed or diverted the flow of seawater. The ocean ridges themselves tend to prevent east–west movement of bottom water. The past position of continents and their change must have affected global climatic patterns.

Changes in sea level have occurred in the past, but in many instances these were not related to changes in the volume of water (melting or freezing of glaciers) but rather to changes in the shape of the ocean basin itself. A reduction in the dimensions of the ocean basin from relative changes in the rates of subduction could result in a worldwide rise in sea level, while an increase in the size of the ocean basin from relative changes in the rates of spreading could cause a drop in sea level.

Unanswered Questions

There are still some points that can be raised against sea-floor spreading and plate tectonics. But again, one of the principal advantages of these concepts is that they can be tested, for example, by deep-ocean drilling. Early results from the Deep Sea Drilling Project in the Atlantic indicate that the lowest, or first formed, sediments immediately above the volcanic rocks become systematically younger as one nears the crest of the Mid-Atlantic Ridge. The systematic change in the age of the sediments indicates a spreading rate similar to that determined from the magnetic pattern. These data strongly support the sea-floor spreading idea. Perhaps when more sophisticated instruments are available, the concept can be directly tested by measuring the actual movement across an area of spreading like an oceanic ridge.

The exact mechanism for sea-floor spreading cannot be unequivocally determined but it seems to be clearly related to convective movements within the earth's crust and mantle. Hot material rises along the ridges, moves along away from the ridge, cools, and eventually sinks. Some feel that the subduction is actually the result of the sinking, with the leading edge of the oceanic dense plate pulling along the remainder of the plate.

Another question concerning the concepts is, Do the plates move in a continuous manner or do relatively large movements occur at irregular intervals? In places where sea-floor spreading processes can be directly observed, such as Iceland, which sits above the Mid-Atlantic Ridge, volcanic activity indeed comes in spurts. A typical pattern seems to be about 100 to 150 years of quiet and then a 5-to-20-year period of spreading and volcanic activity. The type of forces necessary to move or drive the plates is also still debatable as is the actual strength or rigidity of the plates. Numerous important aspects remain concerning mineralization and economic mineral potential associated with spreading and subduction. Little is presently known about past plate movements, if there were any, prior to the formation of Pangaea (about 225 million years ago). An especially interesting question is whether sea-floor spreading was an important process during the early periods of the earth's history. It appears that heating from radioactive decay could have been as much as three times higher in this early period. What effect this might have had—for example, perhaps causing more rapid spreading or thinner plates—is unknown. Some answers to these questions may come from future results of the Ocean Margin Drilling Program.

(a)

(b)

FIGURE 6-20 (a) Dr. Robert D. Ballard of the Woods Hole Oceanographic Institution examines a giant sea worm recovered from hot spring vents in the Galapagos rift. (Photograph courtesy of Mr. Jack Donnelly, Woods Hole Oceanographic Institution.) (b) These stalklike creatures found near a living vent are tubeworms. Marine biologists are familiar with tubeworms but usually in much smaller forms than these, which are 36 to 46 cm (14 to 18 in.) long. This photo was taken at about 2,800 m (about 9,000 ft) with a handheld camera from the submersible *Alvin* by Dr. John Edmond of the Massachusetts Institute of Technology. Limpets, crabs, seaworms, and an unknown variety of fish are also seen in the photograph. (Photograph courtesy of Woods Hole Oceanographic Institution.)

Recent Discoveries

Over the past few years marine scientists have been conducting extensive studies of ocean ridges to learn more about the many aspects of plate tectonics. Some of these studies have used submersibles such as *Alvin* and extensive photographic systems. The results in some instances have been startling.

A series of dives made in 1977 along the Galapagos rift between the Cocos and Nazca plates of the Pacific Ocean (Figure 6-11) resulted in the discovery of many new species of life. These included tubeworms up to almost 2 m (6.5 ft) in length (Figure 6-20) and clams over 30 cm (12 in.) in length (Figure 6-21). The unique creatures were found concentrated around vents along the oceanic ridge from which hot water was being discharged. These discharges are associated with cracks or rifts on the sea floor (Figure 6-22). The hot water (maximum temperature about 22°C, or 71°F) apparently has sufficient concentrations of food (probably bacteria) to support the life. Radioactive studies done at Yale University show that these clams may grow at a rate of 4 cm per year (1.5 in.), or about 500 times more rapidly than other similar deep-sea clams.

These discoveries were followed by even more amazing ones in 1979 from near the Gulf of California along the East Pacific Rise (at about 21° N). Similar, although somewhat larger, animals were found, but the highlight was the discovery of vents discharging, rather dramatically, black clouds of minerals. These vents, as much as 10 m (about 33 ft) high, were discharging water perhaps as hot as 350°C (about 662°F) and were depositing minerals enriched in copper, zinc, cobalt, lead, silver, and other metals. When these hot waters came in contact with surrounding seawater (temperature about 2°C (36°F) at a depth of about 2,800 m or 9,186 ft) the minerals precipitated, forming a chimneylike structure (Figure 7-18).

FIGURE 6-21 *Alvin*'s mechanical arm picks up a large clam specimen from area called Clambake 1 by the scientists. This clam is about 30 cm (12 in.) long. (Photograph courtesy of Dr. Robert D. Ballard, Woods Hole Oceanographic Institution.)

FIGURE 6-22 Tight clusters of clams, mussels, crabs, and other organisms around the living vents on the Galapagos rift contrast with sparse life found elsewhere in the area. These areas of abundant life were found to be about 50 m (164 ft) across. Some of the inhabitants appear to be related to known species, although the rift animals are, in many cases, much larger. Analysis of samples is still in preliminary stages, and some of the rift animals are so far unidentified. This photograph was taken at about 2,800 m (9,000 ft) by a towed camera called ANGUS for Acoustically Navigated Geophysical Underwater Survey system. Five vent communities were found on the Galapagos rift and investigated with the submersible *Alvin* operated by the Woods Hole Oceanographic Institution. Four were active and one vent no longer spouted warm water, as evidenced by dead clams and mussels. (Photograph courtesy of Woods Hole Oceanographic Institution.)

It is premature at this time to speculate if these deposits have economic implications, but it certainly is an interesting series of oceanographic discoveries. It should be emphasized that only about 10 km (about 6 mi) of the 60,000 km (over 37,000 mi) of oceanic ridge have been studied. Certainly many other areas of the oceanwide ocean ridge system will be intensely explored in the coming years.

Summary

The combined concept of sea-floor spreading and plate tectonics is one of the more exciting scientific ideas of this century. Its simplicity and confirming geological and geophysical data argue very strongly for its acceptance. Basically, the upper part of the earth's surface is composed of a series of plates that are in relative motion to each other. Along oceanic ridges the sea floor is slowly spreading and new crust, composed of volcanic material, is slowly being added to the ocean floor (that is, two plates are moving apart from each other). In other areas, plates are colliding; if one is a continental plate and the other oceanic, the lighter continental plate overrides the denser oceanic plate. The result in this case is usually a deep-sea trench or island–arc system. The process is called subduction. If plates of similar composition collide, a mountain range may result. A third type of motion, called strike slip, can occur when plates move parallel to each other.

Sea-floor spreading and plate tectonics can answer almost all questions about the origin and evolution of the ocean floor, but some points such as the type and size of the moving forces remain unanswered. Aspects of the concept can also be used to ascertain areas favorable for some marine resources. Recent discoveries in the Galapagos and East Pacific Rise areas show how dynamic and spectacular the sea-floor spreading processes can actually be, as well as underline the fact that much still remains to be learned about the ocean.

Suggested Further Readings

BALLARD, R. D., "Notes on a Major Oceanographic Find," *Oceanus*, **20,** no. 3 (1977), pp. 35–44.

BALLARD, R. D., and F. J. GRASSLE, "Return to Oases of the Deep," *National Geographic*, **156,** no. 5 (1979), pp. 686–703.

BIRD, J. N., ed., *Plate Tectonics: Selected Papers*. Washington: American Geophysical Union (1980), 992 pp.

BULLARD, EDWARD, "The Origin of the Oceans," *Scientific American*, **22,** no. 3 (1969), pp. 66–75.

BURKE, K. C., and J. T. WILSON, "Hot Spots on the Earth's Surface," *Scientific American*, **235,** no. 2 (1976), pp. 46–59.

Continents Adrift: Readings from Scientific American, Introduction, J. Tuzo Wilson. San Francisco: W. H. Freeman and Company Publishers, 1972.

Cox, A., *Plate Tectonics and Geomagnetic Reversals*. San Francisco: W. H. Freeman and Company Publishers, 1973.

DEWEY, J. F., "Plate Tectonics," *Scientific American*, **226,** no. 5 (1972), pp. 56–72.

DIETZ, R. S., "Geosynclines, Mountains and Continent-Building," *Scientific American*, **226,** no. 3 (1972), pp. 30–45.

FRANKEL, H., "Why Drift Theory Was Accepted with the Confirmation of Henry Hess' Concept of Sea-Floor Spreading," in *Two Hundred Years of Geology in America (Proceedings of the New Hampshire Bicentennial Conference on the History of Geology*, ed., Cecil J. Schneer, ed., Hanover, N.H., University Press of New England, 1979, pp. 337–53.

GASS, I. G., *et al., Understanding the Earth: A Reader in the Earth Sciences*. Cambridge, Mass.: The M.I.T. Press, 1971.

HEIRTZLER, J. R., and W. B. BRYAN, "The Floor of the Mid-Atlantic Rift, *Scientific American*, **233,** no. 2 (1975), pp. 78–91.

JORDAN, T. H., "The Deep Structures of the Continents," *Scientific American*, **240,** no. 1 (1979), pp. 92–107.

McKENZIE, D. P., and J. G. SCLATER, "The Evolution of the Indian Ocean," *Scientific American*, **228,** no. 5 (1973), pp. 62–74.

McKENZIE, D. P., and F. RICHTER, "Convection Currents in the Earth's Mantle," *Scientific American*, **235,** no. 5 (1976), pp. 72–89.

MENARD, H. W., "The Deep-Ocean Floor," *Scientific American*, **221,** no. 3 (1969), pp. 127–42.

MOTTL, M. J., "Submarine Hydrothermal Ore Deposits," *Oceanus*, **23,** no. 2 (1980), pp. 18–27.

RONA, P. A., "Plate Tectonics and Mineral Resources," *Scientific American*, **229,** no. 1 (1973), pp. 2–11.

VALENTINE, J. W., and E. M. MOORES, "Plate Tectonics and the History of Life in the Oceans," *Scientific American*, **230,** no. 5 (1974), pp. 62–77.

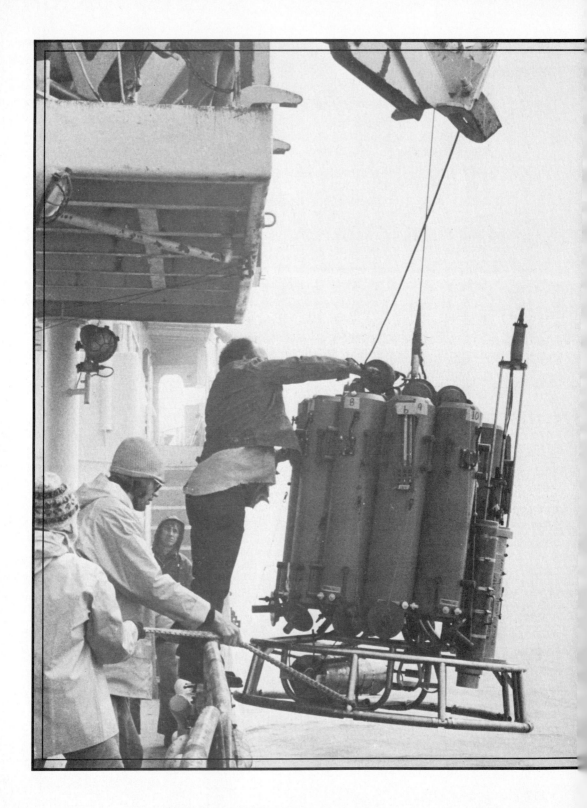

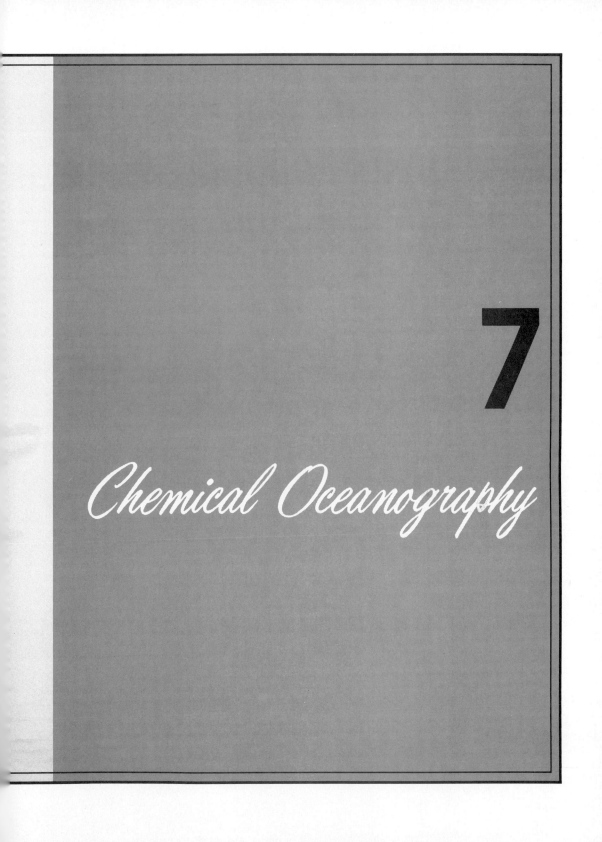

7

Chemical Oceanography

CHEMICAL OCEANOGRAPHERS ARE primarily concerned with the chemical and, in part, biological aspects of the ocean. These include analysis of the different elements and compounds in seawater and their distribution throughout the ocean and an evaluation of the processes that control the presence of these elements and compounds. The distribution of many elements and compounds is strongly influenced by life processes in the ocean and, in some instances, by people's activities. The effects of pollution in the ocean are deeply entwined with the chemical, geological, physical, and biological processes acting within and between the ocean, atmosphere, and ocean floor. A detailed discussion of marine pollution is deferred until Chapter 13. Some chemical oceanographers, often called geochemists, are more interested in the chemical processes that involve seawater and the underlying sediments (including manganese nodules) and rocks.

History of Chemical Oceanography

The sea and its saltiness have been puzzling since prehistoric times. Early people learned to obtain salt from seawater by solar evaporation of the water, a technique that is still used in many parts of the world. Many early Greek and Roman philosophers speculated, usually incorrectly, on the origin of the salt in the water.

The first scientific study of the chemistry of seawater is usually attributed to Robert Boyle, who published a book in 1670 describing his work. Near the end of the eighteenth century, Lavoisier discovered that water is a mixture of oxygen and hydrogen. He also devised techniques for analyzing some of the dissolved materials in seawater. Lavoisier and another chemist, Bergman, also made the first chemical determinations of seawater. Bergman collected and analyzed water obtained from a depth of about 100 m (328 ft). Both men evaporated seawater and then tried to extract different compounds from the residue. The results varied with the techniques and none were accurate.

By 1819, only chloride, sulfate, calcium, potassium, magnesium, and sodium had been detected in seawater (seawater contains over 80 elements and hundreds of different compounds). Boron, iodine, strontium, silver, lithium, arsenic, and fluorine were discovered during the following 50 years.

Although these early chemists were not making accurate estimates of the quantity of particular elements in seawater, they were developing a basic understanding of how eroded material from land was eventually transported into the ocean. For example, in 1822, a chemist named Marcet noted: "For the ocean having communication with every part of the earth through the rivers, all of which ultimately pour their waters into it; . . . I see no reason why the ocean should not be a general receptacle of all bodies which can be held in solution." This idea is not far from the truth since almost all elements are probably to be found in the ocean.

A major advance in the understanding of the chemistry of seawater was made in 1865 by Forchhammer, who noted that although there may be marked differences in total salt content among samples of seawater taken from any area of the ocean, the ratio of the major dissolved components is essentially constant. This concept is known as *Forchhammer's Principle*. Forchhammer also observed that silica and calcium were abundant in river water but nearly depleted in seawater; he concluded that this depletion in seawater was due to the action of marine organisms that were absorbing these elements into their shells. Thus he recognized the fact that biological activity plays an extremely important part in the chemistry of the oceans.

The constancy of composition over the ocean is perhaps not so surprising when one considers that the age of the ocean is probably several billion years. If it is mixed only once every 1,000 or 2,000 years it will have been mixed at least 1 million times during its entire history.

The Challenger Expedition (1872–1876) was a major advance in the chemical study of the ocean. Dittmar, an excellent chemist, analyzed many of the *Challenger*'s water samples and made determinations of dissolved gases (Figure 7-1). Dittmar's work confirmed many of Forchhammer's ideas. He also suggested that using the Forchhammer's Principle, the determination of the quantity of just a single major component in seawater could be used to determine the total salinity of seawater (approximately the weight in grams of dry salts in 1,000 g of seawater). Dittmar also noted a general decrease in oxygen content with water depth (about 1,500 m or about 4,921 ft was the deepest sample) and an increase in carbon dioxide in the surface waters as compared with deeper water. The importance of these differences, which are due to the processes of photosynthesis and respiration, was not realized at that time.

In the late nineteenth and early twentieth centuries, chemists studied the relationship of salinity, density, and chloride content and established in 1902 that salinity equals 1.8050 chlorinity[1] plus 0.03. In 1967 the relationship was redefined

---

[1] Chlorinity (Cl ‰) is defined as the total weight of chlorine, bromine, and iodine in grams in a kilogram (1,000 g) of seawater, assuming that the bromine and iodine are converted to their equivalent weight of chloride. This definition was changed, because of its dependence on the value of the atomic weight of silver, to the weight of silver (Ag) precipitated by 1 kg of seawater; that is, the chlorinity, Cl ‰ = 0.3285234 Ag.

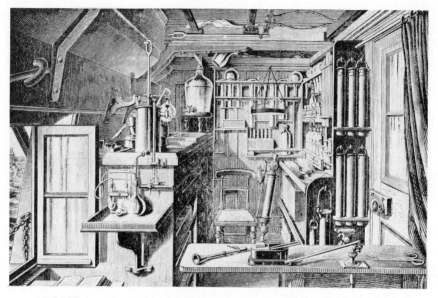

FIGURE 7-1 Chemical laboratory aboard the *Challenger* (1872–1876).

as salinity equals 1.8066 chlorinity. Average values (expressed in parts per thousand) are 35 ‰ for salinity and 19 ‰ for chlorinity. A standard seawater of known chlorinity is available from the Hydrographic Laboratory in Copenhagen, so methods and results of chlorinity determinations of different laboratories can be compared. Within recent years the determination of salinity by chemical precipitation has been replaced by new methods, such as electrical conductivity and devices that can measure salinity and other parameters directly as the instrument is moved through the water (Figure 7-2). Besides being faster, these new techniques are usually more accurate.

By the early twentieth century, chemical oceanographers began to associate the variations of oxygen in the upper parts of the ocean with the biological activity of the microscopic plants floating in the water. Other elements were also believed to be involved in the biological processes, especially nutrients such as nitrate, phosphate, silica, iron, and manganese. Many of these nutrients show distinct vertical and seasonal changes, which is suggestive of biological effects. The study of nutrients continues, but emphasis has shifted to studies of vitamins, trace elements (elements present in very small quantities), and organic compounds, especially pollutants, and their influence on biological growth and development.

As analytical techniques have improved, more information has been obtained about elements that occur in trace concentrations in the sea. Measurements, sometimes accurate to values as small as one part per billion, have shown that more than 80 different elements are present in seawater in measurable quantities. Many elements were actually first found in marine organisms before they were found in seawater. It is probable that as our measurement techniques improve

FIGURE 7-2 The lowering of an STD device into the ocean. This instrument electronically measures the salinity, temperature, and depth (by determining pressure) of the ocean as it is lowered or raised through the water. It then transmits the data back to the ship through the attached wires. (Photograph courtesy of Woods Hole Oceanographic Institution.)

still further, traces of every naturally occurring element, as well as some of the artificially produced radioactive isotopes, will be found in the sea.

Present Objectives

An especially important problem for chemical oceanographers is the increasing presence of pollutants in the ocean. These include lead from internal combustion engines, mercury from various sources, DDT from pesticides, radioactive materials, and petroleum products. Another important problem is acid rain, which has caused much ecological damage to lakes, especially downwind of heavily industrialized areas. Much scientific effort is being used to ascertain the amounts

(a)

(b)

FIGURE 7-3 (a) Lowering of a large-volume water sampler during the GEOSECS project. (Photograph courtesy of Phyllis Laking, Woods Hole Oceanographic Institution.) (b) Part of the "water library" where water samples collected during the GEOSECS program are stored and made available to members of the scientific community.

of these materials in the ocean, their methods of input, and the degree of harm they are causing to the environment (see Chapter 13). Because of this interest in pollution, many marine chemists are turning their attention toward coastal and river processes since these are the source of many pollutants in the ocean.

Another important problem is the effect of animals and plants in the ocean and their life processes on the composition of seawater. It is thus necessary to assay the ocean accurately for those elements, especially nutrients and heavy metals, that have biological implications. A better understanding of these processes could lead to increased biological productivity from the ocean. One such large-scale assay of the ocean was the so-called Geochemical Ocean Sections Study, or GEOSECS, program that was part of the IDOE effort of the 1970s. This successful project involved several countries, and had as a principal goal the study of various chemical properties of the ocean in an attempt to better understand large-scale oceanographic processes. At many stations in the Atlantic, Pacific, and Indian Oceans large volumes of water were collected and analyzed, and part of the sample was stored for further study (Figure 7-3). The data from these stations (Figure 7-4) give a good picture of the general chemical composition of the world's oceans.

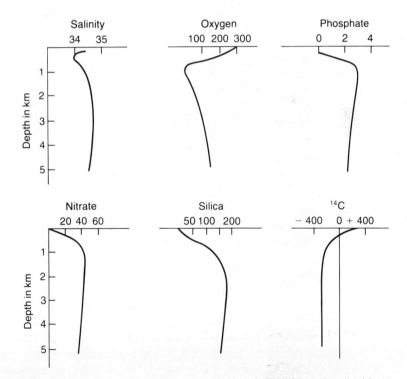

FIGURE 7-4   Part of the data obtained at a typical GEOSECS station (in this case one between San Diego and Hawaii). Units are salinity (‰), oxygen, phosphate, nitrate, and silica (μmol/kg), ^{14}C (Δ‰). (Adapted from Edmond, 1980.)

The GEOSECS program and others have shown the value of stable and unstable radioactive isotopes to determine the rates and importance of some of the chemical and biological reactions in the ocean. These techniques are especially useful in studying marine sediments and the mixing processes of seawater. A new project, Transient Tracers in the Ocean (TTO), has expanded on this idea by studying several man-made compounds (like tritium from atomic bomb blasts) that have been inadvertently introduced into the environment. Another exciting idea concerns some of the recent finds from the recent dives of *Alvin* in the Galapagos rift area (see pages 174–176). Preliminary observations suggest that the source for many of the metals in the ocean may be related to sea-floor spreading processes. Some estimates now suggest that in about 8 million years a volume of water equal to about all that present in the ocean will cycle through these vents and experience some, perhaps dramatic, chemical alteration. Just a few years ago most chemical oceanographers would have thought that nothing major could happen to the chemistry of the ocean in such a short period of time. The implications of this phenomenon are barely understood. Both this and TTO are discussed further in later sections.

Another area of research for marine chemists has been the study of the effects of the increased amounts of CO_2 released to the atmosphere and the ocean from the burning of fossil fuels. According to some the oceans may take up much of this increase, but perhaps not enough to prevent future changes in climate (this interesting and relatively new aspect of oceanography is the subject of Chapter 10).

Other important objectives of chemical oceanographers include those related to the mineral deposits of the sea floor, such as the vast manganese nodule deposits (see Chapter 12 for more details) that cover over 50 percent of the ocean (Figure 7-5). Many potentially important economic materials such as manganese, nickel, lead, cobalt, copper, and zinc are concentrated in some manganese nodule deposits. In other regions of the ocean some deposits are found that contain large quantities of phosphorite that in some instances could be mined and used as fertilizer.

Coastal regions are receiving increased attention from chemical oceanographers. This area is the site of many accidents that cause pollution, discharge from rivers, and localities for the dumping of sewage and wastes. Some of these aspects are discussed further in Chapters 11 and 13.

Chemical Oceanography Instruments

Since chemical oceanographers are mainly interested in the distribution of the various chemical components in seawater and the causes of this distribution, two of their more important needs are a device for sampling seawater and a technique for detecting the elements or compounds. The sampling device should have certain important characteristics:

FIGURE 7-5 Manganese nodules photographed on the ocean floor. The nodules are about the size of baseballs. (Photograph courtesy of Woods Hole Oceanographic Institution.)

1. It must obtain a sufficient volume of water for analysis (Figure 7-3).
2. It must be easily and accurately located according to depth or any other property.
3. It must not allow or introduce contamination.

The volume of water needed depends, of course, on the type of analysis. Salinity determinations use less than a liter (2.1 pints) of water. Carbon-14 analysis can require as much as 150 L (almost 40 gallons) of water, and enough of these samples could quickly fill up the storage area on a ship. Thus in many instances, chemical analyses are done aboard ship or a method of concentration is used.

To obtain samples from a specific water depth a device that can be opened or closed from the ship is required; an excellent instrument for this is the Nansen bottle (Figures 7-6 and 9-2). This old and trusted instrument comes with thermometers that record the water temperature at the time of the sample collection. The device uses two thermometers, one is protected against the effects of pressure, the other is not. When the device is tripped, the temperature values are locked; the temperature difference between the two thermometers can be used to calculate the depth from which the water was collected.

To obtain uncontaminated samples, chemical oceanographers may use stainless steel or Teflon-coated samplers. For some research, particular care has to be taken to avoid chemical alteration after collection, and refrigeration, preservation, or both may be necessary.

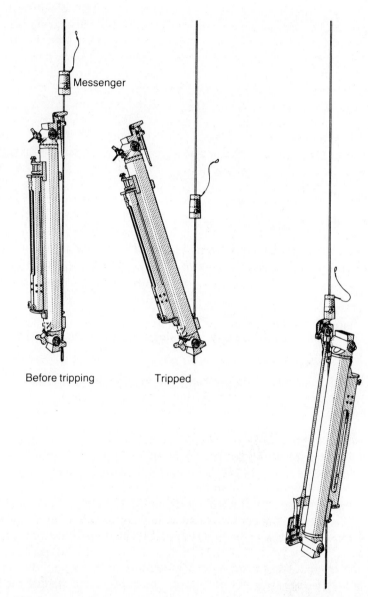

Messenger

Before tripping   Tripped

After tripping

FIGURE 7-6 Nansen bottle in three positions— before tripping, during tripping, and after tripping. When the messenger hits the bottle, it overturns, trapping the water inside and, if desired, can release another messenger to trip a lower bottle (see also Figure 9-2). (From U.S. Naval Hydrographic Office Publication No. 607.)

Once a sample is collected, there are various chemical techniques that may be applied for the determination of different elements or compounds. Among the more common shipboard analyses are those for salinity or chlorinity, for oxygen content, and for one or more nutrients. In the past most analyses were done by chemical titration, a slow and tedious operation at sea. Salinity is now determined by the faster and more accurate technique of measuring the electrical conductivity of the water. New instrumentation has been developed that can make *in situ*, or in place, measurements as the device is lowered through the ocean (Figure 7-2). An *in situ* measurement can avoid many sources of error, such as pressure effects on gas content. Some of the equipment designed for the GEOSECS program uses modern electronic techniques of measuring, recording, and transmitting of *in situ* characteristics of the ocean (Figure 7-7). As the equipment is lowered through the water, data can be transmitted through the cable to the ship and be observed on a computer console (Figure 7-8). When an interesting phenomenon is noticed on the console, a sample bottle can be "triggered" to collect water at that depth.

Eventually electronic systems may be developed to measure almost any element in the ocean. These instruments could be lowered or towed through the ocean with the data transmitted to the ship, producing an instantaneous chemical

FIGURE 7-7 Instrument system used during GEOSECS operation. A sample "rosette" contains 10 plastic sampling bottles of 30 liters (a liter, L, is equal to 0.26 gallons) capacity, a laser nephelometer that measures suspended particulate matter by light scattering, a probe to measure dissolved oxygen, a temperature–salinity measuring probe, and a bottom pinger. Sample bottles can be triggered from the surface by observing data presented on an electronic console (Figure 7-8). (Photograph courtesy of Woods Hole Oceanographic Institution.)

FIGURE 7-8 Scientist watching computer console during GEOSECS operation. It is possible with this system to observe data being measured by the underwater instruments. (Photograph courtesy of Woods Hole Oceanographic Institution.)

profile of the ocean. Long-term unattended monitoring systems could be installed on the sea floor, or suspended from buoys, to measure subtle changes in pollutants or other substances.

Many chemical oceanographers and geochemists are concerned with the sediments of the ocean floor. Their sediment sampling techniques are similar to those employed by geological oceanographers (see Chapter 5). Chemical oceanographers have also become interested in some of the processes of sea-floor spreading, especially those occurring along spreading centers such as the Galapagos. It appears (as discussed in a later section) that a significant percentage of some elements may be entering the ocean via such processes.

Properties of Water

One of the things that makes Earth unique among the planets is its vast quantities of liquid water, without which life as we know it would be impossible. There are over 1,349 million km^3 (326 million mi^3) of water on the surface of the earth's crust, an amount that would make a layer 144 km (or 90 mi) thick if distributed over the United States. Almost all, 97.2 percent to be exact, of this water is contained in the ocean (Table 7-1); if glaciers and ice caps were added to the ocean it would then contain 99.35 percent of the water. The exchange of water between the ocean (mainly surface waters), atmosphere, and land is called the *hydrologic cycle* (Figure 7-9). Some water is continuously moving from one environment to another. This movement occurs mainly by evaporation, precipitation, or river runoff. Evaporation is generally higher over the ocean than is precipitation, while the opposite is true for land. The evaporation loss of the ocean and precipitation gain by land is balanced by river inflow into the ocean (Table 7-2).

Pure water is an extremely simple compound, yet its properties and behavior are remarkably complex. At one time, water was believed to be a separate and indivisible element rather than a chemical compound consisting of two different elements. The chemical composition of water is simply H_2O: two parts hydrogen to one part oxygen. Water has certain unique properties; for example, its dissolving power and its surface tension exceed that of any other liquid. Water also has an exceptionally large capacity for absorbing heat; it will heat or cool more

TABLE 7-1   The World's Water Supply

Area	Water volume (mi^3)	Percentage of total
Surface water		
Freshwater lakes	30,000	0.009
Saline lakes and inland seas	25,000	0.008
Rivers and streams	300	0.0001
Total	55,300	0.017
Subsurface water		
Soil moisture	16,000	0.005
Groundwater	2,000,000	0.62
Total	2,016,000	0.625
Ice caps and glaciers	7,000,000	2.15
Atmosphere	3,100	0.001
Oceans	317,000,000	97.2
Total (approx.)	326,000,000	100

SOURCE: Data from Leopold and Davis, 1966.

191

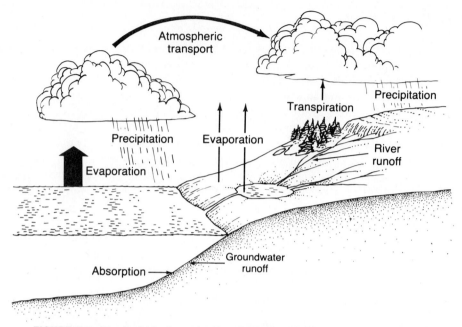

FIGURE 7-9 The hydrologic cycle. For simplicity, biological utilization by plants and animals is not shown.

TABLE 7-2 Evaporation, Precipitation, and River Inflow for the Ocean

Ocean	Evaporation (cm/yr)	Precipitation (cm/yr)	River inflow (cm/yr)
Atlantic	124	89	23
Pacific	132	133	7
Indian	132	117	9
World ocean average	126	114	12

SOURCE: Data from Budyko, 1974.

slowly than other liquids or the atmosphere and thus has a modifying effect on the earth's temperature. Pure water boils at 100°C (212°F) and freezes at 0°C (32°F).

A water molecule is very strong and considerable energy must be applied to break the bonds of the one oxygen atom with the two hydrogen atoms. The bonding between oxygen and hydrogen atoms is called a *covalent bond*. Hydrogen has one electron in its outer shell and space for one more; the outer shell of oxygen has six electrons with space for two more. Unfilled, these electron shells are not completely stable, but become very stable when filled, such as when a water molecule is formed (Figure 7-10).

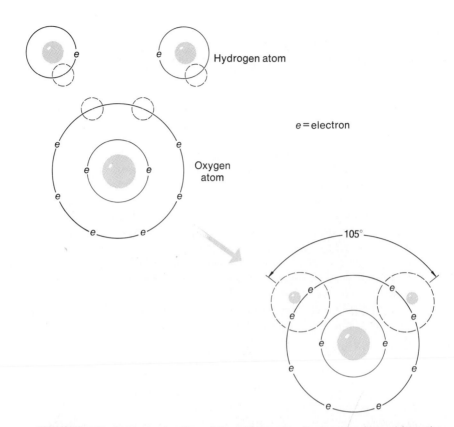

Hydrogen atom

e = electron

Oxygen atom

105°

FIGURE 7-10 An approximation of the water molecule structure formed from the joining of an oxygen atom and two hydrogen atoms.

When the oxygen and hydrogen atoms are united, they produce a somewhat lopsided molecule with the oxygen and hydrogen at an angle of 105° to each other (Figure 7-10). This causes an unequal distribution of electrical charges; the hydrogen side of the molecule is slightly positively charged, the oxygen side negatively charged. A molecule of this type is called a dipole, and it behaves somewhat like a magnet. In other words, its positive side will be attracted to (or will attract) particles having a negative charge and, likewise, its negative side will be attracted to positively charged particles. This attractive action is important when water molecules come in contact with the numerous compounds whose atoms are held together not by covalent bonds but by electrical charges. If a water molecule comes between two atoms held together by an electrical charge it will cancel some of the electrical attraction between the two atoms because water is a dipole. As some of this attraction is canceled, the two atoms will move farther apart from each other and more water molecules can come between the two atoms, until finally the initial attraction between the two atoms is eliminated and the atoms are separated. The original compound, some or all of whose atoms are now surrounded by water, has in this manner been partly or wholly dissolved by the water. This dipole aspect of water explains why it is an excellent solvent, capable of dissolving many compounds.

Water can occur naturally in three phases: solid (ice), liquid, and gas (water vapor). The dominant liquid phase of water results from its surprisingly high boiling point; other similar hydrogen compounds such as hydrogen sulfide have boiling points considerably lower (and thus are gases at that temperature). The high boiling point of water is also a result of the dipole structure of the water molecule. Because of their polarity, individual water molecules can cluster together in aggregates of two to eight molecules. The aggregated molecules are held together by hydrogen bonds; the strength and nature of these bonds determine some of the physical characteristics of water. For example, for water to boil, the bonds must be broken, which requires considerable energy and thus causes the high boiling point. If water consisted of single unclustered and unbound molecules, it would boil at $-80°C$ ($-112°F$) and would be a gas under normal conditions. When water is in the vapor phase, each water molecule is relatively unaffected by any other molecule.

In the ice phase, the water molecules, although tightly bound to each other, have a relatively large separation between each other. In the liquid phase, water molecules are less tightly bound to each other, but the molecules are closer together. Thus ice is bulkier and less dense than water and will therefore float on water. When seawater freezes, the ice contains relatively more freshwater (the salts are left behind) and floats. This is an important point because if ice sank, the ocean or lakes could freeze from the bottom up, rather than from the top down.

The density of freshwater is at a maximum at 4°C, or about 39°F (Figure 7-11). Above this temperature, density decreases with rising temperature; below 4°C, water expands (its density decreases). At 0°C (32°F) the volume of a block of ice is 10 percent greater than that of the same amount of water at 4°C. This phenomenon is important in the weathering and erosion of rock because water,

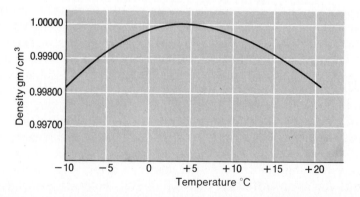

FIGURE 7-11 Density–temperature relationship of freshwater. Note that the maximum density occurs at 4°C (about 39°F).

by freezing and expanding in the cracks of rocks, will cause them to fracture. Another important effect can occur in freshwater lakes in cold climates. If the surface water of the lake is cooled to 4°C, it sinks and is replaced by lighter and warmer bottom water. If these bottom waters lack oxygen, the foul odor of hydrogen sulfide from the bottom might be carried to the surface and could kill most of the organisms living in the surface waters. This anomalous condition does not occur in seawater if the salinity is greater than 24.7 ‰.

Water has a large capacity for storing heat, and in doing so, helps prevent wide variations in temperature on the earth. To raise the temperature of 1 g of liquid water by 1°C requires a calorie (cal) of heat; to reduce it by 1°C, 1 cal of heat (a form of energy) must be removed. Considerably more energy must be used to change ice to water or water to gas; melting 1 g of ice requires 80 cal, and 540 cal are needed to evaporate 1 g of water that is already at 100°C (212°F). Much of this energy is used to break the bonds of the water structure. Going back in the opposite direction, that is, freezing water at 0°C, requires the removal of 80 cal per gram of water.

It is interesting to consider what the ocean would be like if some of its properties were different. For example, if its density were less, ships would have to be much larger to be as efficient as they are now. The high density of water is important for many large marine animals. Water will support the huge weight of animals like whales, whereas on land, animals have to develop special muscles and a strong skeleton to support their own weight. If the surface tension of water were less, it would stick to everything like syrup. Perhaps most important, if water was less transparent to light, then a smaller depth of the ocean could be used by plants for photosynthesis, a process that requires sunlight. Actually, probably most aspects of water are fine as they are, except that some areas of the land could use more freshwater.

Effect of Adding Salt to Water

Seawater, resulting from the addition of numerous elements and compounds to water, is a very complex mixture. It averages about 3.5 percent salt (often expressed in parts per thousand, that is, 35 ‰). Some of the properties of seawater are similar to those of freshwater, such as its capacity to absorb or give off heat. Other properties, especially those related to biological activity, are strongly influenced by the chemical composition of the water.

In several instances the addition of salts to water will reduce the normal properties of water. For example, before the addition of salt, water molecules can react easily with each other and go from the gaseous to the liquid state in a continuous process. Ions, such as sodium and chloride, when added to water will absorb water molecules and become hydrated (Figure 7-12). The bond between the ion and the water molecule has to be overcome for saltwater to freeze or boil. This results in an increase in the boiling point and a decrease in the freezing point

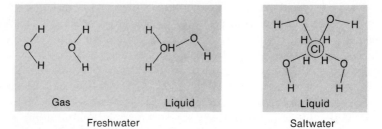

Freshwater   Saltwater

FIGURE 7-12 Diagrammatic illustration of the various states of water and the effect of adding an ion to the water.

of seawater, as compared with those of freshwater. When seawater boils or freezes, most of the salts remain behind in the liquid phase, increasing its salinity.

Some important changes in the properties of water that occur with the addition of salt are the following:

1. The specific heat [amount of heat necessary to raise the temperature of 1 g of water (at constant pressure) 1°C] decreases with increasing salinity. However, the specific heat also increases with increasing temperature in waters of normal salinity. In other words, as the temperature of the water increases, it becomes harder to remove the last few water molecules from a hydrated ion. Thus the boiling point of seawater is increased with increasing salinity.

2. The density increases approximately linearly with increasing salinity. Pure water has a maximum density at 4°C. The addition of salt lowers the temperature of maximum density, and at salinities greater than 20 ‰ the maximum density occurs at a temperature below the normal (0°C) freezing point (Figure 7-13).

3. The freezing point is lowered with the addition of salt (Figure 7-13). This characteristic combined with the temperature and salinity effect on density (the density of seawater increases as the temperature decreases) means that the highest density

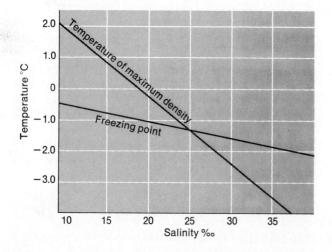

FIGURE 7-13 Relationship of temperature of maximum density and of freezing point to salinity. (After Sverdrup, Johnson, and Fleming, 1942.)

water in the ocean is the coldest, most saline water. Seawater of normal salinity can exist as a liquid when its temperature is as low as about $-2°C$. The lowest density water is that with high temperature and low salinity.

4. The vapor pressure (which is a measure of how easily water molecules escape from the liquid phase into the gaseous phase) is lowered with increasing salinity because the salts tend to make water molecules less available for evaporation. Freshwater evaporates more quickly than seawater.

5. The osmotic pressure of water is increased with increasing salinity. Osmotic pressure relates to the flow of solutions (not elements dissolved in solution) through semi-permeable membranes. The amount of flow increases with increasing salinity. This characteristic is very important to organisms since their cell membranes act as semipermeable membranes through which fluids can flow (Figure 8-7). The direction of flow depends on whether the osmotic pressure of the internal medium (the organism) is higher or lower than that of the external medium (the ocean). The flow is toward the more concentrated medium.

Some of the properties above are called *colligative properties* (properties that vary because of the amount of elements in solution and not because of the type of elements), namely, the boiling-point elevation, freezing-point depression, vapor-pressure lowering, and osmotic pressure. If the magnitude of one of the properties is known under a given set of conditions, the others may be calculated.

An important consequence of the relationships among salinity, temperature of maximum density, and freezing point is that in waters having a salinity greater than 24.7 ‰, the temperature of maximum density is lower than the freezing point. Thus as ocean water continues to cool, it continues to grow denser (Figure 7-13). Since the cooling is from the surface, the surface water becomes heavier than the underlying water and sinks. The underlying warmer and less dense water rises to replace the cooled water and, in turn, is cooled and sinks. In this manner a deep circulation is initiated, and freezing will not occur until the entire body of water is cooled to the freezing point. This generally will not occur in a large body of water. If, however, the salinity of the water is less than 24.7 ‰, the temperature of maximum density is reached before the freezing point is reached. As the surface water is cooled, it reaches its maximum density and then decreases in density. This means that the water remains near the surface and is cooled further; eventually the freezing point is reached and a layer of ice forms on the surface. Thus the relationships of salinity, temperature of maximum density, and freezing point prevent the ocean from freezing over.

Composition of Seawater

The normal salinity of the oceans, away from rivers or melting ice, usually ranges from 33 ‰ to 37 ‰. Six major elements (chlorine, sodium, magnesium, sulfur, calcium, and potassium) constitute more than 90 percent of the total elements in solution. These, and some of the so-called minor elements (strontium, bromine, and boron), have an essentially constant ratio to each other (Forchhammer's Principle) and are the "conservative" constituents of seawater. By knowing the amount of one of these elements in seawater, one can calculate the others. Many of the remaining constituents in seawater, including other elements, dissolved gases, organic compounds, and particulate matter, occur in varying proportions. These variations are due in large part to biological reactions. Actually, as we shall see in later sections of this chapter, most of the important chemical reactions in the ocean are due to biological processes.

Perhaps surprisingly, the composition of seawater is not very accurately known. One reason is that some areas of the ocean have not been adequately sampled; another is that many components are present in extremely small quantities. Furthermore, elements involved in biochemical reactions may vary in concentration by a factor of 1,000 or more.

The chemical composition of the ocean can be divided into four parts: (1) dissolved inorganic matter, (2) dissolved gases, (3) dissolved organic matter, and (4) particulate matter. A later section will be devoted to the biological effects on the chemical composition of the ocean.

DISSOLVED INORGANIC MATTER

Seawater by weight is about 96.5 percent pure water (Figure 7-14) and about 3.5 percent (35 ‰) dissolved inorganic components. The major inorganic elements [those present in quantities greater than 100 parts per million (ppm), or 100 mg per liter] are chlorine, sodium, magnesium, sulfur (usually expressed as sulfate), calcium, and potassium. Minor elements (more than 1 ppm but less than 100 ppm) are bromine, carbon, strontium, boron, silicon, and fluorine. Common trace elements (concentrations less than 1 ppm) are nitrogen, lithium, rubidium, phosphorus, iodine, iron, zinc, and molybdenum. At least 50 other elements (Table 7-3), and possibly all the known naturally occurring elements, are present in quantities of less than 10 parts per billion (ppb). At present about 84 elements have already been detected in the ocean. It should be emphasized that the concentrations of many of these elements can vary considerably with location, time, or season, and especially with biological activity.

Elements in seawater are virtually always present as component parts of a chemical compound. Some compounds, such as those containing sodium and potassium, are very stable; others, such as those containing silicon and manganese, are usually relatively unstable. The relative stability of chemical compounds is important in controlling the general composition of the ocean. Apparently some elements are being concentrated in the ocean while others are quickly passing through the ocean system and being precipitated on the sea floor. In other words, the residence time of different elements in the ocean can be extremely variable. (This interesting concept of residence time is further discussed on page 204.)

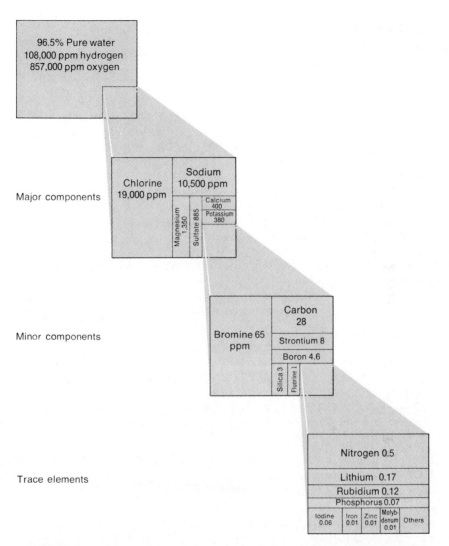

FIGURE 7-14 Dissolved inorganic compounds in seawater. (*Note:* All values in ppm.)

DISSOLVED GASES

The major gases in seawater are nitrogen, oxygen, and carbon dioxide; occurring in lesser quantities are helium and the inert gases neon, argon, krypton, and xenon. Gases present in the ocean generally enter it from the atmosphere; however, some very rare gases come from radioactive decay processes within the sediments on the ocean bottom.

The solubility of a gas, or its ability to go into solution, depends on three

TABLE 7-3   Trace Elements in Seawater (Note all values in parts per billion)

Element	Concentration (ppb)	Element	Concentration (ppb)
Carbon	200–3,000	Cobalt	0.2–0.7
Lithium	170	Mercury	0.15–0.27
Rubidium	120	Silver	0.145
Barium	10–63	Chromium	0.13–0.25
Molybdenum	4.0–12.0	Tungsten	0.12
Selenium	4.0–6.0	Cadmium	0.11
Arsenic	3.0	Manganese	0.1–8.0
Uranium	3.0	Neon	0.1
Vanadium	2.0	Xenon	0.1
Nickel	2.0	Germanium	0.07
Iron	1.7–150	Thorium	0.05
Zinc	1.5–10	Scandium	0.04
Aluminum	1.0–10	Bismuth	0.02
Lead	0.6–1.5	Titanium	0.02
Copper	0.5–3.5	Gold	0.015–0.4
Antimony	0.5	Niobium	0.01–0.02
Cesium	0.5	Gallium	0.007–0.03
Cerium	0.4	Helium	0.005
Krypton	0.3	Beryllium	0.0005
Yttrium	0.3	Protactinium	2×10^{-6}
Tin	0.3	Radium	1×10^{-7}
Lanthanum	0.3	Radon	0.6×10^{-12}

SOURCE: Data from Goldberg, 1963; Hood, 1963 and 1966.

factors: (1) temperature of the gas and the solution, (2) atmospheric partial pressure of the gas, and (3) salt content of the solution. The quantity of gases in seawater, with the notable exceptions of oxygen and carbon dioxide, is mainly determined by these factors. Gases, in general, are relatively unreactive once in the marine environment. If the quantity of a gas is higher or lower than indicated by these three factors, it would suggest that something in the marine environment is causing the variation. Concentrations of oxygen and carbon dioxide can vary independently of the above factors.

Oxygen concentration in seawater varies with depth. In surface waters, it is related to temperature: the higher the temperature, the lower the solubility of a gas. A few hundred meters below the surface, however, an oxygen-minimum or oxygen-poor zone is usually found (Figures 7-21 and 9-18).

Seawater has two sources of oxygen: the atmosphere and the plants that live in the ocean. Surface waters, because of their contact with the atmosphere, generally contain their expected amount of oxygen. In some instances a supersaturation, or extremely large amount, of oxygen is observed. This is usually due to photosynthesis, the process whereby plants use carbon dioxide (CO_2), water

(H_2O), nutrients, and solar energy and produce organic matter and oxygen (O_2). This photosynthetic reaction, because it is dependent on sunlight, takes place only in the upper layers of the ocean, usually above 200 m depth. It can result in an increase in the oxygen content in the surface waters. Below and in these upper layers organic matter and oxygen are utilized by organisms including bacteria. This process, called *respiration*, results in the consumption of oxygen and its removal as a gas from the water. The two reactions can be expressed as follows:

Photosynthesis (plants)—occurs in upper layers of ocean
$$CO_2 + H_2O + \text{nutrients} + \text{solar energy}$$
$$\longrightarrow \text{organic matter } (CH_2O) + O_2$$

Respiration (plants and animals)—occurs throughout ocean
$$\text{Organic matter } (CH_2O) + O_2 \longrightarrow CO_2 + H_2O$$

The oxygen-minimum zone results from the respiration of animals and plants and from the bacterial oxidation of organic debris. The presence or absence of the oxygen-minimum zone depends on whether the depletion of oxygen by respiration exceeds the renewal of oxygen by mixing of surface and deeper waters. The increase of oxygen in depths below the oxygen-minimum zone is believed to be due to the influx of oxygen-rich waters from the polar regions into the deeper parts of the ocean (Figure 9-18).

If an area is isolated from a potential oxygen source such as deep-polar waters, then it is possible for most or all of the oxygen at depth to be used up. One such isolated area is the Black Sea. Waters devoid of oxygen are called *anaerobic*; however, organic material can be decomposed there by sulfate-reducing bacteria. The sulfide formed can combine with hydrogen to form hydrogen sulfide, which is odorous and lethal to many organisms. If oxygen-deficient deep waters come to the surface by some form of water movement, mass mortality of the animal life in the surface waters usually occurs.

Carbon dioxide is present in seawater at considerably higher concentrations than in the atmosphere. One reason for this is that water, volume for volume, can absorb a larger quantity of carbon dioxide than can air. A more important reason is that seawater is slightly alkaline and contains certain cations such as magnesium and calcium in excess of equivalent anions. This allows carbon dioxide, as shown in the equation below, to combine with seawater to form carbonates and bicarbonates.

$$CO_2 + H_2O \rightleftharpoons H_2CO_3$$
$$\uparrow \quad \downarrow$$
$$2H^+ + CO_3^{2-} \rightleftharpoons HCO_3^- + H^+$$

If carbon dioxide is removed from seawater, for example by growing plants during photosynthesis, the bicarbonates (HCO_3^- and H_2CO_3) and carbonates (CO_3^{2-}) will give off carbon dioxide. This mechanism provides a large reservoir of carbon dioxide for photosynthetic reactions. At night when carbon dioxide is being produced by respiration, it is again chemically combined and stored.

Carbon dioxide in seawater has a complex relationship with pH[2], temperature, and salinity. With pH constant, the total carbon dioxide content increases with rising salinity and decreasing temperature. However, pH depends in part on the amount of carbon dioxide present and also is influenced by water temperature and pressure. An understanding of the complex dynamic relationship of carbon dioxide in the air, sea, and sediments has been one of the most difficult problems in chemical oceanography.

Three radioactive isotopic gases are found in the ocean: tritium (^{3}H), carbon-14 (^{14}C), and argon-39 (^{39}A). They are formed in the atmosphere by cosmic rays; the gases ^{3}H and ^{14}C are also formed by the detonation of nuclear bombs. Once in the ocean, radioactive isotopes continue their decay process and can be used to date changes or marine processes (discussed in a later section).

DISSOLVED ORGANIC MATTER

Dissolved organic matter is present in seawater in moderately small and usually variable amounts [between 0 and 6 mg (0.006 g) per liter or 0 to 6 parts per million]. The sources of this material are excreta and dead organisms. Included as dissolved organic material are the nitrogen and phosphorus that are chemically combined in organic compounds and are eventually oxidized, in some instances by bacteria, to nitrate and phosphate.

Other dissolved organic compounds in seawater are organic carbon, carbohydrates, proteins, amino acids, organic acids, and vitamins. Aside from the nutrients, nitrogen, and phosphorus, very little is known about the vertical and horizontal distribution of dissolved organic material. It is thought that only about 10 percent of the organic compounds present in seawater have been identified. The chemistry and distribution of nutrients are discussed in a following section.

PARTICULATE MATTER

Particulate matter, excluding living organisms, in seawater includes organic detritus, some complexes of organic and nonorganic material, and fine-grained minerals. The complexes of organic and nonorganic material may account for local variations in concentration of some elements; the high iron abundance in near-shore waters may be due to the formation of ferric–organic complexes. Freshwater diatoms and minerals have been detected in surface ocean waters thousands of miles from their probable source and were probably transported there by winds

---

[2] The unit pH is the negative logarithm of the hydrogen ion activity (or essentially the hydrogen ion concentration). A pH of less than 7 indicates an acidic solution; higher than 7, a basic or alkaline solution.

FIGURE 7-15 (a) A series of sediment traps being prepared for lowering and (b) being lowered. (Photographs courtesy of Dr. Derek Spencer and N. Green, Woods Hole Oceanographic Institution.)

(Figure 5-39). Particulate matter in the ocean, thus, is highly variable in concentration. It may be seen to respond to local geography, biological production, atmospheric conditions, and other unknown conditions.

The movement of particulate matter through the ocean water to the sea floor is a new and important field of study in chemical oceanography. Its importance relates to the movement of pollutants and organic matter as well as other substances from their source to their final resting place (the sea floor), although some studies have shown that the sea floor is often considerably disturbed by bottom currents. The use of sediment traps, placed at or near the sea floor (Figure 7-15) and left for long periods of time, has provided some quantitative data on the movement of particulate matter. These traps have shown that fecal matter (usually in the form of pellets) is an important mechanism for transporting material from the surface of the ocean to its deeper waters.

SUMMARY

In summary, there are many factors that control or influence the chemical composition of the sea. These are

1. Exchange with the atmosphere.
2. Solubility of different compounds.
3. Reduction by anaerobic bacteria.
4. Precipitation and exchange with the ocean bottom.

5. Inflow of fresh water.
6. Freezing and melting of sea ice.
7. Chemical reactions that control or influence the concentrations of different elements.
8. Biological processes, including life processes and decomposition of organic matter.

Because of their importance, some specific factors are discussed further in the next section.

Factors that Influence the Ocean's Chemical Composition

CHEMICAL REACTIONS

A large portion of the elements in the ocean are carried there by rivers as erosion products from land. At present, rivers carry about 4 billion tons of dissolved materials to the ocean each year. Some of this material is recycled (Figure 7-9), that is, evaporated from the ocean into the atmosphere, precipitated out as rain on land, and carried back into the ocean by rivers. Once in the ocean, different elements have different abilities to react in the marine environment. These differences can be expressed by a simple model of input of elements to the ocean and their subsequent removal (Figure 7-16). The concept assumes that the amount of an element introduced into the ocean per unit time equals the amount deposited out as sediment per unit time on the sea floor—in other words, that a steady state condition exists. This assumption seems reasonable since the salinity of seawater seems to have been fairly constant for long periods of time in spite of the large volume of salts carried into the ocean by rivers. Another assumption is that the elements are uniformly and quickly mixed in the ocean. The concept does not state that the different elements are saturated in the ocean. Using these conditions, the residence time of an element in the ocean can be defined as the total amount of a specific element in the ocean divided by its rate of introduction (or rate of precipitation out as sediment). In other words, residence time is the time needed to replace the amount of a specific element in the ocean. Early estimates of

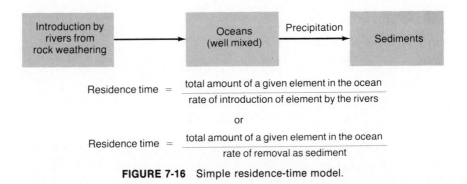

$$\text{Residence time} = \frac{\text{total amount of a given element in the ocean}}{\text{rate of introduction of element by the rivers}}$$

or

$$\text{Residence time} = \frac{\text{total amount of a given element in the ocean}}{\text{rate of removal as sediment}}$$

FIGURE 7-16 Simple residence-time model.

TABLE 7-4 Residence Time of Some
Elements in Seawater

Element	Amount in ocean (g)	Residence time (yr)
Sodium	147×10^{20}	260,000,000
Magnesium	18×10^{20}	12,000,000
Potassium	5.3×10^{20}	11,000,000
Calcium	5.6×10^{20}	1,000,000
Silicon	5.2×10^{18}	8,000
Manganese	1.4×10^{15}	700
Iron	1.4×10^{16}	140
Aluminum	1.4×10^{16}	100

residence time considered the amount of the elements introduced by rivers. Later scientists examined the same problem but emphasized the amount of the elements in the sediments (Table 7-4). Unquestionably, the model is an oversimplified picture of the ocean but it is significant that the results of two different investigations are similar; the only major difference is the result for calcium. What is more important to the understanding of the ocean's composition is the wide range in residence time between some elements; for example, aluminum and iron remain in seawater only hundreds of years while others, such as sodium and magnesium, remain for millions of years.

The significance of these calculations is that the elements with long residence times, such as sodium, are also those elements that are not very reactive in the marine environment and therefore remain in the ocean for long periods of time. The low residence times of silicon, manganese, iron, and aluminum are related to their biological activity; in addition, significant quantities of silicon, iron, and aluminum enter the ocean in the particulate phase and can quickly precipitate out to the sediments. These particulate phases are often minerals such as quartz, feldspar, or material from volcanic activity that settle to the ocean bottom. The high chemical reactivities of manganese, iron, and aluminum are also due to their ability to form mineral deposits on the sea floor, such as manganese minerals (Figures 7-5 and 7-17), glauconite, and zeolites.

Manganese nodules (sometimes called iron manganese or ferromanganese deposits) are one of the more interesting and potentially valuable deposits on the ocean floor (see Chapter 12). These can occur as round spheres from about 1 to 20 cm (about 0.4 to 7.8 in.) in diameter (Figure 7-17), as coatings on rocks and other objects, or as long slabs commonly called *manganese pavements*. Regardless of their form, manganese nodules are mainly composed of hydrated oxides of iron and manganese and often form around a nucleus like a small shell, a rock, or a shark's tooth and slowly grow outward in concentric rings. Ferromanganese deposits are commonly found throughout the deep-sea environment of all oceans as well as in some lakes and other shallow-water bodies. The economic interest in the nodules is principally because of their accessory trace elements—copper,

Unetched

Etched

0   2   4 cm

FIGURE 7-17 Cross section of a manganese nodule. Deposits of these nodules cover vast areas of the sea floor. The lower nodule has been etched with acid and shows its growth rings. The white material is calcium carbonate that is deposited within small cracks. (Photograph courtesy of F. T. Manheim.)

nickel, and cobalt—as well as their main component, manganese, and their supposed ease of recovery from the sea floor (see Chapter 12).

A summary of the classification and genesis of manganese nodules by Bonatti, Kraemer, and Rydell (1972) suggests four main mechanisms for their origin.

1. They form by slow precipitation from seawater (hydrogenous deposits).
2. They form by precipitation from submarine thermal solutions, often associated with volcanism (hydrothermal deposits), and some are found along oceanic ridges and rifts.
3. They form where the manganese comes from weathering of basaltic rock (halmyrolytic deposits).
4. They form under reducing conditions in sediments rich in organic material (diagenetic deposits).

A combination of these processes is possible. Nodules are especially common in areas of low sedimentation rates, such as some abyssal plains. This occurrence is due, in part, to the extremely slow rate of formation of nodules, which can be of the order of 1 to 3 mm (0.039 to 0.117 in.) per million years; thus they could easily be buried in places where the sedimentation rate is high. Nodules appear to form directly on the ocean bottom, as few are found buried. The fact that they are round may suggest that they are rolled around during their growth. The concentration of nodules on the sea floor can be impressive. One of the highest concentrations observed (from bottom photos) is 100 kg of nodules per square meter, or about 300,000 tons per square mile. Calculations of the total rate of formation of nodules indicate a production rate of about 10 million tons per year. This means that some elements are being incorporated into the nodules in the ocean at a rate faster than their present industrial consumption on land.

Some areas of nodules on the sea floor have different metal contents than do other areas, a fact that is not yet understood. It has been suggested by some scientists that organisms, such as plankton, may concentrate metals within their organic matter, and after death these metals become incorporated into the bottom sediments. After the eventual decay of the organism the metals could be released and move up to the sediment surface where they may become incorporated into the nodule. It has been observed that the sediments in areas of nickel-rich nodules are depleted in nickel, which suggests that this biological mechanism has occurred. The different metal contents of different sediment types may explain some of the variations in metal content of the nodules. The interesting discovery of metals' being released along spreading centers suggests that volcanic activity may have a role in the variation of chemical composition over different areas.

The recent discovery along the Galapagos region of the East Pacific Rise [about 1,000 km (about 621 mi) west of Ecuador] of underwater geysers may lead to some rethinking of the chemistry of the ocean. Discharges of hot (hydrothermal), salty water from the Red Sea and Pacific Ocean (see pages 408–410) have been known since the mid-1960s. However, the findings in the Galapagos region (using *Alvin*) are especially impressive and may indicate a process active along many areas of sea-floor spreading. In one region almost 30 vents or "chimneys" were found erupting extremely hot water and mineral material (Figure 7-18). The latter may represent the early stages of a mineral deposit (discussed later in Chapter 12). The water temperature within the vents may be as high as 300°C (about 570°F). Normal seawater at these depths (about 3,000 m or 9,843 ft) is only a few degrees centigrade. Associated with these and smaller mound areas is a unique biological assemblage consisting of large clams, worms, crabs, and so on (Figure 6-22). These "eruptions," or vents, are occurring along an oceanic rift where two crustal plates are moving apart, with volcanic rock filling the resulting gap (Figure 6-3). Among the interesting chemical aspects of these discoveries is that the water coming out appears to be recycled seawater that has sunk and moved through parts of the ridge crests. In doing so it has picked up relatively large quantities of some elements and heat that are ultimately released to the ocean (Figure 7-19). The rates of the water motions are still not well known but some scientists have suggested that the movement of dissolved material by the water from the ridge axis may be comparable to the amount of material contributed to the ocean from weathering on land. If this premise is correct, and we should know in a few years, it will necessitate a revision of the geochemical cycle of the ocean.

FIGURE 7-18 An erupting undersea vent spews a black plume of mineral-laden matter onto the sea floor some 2,800 m (about 9,186 ft) deep in the East Pacific Rise off the coast of Mexico. Scientists in the Woods Hole Oceanographic Institution's deep-diving submersible *Alvin* observed for the first time how deposits of silver, zinc, cobalt, lead, and other metals are formed at the bottom of the sea. (Photograph courtesy of Dr. Robert D. Ballard, Woods Hole Oceanographic Institution.)

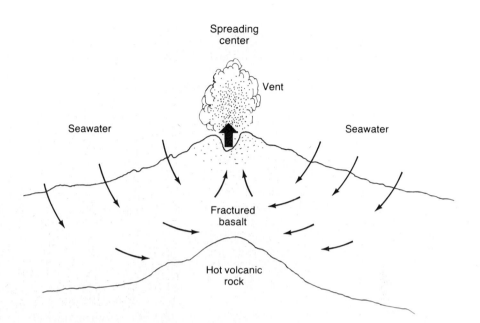

FIGURE 7-19 Possible mechanism of recycling of seawater through spreading center areas.

Another important reaction that can influence the chemical composition of the ocean is exchange processes between the sediments on the sea floor and the overlying water. There can be movement of certain elements to or from the sediments (from the water) as well as from the water buried with the sediments. For some elements the amounts involved are similar to the amounts carried into the ocean by rivers.

BIOCHEMICAL REACTIONS

Probably the most important chemical reactions in the ocean are those due to life processes. These processes, often collectively called the biochemical cycle, result in a flow of nutrients from the organic material formed by plants during photosynthesis, through the life history of the organisms in the ocean, and eventually back into the photosynthetic zone (Figure 7-20). It is clear that the animals and plants of the ocean have a great influence on the composition of the seawater. The effects of photosynthesis and respiration on the oxygen and carbon dioxide content of the surface waters have already been described. Other elements, such as nitrogen, carbon, and phosphorus, and trace elements, such as silicon and iron, are involved in the biochemical cycle. These elements are withdrawn from seawater in the formation of organic material during the photosynthetic growth phases of marine plants and are later returned to seawater as waste and decomposition products of the organic material.

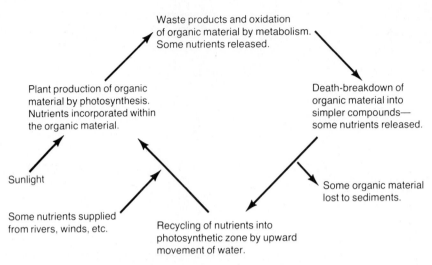

FIGURE 7-20 The biochemical cycle.

Generally there is not much accumulation of organic matter on the sea floor, because the organic material has either been decayed or been consumed while falling through the water or, once on the bottom, been digested by bottom-dwelling organisms. When the organic matter decays, or is oxidized, the nutrients it contains can be released to the environment. There are certain areas, such as the Black Sea, where organic matter can accumulate because of the absence of oxygen in the bottom waters. The use of oxygen in the oxidation of organic material helps explain the relative absence of oxygen at intermediate depths of the ocean (Figure 7-21). Oxidation, or respiration, unlike photosynthesis, is independent of light and may take place at any depth.

The distribution of nutrients in the ocean waters can be characterized by four layers (see Figure 7-21 for phosphorus).

1. A surface layer, usually about 100 or 200 m (about 328 to 656 ft) thick, where nutrient concentration is low because it is utilized during photosynthesis and incorporated within the organic material.

2. A second layer, several hundred meters thick, where nutrient concentration increases very rapidly due to its release from organic material by oxidation processes.

3. A layer of maximum nutrient concentration usually between 500 and 1,000 m (about 1,640 to 3,281 ft) depth.

4. A layer, usually extending to the bottom, where the nutrient concentration is uniform.

In some instances the maximum concentration of the nutrients does not coincide with the minimum concentration of oxygen because of either the presence of nutrients released during a previous cycle or a different source for oxygen.

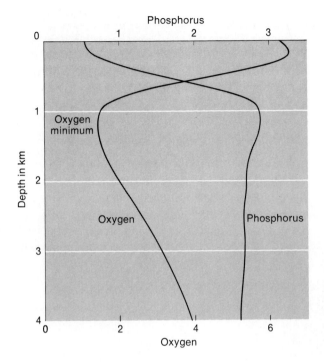

FIGURE 7-21 Vertical distribution of oxygen and phosphorus in the ocean. The units for oxygen are milliliters per liter or parts per million while each unit for phosphorus is equivalent to about 30×10^{-6} g (or 1/300,000 g) per liter of seawater.

The oxidation and subsequent return of the nutrients in the second layer results in a large reservoir of nutrient-rich water below the photosynthetic zone (upper 100 to 200 m). These nutrients can return to the photosynthetic zone (where they will become reincorporated into organic material) only by the physical circulation or movement of the water. This movement takes place in several ways: by the worldwide circulation of the oceans; by upwelling (a type of vertical mixing; Figures 9-15 and 9-16) in coastal, offshore, and equatorial regions; and by annual vertical mixing in temperate and high latitude areas. How these movements occur is described in Chapter 9 (pages 308–316). The total process results in a downward movement of nutrients due to biochemical reactions, compensated by an upward movement of these same elements as a result of water circulation. The organic material that is deposited in the sediments is removed from the system, but this can be balanced by material that is added from rivers and other sources (Figure 7-20).

In addition to these vertical differences in nutrient concentration, there are also oceanwide differences and seasonal variations. The deep waters of the Pacific and Indian Oceans, for example, contain more phosphorus and less oxygen than do the deep waters of the Atlantic and Arctic Oceans because the source of the waters, their composition, and their subsequent modification by biological and physical factors are different.

Seasonal changes in nutrient content are most evident in temperate areas, where phytoplankton have two growth periods each year—the spring and early autumn. Studies made in the English Channel, which analyzed the animals and plants living in the water and the water itself, showed seasonal patterns and changes in the form of phosphorus (Figure 7-22). The result shows two annual plankton growth periods, each accompanied by an increase in dissolved organic phosphorus.

The preceding discussion has centered on the biochemical reactions affecting the nutrients. Many other elements, however, are concentrated by organisms either during photosynthesis or by their eating organic material and eventually liberating it either as waste products, or by the death and decomposition of the organism. Calcium and silicon, for example, are used by some organisms to form their shells. Extensive deposits of these shells are found on the sea floor (Figures 5-41 and 5-42). Several crustacean species concentrate copper; and other organisms, such as tunicates, concentrate large quantities of vanadium. In some marine organisms, trace elements occur in greater concentrations than in the seawater.

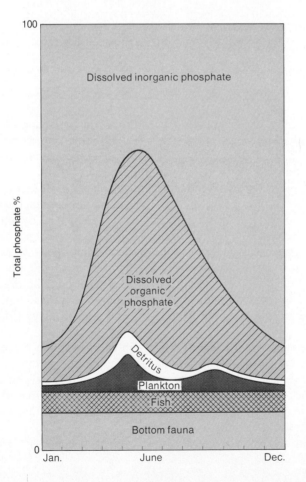

FIGURE 7-22 Seasonal variation of phosphate in waters of the English Channel. (After Harvey, 1950, 1960.)

When one considers the distribution of the reactive elements in seawater, it should be remembered that the dynamics of the water must also be evaluated. For example, if the movement or flow of water is sufficiently large, it can overwhelm or obscure differences in chemical composition caused by biological processes, evaporation, heating, or any other phenomena.

PHYSICAL PROCESSES

Changes in the salt content of ocean waters are mainly caused by differences in evaporation and precipitation (rainfall), and melting of ice. In the polar region, surface waters of relatively low salinity are common because precipitation is high (diluting the seawater), whereas waters of relatively high salinity occur in subtropical areas where evaporation is high.

Currents, waves, and other forms of turbulence usually keep the upper parts of the ocean moderately well mixed. Near land, warm surface water may be blown away from the coast by strong offshore winds and be replaced by cooler, nutrient-rich subsurface water by upwelling. This process can lead to large amounts of nutrient-rich water and in turn to high levels of photosynthesis, and often areas of rich biological resources.

HUMAN INPUT

Human input on the composition of the ocean, although small, can be seen, for example, by the distribution of lead. The development and almost universal acceptance of the internal combustion engine and the use of lead compounds in its fuel have caused a significant increase in lead concentration in the atmosphere and subsequently in the ocean. More than 160,000 tons of lead per year are put into the atmosphere just by the United States. Much of this lead is carried to the ocean by rain or ends up in rivers that reach the ocean. Worldwide more than 250,000 tons of lead gets into the ocean each year. This input can be seen near industrialized regions in surface layers of the ocean, where the lead content can be 5 to 50 times that of the deeper waters.

All these processes are part of the large and complex geochemical cycle that is affecting the chemistry of the ocean (Figure 7-23). Many of these processes have been active for over a billion years, and have resulted in the present composition of the ocean.

Isotope Chemical Oceanography

Atoms of elements may occur in several forms called *isotopes*. Relatively rare isotopes differ from the more typical forms mainly in their atomic weight; in most other respects they have chemical properties similar to those of the common form

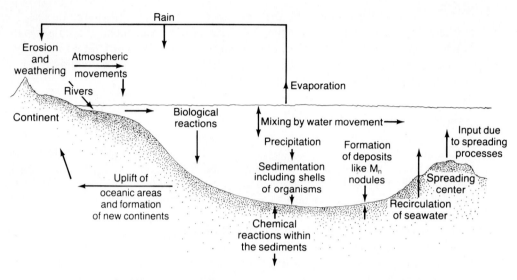

FIGURE 7-23 A model of the geochemical cycle of the ocean.

of the element. The differences in atomic weight, which are slight, are mainly evident as small changes of boiling point, freezing point, and rates of diffusion of the rare isotope compared with the normal form. These variations can be used to study certain oceanographic processes. Some isotopes are radioactive and can change, or decay, by losing part of their atomic structure to become a different isotope or element. For example, carbon occurs in three forms: carbon-12 (^{12}C), carbon-13 (^{13}C), and carbon-14 (^{14}C). The numbers 12, 13, and 14 indicate the different atomic weights of the element. These three forms are chemically similar; however, ^{14}C is radioactive and will decay to nitrogen-14 (^{14}N). The original radioactive isotope is known as the parent and the resulting form or forms (which, if radioactive, will also decay or change) are called the *daughters*.

The rate at which radioactive isotopes decay is constant for each isotope (Table 7-5). The term *half-life* indicates how long it takes for half of the original quantity of a radioactive material to decay into another isotope. Therefore, by measuring the relative amounts of the parent and daughter, we can determine the

TABLE 7-5 Some Important Radioactive Isotopes
and Their Half-lives

Parent	Stable daughter	Half-life in years
Thorium-232 (^{232}Th)	Lead-208 (^{208}Pb)	1.4 billion
Uranium-238 (^{238}U)	Lead-206 (^{206}Pb)	4.5 billion
Potassium-40 (^{40}K)	Argon-40 (^{40}Ar)	1.3 billion
Uranium-235 (^{235}U)	Lead-207 (^{207}Pb)	700 million
Carbon-14 (^{14}C)	Nitrogen-14 (^{14}N)	5,560

age of a rock. For example, if the ratio of potassium-40 (^{40}K) to argon-40 (^{40}Ar) in a rock is 1 to 1, the age of the rock is 1.3 billion years because that figure is the half-life of ^{40}K or the amount of time necessary for half of the ^{40}K to decay to ^{40}Ar. If the ratio was 1 to 3, the age of the rock would be 2.6 billion years (three-fourths of the original ^{40}K will have been converted to ^{40}Ar).

Actually, the age obtained only indicates when the rock last solidified. The reliability of this technique is based on two assumptions.

1. No daughter isotope was originally present.
2. There has been no addition of parent or daughter isotope during the aging of the rock.

It is usually difficult to test these assumptions, but if more than one parent–daughter isotope pair is used, they can provide an independent check on each other. In other words, two different radioactive isotopes could produce similar results only if the assumptions were correct or, alternatively, if they were incorrect in exactly the right proportions—an unlikely situation.

Isotopes that have half-lives on the orders of billions of years are best suited for evaluation of the age of the earth, since sufficient amounts of the parent will still be present and measurable after long periods of time. Isotopes with relatively short half-lives, like ^{14}C, are used for measuring recent events of shorter duration, such as dating of recent sediments, or archaeological events.

The natural fractionation or separation of stable isotopes has been used in interpreting some chemical and physical oceanographic processes, such as evaporation and changes in temperature. Radioactive isotopes, both naturally occurring and man-made by atomic bomb explosions, can be used to study and date the history of the ocean and the presently active dynamic processes, such as circulation and ocean mixing. Radioactive isotopes are also useful in determining rates of accumulation of sediments and, by dating of the sediments, in comparing past events in the ocean with past events on the continents.

STABLE ISOTOPES

The stable hydrogen and oxygen isotopes that occur in a water molecule are good examples to use to show the value of stable isotopic studies. Hydrogen is present in two forms: the common hydrogen atom (^{1}H), which constitutes more than 99.9 percent of all the hydrogen atoms, and deuterium (^{2}H), which is heavier and much less abundant. Tritium, ^{3}H, is a very rare radioactive isotope of hydrogen. Oxygen occurs in three forms: Oxygen-16, -17, and -18 (^{16}O, ^{17}O, and ^{18}O). The lightest and most abundant isotope is ^{16}O, making up over 99.7 percent of all oxygen atoms.

The vapor pressure of water molecules composed of the lighter isotopes ^{1}H and ^{16}O is higher than it would be if one of the heavier isotopes (^{2}H, ^{17}O, ^{18}O) were substituted in the molecules. Because of the higher vapor pressure, it is easier for molecules containing lighter isotopes to be evaporated from the surface waters of the ocean. Therefore when there is evaporation, surface waters have a decrease in their lighter isotope components and an increase in the heavier isotopes. Latitude is also important since there is more evaporation in equatorial areas than in the high latitude areas. In addition, in high latitudes there is less melting ice, which has a relatively small quantity of the heavier isotopes. Using these facts and measuring the hydrogen and oxygen isotopic composition, it is possible to determine some aspects of the history or mixing of ocean water.

Shell-forming animals and plants that live in the surface waters also reflect the conditions of the water by their ratio of ^{18}O to ^{16}O within their shells. After death, the shells settle to the ocean floor and become incorporated into the bottom sediments. The analyses of these shells with depth in the sediment can produce a continuous record of the temperature changes that has occurred in the surface waters. If radioactive isotopes are also used, the record can be dated and compared with events on land.

There is also a variation of isotopes within the biological environment; for example, there are more heavier isotopes of oxygen in the oxygen-minimum zone. Apparently marine plankton and other organisms that utilize organic material in their oxidation processes selectively remove the lighter isotopes of oxygen, leaving the heavier isotopes behind.

RADIOACTIVE ISOTOPES

Radioactive isotope analysis is equally valuable to the oceanographer as it allows understanding of time effects on the rates of chemical reactions in the ocean. These isotopes are especially useful in the study of deep-sea sediments, which are deposited at extremely slow rates. Besides being able to determine the rate of deposition, marine geochemists can establish correlations or time equivalencies of events, expressed in the sediments, over large areas of the ocean and perhaps even correlate them with events on land.

Radioactive isotopes in the ocean and sediment have three origins.

1. Material formed when the earth was created.
2. Cosmic and solar reactions with the atmosphere.
3. Atomic reactions produced by man.

The first group are generally used for dating extremely slow processes, such as sedimentation rates of the very slowly deposited deep-sea sediments. This is because radioactive elements have extremely long half-lives and therefore are still present after about 4.5 billion years of earth history. Potentially valuable for dating are some members of the uranium decay series, protactinium (^{231}Pa) and (^{230}Th). There are two sources of these isotopes in sediments: precipitation from seawater (most important) and decay of uranium in the sediment. Although the content of these isotopes is very small in sediments (generally measured in parts per billion or less), the concentrations appear to be closely related to sed-

TABLE 7-6 Some Radioactive Isotopes that Are Produced by
Cosmic and Solar Reactions, and Their Half-lives

Isotope	Half-life in years	Source
Carbon-14 (^{14}C)	5,560	Cosmic rays and nuclear bombs
Silicon-32 (^{32}Si)	500	Cosmic rays
Tritium (^{3}H)	12	Cosmic rays and nuclear bombs
Strontium-90 (^{90}Sr)	28	Nuclear bombs
Cesium-137 (^{137}Cs)	30	Nuclear bombs

imentation rates as determined from other techniques. The uranium decay series is useful mainly for dating material younger than about 300,000 years.

Isotopes produced by cosmic and solar reactions with the atmosphere are sometimes of value to oceanographers because of their relatively short half-lives (Table 7-6). Short half-lives are better for the measurement of short-term events, such as water circulation and ages of relatively quickly deposited sediments.

Probably the most useful of these isotopes is ^{14}C. It is formed by the interaction of cosmic rays with the atmosphere, which produce higher-energy neutrons, most of which in turn are captured by ^{14}N to form ^{14}C. The ^{14}C combines with oxygen, producing carbon dioxide, which then enters the ocean by exchange with the atmosphere to be utilized in life processes. The ^{14}C decays at a constant rate; its half-life is 5,560 years.

The ^{14}C in organisms or surface waters is in equilibrium with the environment. After death of the organisms or sinking of the water, however, the ^{14}C starts to decay (50 percent decrease every 5,560 years). Therefore the measurement of its content indicates how long the material has ceased to be in contact with the surface waters. For water, the resulting age can also reflect the degree of mixing with waters of differing ^{14}C ages.

In general ^{14}C is used to date only material having an age of less than about 100,000 years. This is due to its relatively short half-life and the difficulty of detecting the small amounts present after this period of time. Two corrections must be made to the ^{14}C to stable carbon ratio before it can be used: one for the recent increase in carbon in the atmosphere due to the burning of fossil fuels and the other for the ^{14}C produced by nuclear explosions.

An application of ^{14}C dating to sediments is an especially valuable technique in understanding the recent glacial history of the ocean. Carbon-14 has also been used to calculate the age (or how long the water has been away from the surface) of various water masses. These calculations are based on numerous assumptions and usually depend on models of differing complexities. Ages will vary with the model, but it is generally agreed that the age of deep Pacific water is between 1,000 and 1,600 years and of deep Atlantic water about half that.

An important problem in oceanography is to understand the motion of water and how it mixes, both in a horizontal and vertical sense. Such information is needed to determine the fate of pollutants in the oceans as well as to better

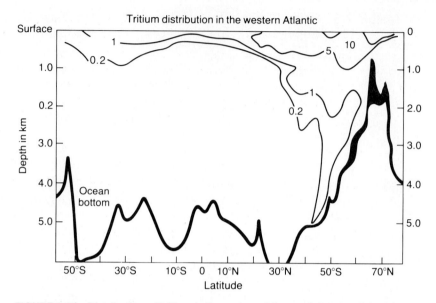

FIGURE 7-24 Distribution of tritium in the waters of the western part of the Atlantic Ocean. Note how the tritium has penetrated the waters of the North Atlantic. Units are tritium units or number of tritium atoms in 10^{18} atoms of hydrogen. (Data from numerous GEOSECS cruises.)

understand the biological and geochemical cycles. A new program, called Transient Tracers in the Ocean (TTO), is using several man-made compounds that have found their way into the ocean to trace water motion. These substances include ^3H and ^{14}C produced from atomic bomb blasts and carbon dioxide from the burning of fossil fuels. A key point to this program came from GEOSECS studies in which a deep penetration of ^3H was noted in the deep waters of the North Atlantic (Figure 7-24). Since ^3H has a half-life of only about 12 years (and decays to helium), it was surprising to see it penetrate so far into the ocean. In general, ^3H, the heaviest isotope of hydrogen, is relatively rare. However, there was a major input from nuclear testing in the 1950s and 1960s. This input into the atmosphere eventually reached the surface waters of the ocean and, by mixing, the deeper waters. By measuring the helium/tritium ratio an evaluation of mixing can be made. Without the man-made bomb input this analysis would have been impossible because of the small quantities of naturally-occurring ^3H.

Summary

Chemical oceanographers principally study the chemical aspects of the ocean, which in many instances are intimately involved with biological activities. Early studies in this field stressed cataloging what was present in seawater. More recent efforts have focused on the processes that control the composition of the ocean and influence things like pollution, economic deposits (like manganese nodules), and rates of reactions (radioactive isotopes are especially useful in this category).

Water, besides being a very common compound, is unique in that over 97 percent of the water on the earth's surface is in the ocean. Although water is a simple compound it has rather complex properties that, to a large part, are due to its dipole aspect. If this aspect did not exist, life as we know it probably would not be possible. The addition of salts to freshwater increases the complexity of water properties.

The chemical composition of seawater is a mixture of dissolved organic matter, dissolved gases, particulate matter and, of course, water. The amount of many elements in sea water can be estimated by the Forchhammer Principle if the concentration of one element (not involved in biological reactions) is known. Biological processes, especially photosynthesis, can strongly influence the concentrations of some elements (especially nutrients) and some gases (especially oxygen and carbon dioxide). The key factors that influence the chemical composition of the ocean are exchange with the atmosphere, solubility of compounds, reduction of organic matter by anaerobic bacteria, precipitation to and exchange with the ocean bottom, inflow of fresh water, freezing and melting of sea ice, and chemical and biological reactions and processes. The formation of manganese nodules and hydrothermal activity along spreading centers are two examples of processes that can influence the chemical composition of the ocean—although neither is yet well understood. Better understood are some biochemical aspects and their seasonal variations and influences on the ocean. Elements that are unreactive in seawater tend to have a long residence time in sea water.

Isotopes, especially some radioactive ones, can be used to introduce time in the assessment of oceanographic and chemical processes.

Suggested Further Readings

BOLIN, BERT, "The Carbon Cycle," *Scientific American,* **223,** no. 3 (1980), pp. 125–32.

CLOUD, PRESTON, and AHARON GIBOR, "The Oxygen Cycle," *Scientific American,* **223,** no. 3 (1970), pp. 111–23.

EDMOND, J. M., "Ridge Crest Hot Springs: The Story So Far," *EOS,* **61,** no. 12 (1980), pp. 129–31.

GOLDBERG, E. D., "Chemical Descriptions of the Oceans," *Technology Review,* **72** (1970), pp. 25–29.

HEATH, G. R., "Deep-Sea Manganese Nodules," *Oceanus,* **21,** no. 1 (1978), pp. 60–68.

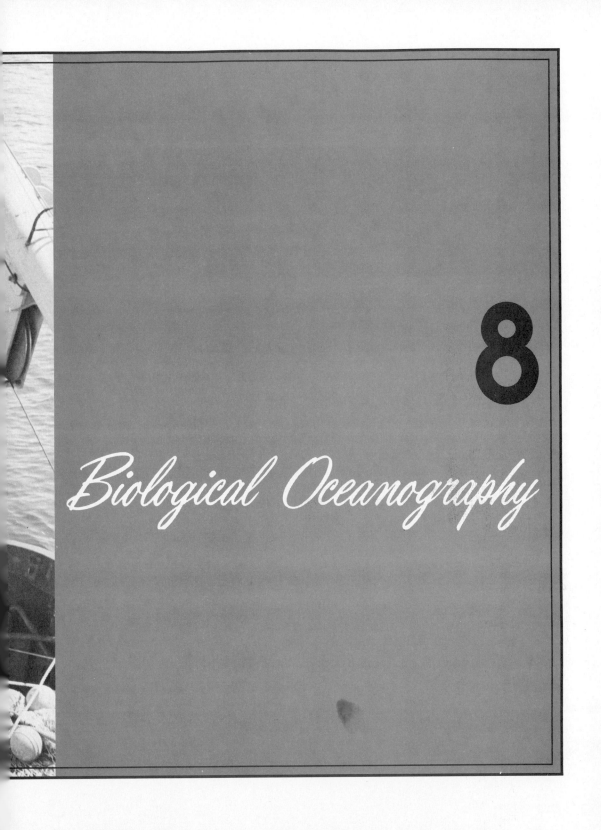

8

Biological Oceanography

BIOLOGICAL OCEANOGRAPHY IS one of the most intriguing fields of oceanography. Many animals and plants in the ocean remain to be discovered, as evidenced by the recent discoveries along the East Pacific Rise (Figures 8-1, 6-20, and 6-21). Marine biologists, in many instances, are just beginning to comprehend the complex ocean ecology, or relationships of the organisms with the ocean. The danger of pollution to the biological environment is clear, but little is known about its actual or permanent effects. Many feel that the ocean has the potential for a much larger supply of food. This potential is obvious when one realizes that land, which covers less than 30 percent of the earth's surface, presently yields about 1,000 times more food than the ocean. Some scientists, however, feel that the oceans are already being fished close to their maximum.

Much marine biological research is focusing on how marine animals and plants interact with each other and the environment. This often involves a detailed knowledge of water chemistry and sometimes physical oceanography and marine geology as well as an understanding of basic biological processes. Applied research often emphasizes pollution and its effects, detection, and remedies, as well as ways to increase food production from the sea. Underlying both these types of research are four questions important to the field of biological oceanography.

What are the plants and animals that live in the ocean?
How are these organisms distributed in time and space?
What are the factors that control their distribution?
How do the organisms in the ocean live and survive?

This chapter emphasizes the sea as a biological environment, the animals and plants of the sea and their interaction with each other and the ocean, and the processes involved in the production of organic matter (food) in the sea. The effects of pollution are discussed in Chapter 13 and the biological resources in Chapter 12.

FIGURE 8-1 Photograph taken from an exterior camera on *Alvin* showing a cluster of tube worms, mussels, and white crabs growing in a portion of the Galapagos spreading center. These creatures were unknown until the recent exploration of this region (see Figures 6-20, 6-21, and 6-22 and discussion in Chapter 6). A portion of *Alvin*'s sampling system is seen in the foreground of the picture. (Photograph courtesy of Woods Hole Oceanographic Institution.)

History of Biological Oceanography

Ancient people clearly knew something of the food that could be obtained from the sea, since ancient dwellings frequently have large quantities of shells of marine organisms and other evidence that "sea food" was then popular. Drawings in ancient Egyptian tombs occasionally show the various fish that were caught in the Nile River, Mediterranean Sea, or Red Sea.

Early marine scientists were mainly concerned with the depth and shape of the ocean but some biologists were interested in the distribution of life, especially at great depths in the ocean. Samples of bottom life were occasionally obtained with the devices used in deep-sea soundings and from dredgings.

An important early marine biologist was C. G. Ehrenberg (1795–1876). Ehrenberg noted that many of the rocks he collected on land were composed of skeletons of microscopic organisms such as diatoms, radiolarians, and sponges and that similar organisms were found living in the ocean. He thus concluded that many of the rocks found on land must have been formed on the bottom of the sea.

Edward Forbes, considered to be the father of marine biology, in 1844 divided the ocean into eight zones on the basis of their marine organisms. He observed that animal life existed, although in decreasing amounts relative to the surface water, to a depth of about 550 m or 300 fathoms; below this depth he thought no life existed. Forbes's observation was somewhat surprising, considering that animal life had already been brought to the surface from greater depths. The controversy that resulted, concerning the existence of life at depth, was beneficial to the blossoming science of oceanography because it stirred interest in the field and helped launch the Challenger Expedition of 1872 to 1876.

Another incident that helped spark early interest in oceanography was the amusing "*Bathybius* mystery." Samples of calcareous ooze from the sea floor were found to contain a strange gelatinous substance believed by some scientists to be a primitive form of life that was called *Bathybius*. The mystery came to an inglorious end when people associated with the Challenger Expedition found that *Bathybius* was the result of the interaction of the alcohol preservative and the collected sample.

The Challenger Expedition was clearly one of the major advances in the field of biological oceanography. This expedition collected large quantities of data on the animals and plants in the ocean. There were, however, many other important, although not as well-publicized, expeditions that also made significant contributions to biological oceanography. Important studies were made by J. Vaughan Thompson, one of the first to use a net to collect plankton; Johannes Muller, a German naturalist; and Sir John Murray and Victor Hensen, who studied and named plankton. These early scientists were mainly interested in collecting, describing, and studying organisms. In later years, some advances were made in understanding the intimate relationships between the organisms and the seawater. Descriptive work continues but as the different groups of organisms become better known, the need for these studies decreases. Emphasis is shifting to studies of the environment, interrelationships among organisms, and production of organic matter.

Instruments of the Biological Oceanographer

Since the biological oceanographer is primarily concerned with the distribution and relationships of animals and plants in the ocean, he or she needs a way of sampling the population of the sea. A common method is to pull some device, usually a net, through the water. A typical one is the plankton net (Figure 8-2) that is used to catch small floating plants and animals. Larger nets (Figure 8-3) are used to collect larger organisms and slow-swimming fish. One question raised by samples obtained in nets is, Do they really represent the environment? Obviously, large and fast fish can avoid nets, while slower-moving organisms will

FIGURE 8-2 Plankton net being retrieved. The net has been towed through the water, and small plants and animals have been trapped and are being removed by the kneeling technician from the end of the net. These nets can be rigged to close by command and thus only sample selected portions of the water column. (Photograph courtesy of Vicky Cullen, Woods Hole Oceanographic Institution.)

FIGURE 8-3   Isaacs–Kidd net being brought aboard a research vessel. This net can catch slow-swimming organisms. (Photograph courtesy of Woods Hole Oceanographic Institution.)

be more easily caught. In addition, certain species tend for some (usually unknown) reason to be caught more easily than others. Thus it is probable that these types of sampling devices can discriminate against certain organisms and present an unrealistic picture of the biological population. With nets, the biological population is usually sampled in only one direction, such as in the horizontal sense, as the net is dragged through the water at essentially a constant speed. Lowering the net to a certain depth and then bringing it up can give a sample of the vertical distribution of life. It is also possible to open and close nets at selected depths for more specific sampling. Specialized nets have been designed to sample just the surface layers of the ocean (Figure 8-4).

Biological oceanographers want to know both the total amount of organisms collected as well as the number of organisms per volume of water (in other words, the population density). To accomplish this the volume of water passing through the sampler must be measured (perhaps by a small current meter in front of the net) or be reasonably estimated.

The bottom-living fauna can be sampled by devices pulled or dragged along the bottom (Figure 8-5). Additional sources of information concerning bottom fauna are bottom photographs (see later figures in this chapter) and rocks and submarine cables when they are brought up to the surface.

Other methods of observing organisms in the ocean include direct obser-

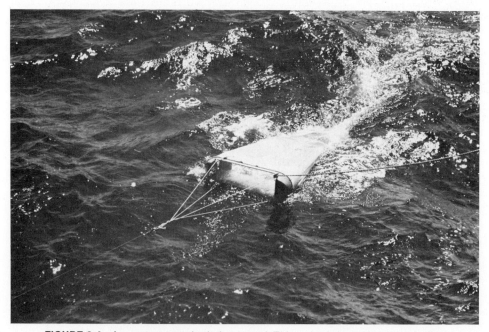

FIGURE 8-4 A neuston sampler being towed. This net is kept at the water surface where it can sample surface living organisms as well as some pollutants, such as floating tar balls. (Photograph courtesy of Dr. George Grice.)

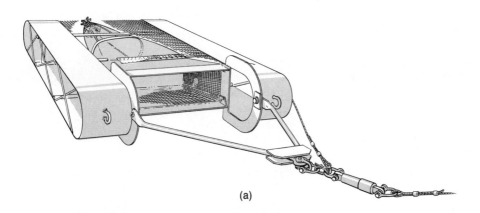

(a)

FIGURE 8-5 A biologic dredge. (a) This device is dragged along the ocean bottom, collecting bottom-living (benthic) organisms and sediment in the net held between the two runners; most of the sediment will pass through the net, but the larger animals will be caught. (b) Emptying out a biologic dredge after it has finished its sampling. (Photograph courtesy of Woods Hole Oceanographic Institution.)

(b)

vations by diving, either in a submersible or by using scuba. Some electronic devices, such as echo sounders, can be used to detect large individuals or groups of animals in the water (Figure 12-18). The sound made by, or the echo pattern from, different organisms can sometimes be used to determine the identity of the species (Figure 8-48). Many modern fishing boats have sonic devices to locate schools of fish. Fishermen and fish-tagging programs have also supplied much information about the migratory patterns of fish.

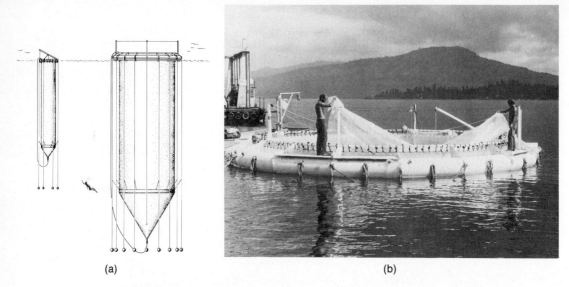

(a) (b)

FIGURE 8-6 (a) An illustration of two types of bags used in the CEPEX work. (b) Technicians preparing part of one of the CEPEX bags. (Both photographs courtesy of Dr. George Grice.)

In recent years oceanographers have developed sophisticated instruments to measure subtle changes of the fine parameters in the marine environment, and these can be studied to determine how they affect biological activity. Automated instruments (Figure 3-24) that measure temperature, salinity, light, and other characteristics of the environment are often used with observing devices such as television and photography. The data obtained from such systems allow scientists to better evaluate the effect of the environment on marine life. Laboratory facilities that can simulate marine conditions are also among the more important instruments of the biological oceanographer. Some more innovative biological oceanographers have used large plastic bags filled with seawater for their experiments (Figure 8-6). One such project called CEPEX (Controlled Ecosystem Populations Experiment) used bags that contained as much as 1,324,750 L or 350,000 gals of seawater. Within such bags the scientists could run controlled experiments and evaluate effects (such as adding pollutants) over a period of time.

The Biological Environment of the Sea

Most marine organisms are constantly immersed in seawater, a fact that has certain advantages and disadvantages. The chemical or physical state of the water will affect the organism. Fortunately, the physical and chemical characteristics of seawater tend to be relatively stable, and most marine organisms are not exposed to sudden environmental changes as are their terrestrial counterparts. Animals living in the ocean have another advantage over land varieties in that they are free from the effects of desiccation, or drying out. Animals of terrestrial environments have had to evolve impervious skins or scales, and plants, extensive root systems to preserve or obtain more water. As discussed in Chapter 7 and subsequent sections of this chapter, marine organisms can also directly influence the chemistry of seawater by various aspects of their life processes. Some of the biologically important physical and chemical characteristics of seawater are discussed in the following sections.

BIOLOGICALLY IMPORTANT PROPERTIES OF SEAWATER

Water is essential for the production of food by plants. In this respect, an important property of water is its solvent or dissolving power. Water can dissolve and carry more different materials than any other common liquid and thus can hold the critical gases and minerals necessary for animal and plant life.

Seawater, because of its density, provides the physical support for many organisms in the ocean and in some instances eliminates the need for skeletal structures. Examples of this are jellyfish and other small animals that float in the ocean. Another effect of this support is to enable extremely large animals, such as whales, to exist in the ocean. (The largest dinosaurs were also aquatic.)

Seawater is a buffered solution, which means that it is difficult to change its pH toward a more alkaline or acidic state. In general, seawater is slightly alkaline, having a pH between 7.5 and 8.4. The acidity or alkalinity in water is determined by the concentration of the hydrogen (H^+) and the hydroxyl (OH^-) ions. If equal amounts are present, the solution is said to be neutral. If the H^+ concentration is greater than the OH^- concentration, the solution is acidic; if less, the solution is alkaline. Conventionally, the concentrations are expressed by pH (see footnote 2, page 202). An alkaline state of seawater is necessary for the organisms that secrete calcium carbonate shells; if it were acidic, the carbonate could be dissolved. An additional advantage of seawater's being buffered is that abundant carbon, in the form of carbon dioxide, can be present in the water without changing the pH. The carbon is necessary for plants in the production of organic matter.

The transparency of water is another biologically important property. Light can penetrate seawater to a considerable depth, and since photosynthesis is light dependent, it means that this important process is not restricted to the upper few meters of the ocean but can take place as deep as 100 m (328 ft) or more, depending on the clearness of the water.

Seawater also has a high heat capacity (a large amount of heat is necessary to raise its temperature) and a high latent heat of evaporation (heat is released during condensation and absorbed during evaporation). Both of these characteristics prevent rapid changes in temperature, which could be very harmful to many marine organisms.

Many of the elements dissolved in seawater are very important biologically. The ratio of some of these elements in seawater is amazingly similar to the ratio in the body fluids of most marine organisms but less similar to the body fluids of land animals. This similarity between the external medium (the ocean) and the

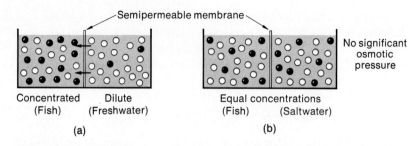

Semipermeable membrane

Concentrated (Fish)   Dilute (Freshwater)

(a)

Equal concentrations (Fish)   (Saltwater)

No significant osmotic pressure

(b)

FIGURE 8-7 (a) Diagram illustrating the osmotic effect of an animal living in a fluid that has a different concentration than its internal fluids. This difference creates an osmotic pressure that the animal has to work against to maintain its internal composition. (b) The concentration of the body fluid and the external fluid is similar, and no significant osmotic pressure results. Situation (a) is similar to that of a fish in freshwater, situation (b) of a fish in saltwater.

internal medium (the body fluids) is critical for osmosis. Osmosis occurs when two solutions of different concentrations are separated by a semipermeable membrane. An osmotic pressure is created when there is a difference in concentration on either side of the membrane. Water will then move through the membrane into the more concentrated solution (Figure 8-7). The greater the difference in concentration, the greater the osmotic pressure.

Organisms must work against osmotic pressure to maintain the composition of their internal fluids. In the marine environment, the similarity of body fluids in marine organisms to the external medium means that only a small osmotic pressure exists. This in turn means that marine organisms need not use as much energy as freshwater organisms in maintaining their body fluids. The easier control of osmotic balance in saltwater than in freshwater may explain why some saltwater animals are larger than the largest freshwater animals.

GENERAL CHARACTERISTICS OF THE OCEAN AS A BIOLOGICAL ENVIRONMENT

Ocean water temperature ranges from almost −2°C (28°F) (at great depths) to more than 40°C (104°F) in places like the Persian Gulf. Salinity can vary from near zero in estuary and nearshore conditions to about 40 ‰ in areas having high evaporation, such as the Red Sea. Ocean depths can reach over 10,000 m (over 32,810 ft). Pressure, due to the weight of the overlying water, ranges from 1 atm (1 kg per square centimeter or about 14.7 lb per square inch) at the surface to over 1,000 atm at great depths. There is a 1 atm increase with each 10 m (32.8 ft) of depth. Light penetration can also vary; hardly a trace of sunlight penetrates deeper than about 1,000 m (3,281 ft). Even though the range of these characteristics is extensive, many large areas of the ocean have uniform conditions. Temperature, for example, is fairly similar over large parts of the ocean (Figure 8-8). Salinity is very constant in the surface waters of the open ocean ranging between

33 to 37 ‰, and only in isolated areas does it get higher. The salinity of the deeper water is even more uniform, having a normal range of 34.6 to 35 ‰. The supply of oxygen could be a limiting factor for most forms of life in the deep sea if it were not for the sinking of cold surface waters at high latitudes (Figures 9-17 and 9-18). These waters carry oxygen-rich waters to the bottom and near-bottom, making life possible at all depths. Even though pressure clearly does have an effect on animal life, many organisms, through various adaptations, are able to range over considerable depths; whales, for example, can dive from the surface to depths of 1,000 m (3,281 ft) (a pressure of 100 atm).

A common misconception is that a fish brought to the surface with one of its internal bladders protruding from its mouth and found to be dead, died from pressure change. In most instances, however, death was due to the large changes in temperature the fish experienced in being brought to the surface.

The movement of seawater can be biologically very important. Motion can move nutrients necessary for plant growth from deep water to the surface water where they can be utilized by plants. Motion can also disperse water products,

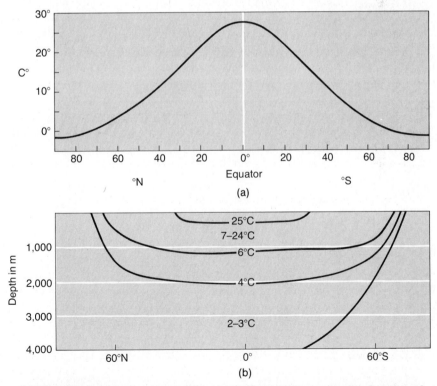

FIGURE 8-8   (a) The average surface temperature of the ocean as a function of latitude. (b) The range of temperature with depth and latitude in an idealized ocean. (After Raymont, 1963.)

eggs, and larvae, or adult forms of life. Water motion may also have some non-beneficial results, such as carrying animals out of their natural environment into an unfavorable one. This can happen where warm water and cold water come in contact.

THE DIVISIONS OF THE MARINE ENVIRONMENT

The marine environment can be divided into two major realms: the *benthic*, which refers to the ocean bottom, and the *pelagic*, which refers to the overlying water (Figure 8-9). The pelagic realm can be subdivided into the neritic environment (the water that overlies the continental shelf) and the oceanic environment (the water of the deep sea). A depth division is also possible for the pelagic realm. The benthic realm is usually divided into a littoral (out to a depth of 200 m or 656 ft) and a deep-sea system. A more detailed division is possible for the nearshore region. Depth ranges of these different environments are shown in Table 8-1. Note that similar terms can be used for both the benthic and pelagic realms and that the depth divisions are not absolute figures but, being dependent on life-forms, can vary somewhat in meaning. Therefore, if one wanted to talk about the water at a depth of 7,000 m (22,960 ft), one would speak of the hadal pelagic environment; the bottom at 4,000 m (13,120 ft) would be the abyssal benthic environment. The oceanic province of the pelagic environment can also be divided

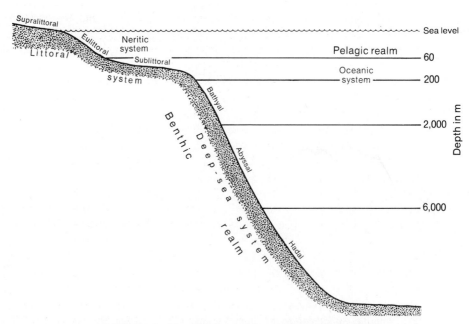

FIGURE 8-9   The divisions of the marine environment.

TABLE 8-1   Depth Ranges of the Different Environments of the Ocean

Depth	Benthic environment		Pelagic environment	
Above high tide High tide to 40–60 m 60 to 200 m	Littoral system	Supralittoral Eulittoral Sublittoral	Neritic system	
200 to 2,000 m 2,000 to 6,000 m Greater than 6,000 m	Deep-sea system	Bathyal Abyssal Hadal	Oceanic system	Bathyl Abyssal Hadal

into two zones on the basis that sufficient light for photosynthesis hardly penetrates deeper than 200 m (656 ft). The upper, lighted area is called the *euphotic zone*, and it is within this region that photosynthesis by plants occurs. The deeper, darker region is called the *aphotic zone*.

BENTHIC ENVIRONMENTS

The benthic environment (Table 8-1 and Figure 8-9) covers a wide range of oceanographic conditions, from the exposed supralittoral areas to the hadal environment. Obviously then, the benthic organisms (those that live on the ocean bottom) will vary in the different environments and no one organism will be found in all environments.

The littoral system can be divided into three divisions, from the essentially exposed *supralittoral* to the sometimes partially exposed *eulittoral* to the deeper *sublittoral*. The supralittoral environment is extremely rugged. Animals living there are almost continuously exposed, being immersed only during periods of extremely high tides and storms and by the spray from breaking waves. Animals of this environment are similar the world over; generally small gastropods and lichens on rocks, and crabs and amphipods on beaches.

The eulittoral environment includes the region that is periodically exposed at low tide and extends out to a depth of 40 to 60 m (131 to 197 ft). The width of the intertidal region depends on the tidal range and the slope of the ocean bottom. Animals living in this environment also must withstand the effect of breaking waves. Many animals accomplish this by burrowing into the bottom, thus also lessening some of the harmful effects of exposure at low tide. The outer edge of the eulittoral is near the depth limit at which most attached plants can grow on the bottom; they cannot grow in deep water because of the absence of sufficient light. Actually only a small portion of the sea floor is available for the growth of attached plants, and even within this small area many parts cannot be utilized because the bottom may be muddy or otherwise unsuitable for plants. The eulittoral may be the best studied marine biological environment, especially since it can be observed by divers. The animals and plants in this environment are very numerous and varied.

233

The outer part of the littoral system is the sublittoral division, the depth limits of which extend to 200 m (656 ft) or more. Proceeding seaward from the eulittoral to the sublittoral there is a decrease in plant life and an increase in animal life. The outer part of the sublittoral division, which generally conforms to the edge of the continental shelf, is an area extensively exploited by commercial fishermen.

The deep-sea system, composed of the *bathyal, abyssal,* and *hadal* divisions, is obviously not as well known as the shallower littoral system. The deep-sea system is devoid of higher plant life (no light) but bacteria can live at this depth.

Oceanographic conditions of the deep parts of the ocean are relatively uniform: Temperature decreases slowly with depth, salinity is essentially constant, and pressure increases 1 atm with each 10 m (32.8 ft) of depth. As most organisms in this environment are composed primarily of water, with few if any air spaces, and because water is not very compressible, deep-sea pressure itself is not an excluding factor for life there. Pressure can, however, have other effects that can affect the life processes of deep-sea animals.

The uniform conditions of the deep sea suggest that seasons would have little biological importance there, as compared with their effect on the numerous phenomena such as breeding in the shallower waters. Oceanographers from Florida State University, however, have recently found that some seasonal effects may extend to deeper waters. They noted that some deep-sea animals were spawning off North Carolina only during August to November and in the Antarctic only during July to October. This presents two interesting problems: How do the animals recognize the seasons? Why are the breeding periods seasonally similar in both hemispheres (remember that when it is winter in the northern hemisphere, it is summer in the southern hemisphere)?

Food in the deep sea is not so abundant as it is in the littoral system. Animals living there are thought to receive most of their food from organic material that falls from near-surface waters to the ocean bottom [Figure 8-34(b)]. The production of organic matter in surface waters is usually higher over the littoral system and decreases with distance from land, becoming rather low over the deep sea. Thus the amount of food reaching the deep-ocean bottom often is small, and this quantity is further reduced by its disintegration and decay while sinking through the water. It follows, therefore, that the animals of the deep sea should be small scavengers rather than large predators.

Many animals of the deep sea are bizarre in appearance (Figures 8-10 and 8-11); however, their size is generally small. Most of our information about the deep sea is derived from underwater photographs, deep-sea dredgings, and occasional glimpses of the bottom from submersibles. In general, relatively little is known about the biological environment of the deep sea.

The density of life in the ocean can be expressed by a measure called *biomass* (the amount of living organisms in grams per square meter of the ocean bottom). Biomass generally shows high values in the littoral region and low values in deep water (Table 8-2).

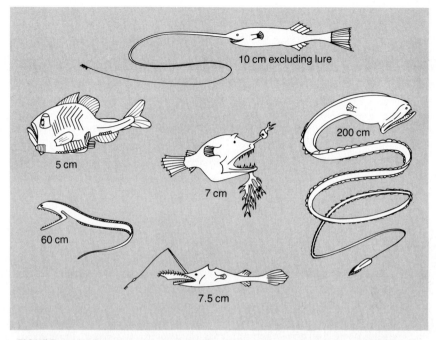

FIGURE 8-10 Some deep-sea fish. Their approximate length in centimeters is indicated. Note that although these fish look very ferocious, they are usually very small. (From R. V. Tait and R. S. DeSanto, *Elements of Marine Ecology*, New York: Springer-Verlag, 1972. Figure courtesy of Professor R. V. Tait.)

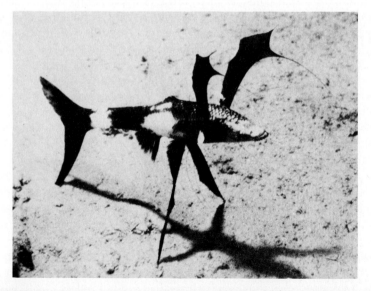

FIGURE 8-11 One of the bizarre fishes of the deep sea. This is commonly called a tripod fish (officially it is *Bathypterois bigelowi*). (Photograph courtesy of C. D. Hollister.)

TABLE 8-2 Average Biomass
in Different Parts
of the Ocean

Area	Biomass (g/m²)
Coastal zone	100–5,000
50–200 m depth	200
about 4,000 m	about 5
Central part of ocean floor	0.01
Kuril–Kamchatka Trench	
about 6,000 m	1.2
about 8,500 m	0.3
Tonga Trench, 10,500 m	0.001

The pelagic environment, which refers to the water, is divided into the nearshore neritic system and the offshore ocean system. The border between the two is not very definitive and is often set at the edge of the continental shelf.

The neritic pelagic environment generally has a large diversity of conditions, especially if there is freshwater discharge from a river. Organisms living in this environment must be able to tolerate a wide range in salinity. Nutrients can enter this environment by upwelling due to coastal winds (Figure 9-15) or from rivers. A good nutrient supply will sustain the abundant growth of plankton, the basic food of the sea. This, in turn, attracts other forms of life, and as a result, the neritic area is generally the most biologically productive area of the sea and is where most fish and other types of food from the sea are caught.

The oceanic system can be divided as shown in Table 8-1 or, as previously mentioned, into a light (euphotic) and dark (aphotic) zone with the boundary at a depth of about 200 m (656 ft) or less. Compared with the salinity of the neritic environment, that of the oceanic area is relatively constant. Temperature decreases with depth, the greatest change occurring in a region called the *thermocline* (Figure 9-7). The surface waters have a temperature variation that is a function of latitude (Figure 8-8). Nutrients are usually low in the surface waters and increase with depth (Figure 7-21).

Currents in the deep waters of the ocean are generally relatively slow. The deep sea is an area of almost complete darkness with very subtle, if any, changes. The abyssal pelagic area is one of the world's largest ecological units, enclosing about three-quarters of the total volume of the ocean.

Little is known about the deep hadal zone mainly because of the difficulty of sampling the animals that live in these waters; the benthic fauna have been better sampled and more than 350 species have been identified.

A general summary of some of the important characteristics of the marine environment is shown in Figure 8-12.

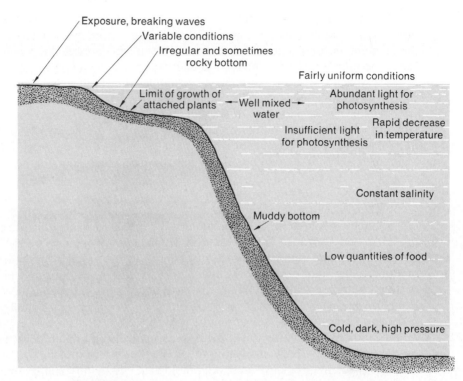

Exposure, breaking waves

Variable conditions

Irregular and sometimes
rocky bottom

Fairly uniform conditions

Limit of growth of
attached plants

Well mixed
water

Abundant light for
photosynthesis

Insufficient light
for photosynthesis

Rapid decrease
in temperature

Constant salinity

Muddy bottom

Low quantities of food

Cold, dark, high pressure

FIGURE 8-12   General characteristics of the marine environment.

THE BIOLOGICAL COMMUNITY

The study of organisms and their environment is called *ecology*. An understanding of the relationships between organisms and environment is fundamental since organisms in the ocean (or on land) do not live as individual creatures but rather are closely controlled and influenced by the physical and chemical environment as well as by the other plants and animals with which they must interact. Such interactions can in turn affect the environment. Thus one must really consider animals and plants as part of a complex system involving physical, chemical, and biological aspects (such a system is commonly called an *ecosystem*). An ecosystem can be defined as an entire range of organisms and the environment in which the organisms interact. There are two broad categories to the organisms: those that produce their own food (autotrophs) and those dependent on autotrophs for food (heterotrophs). Prominent among the autotrophs are green plants that produce organic matter by photosynthesis, and various forms of bacteria.

Marine biologists can also consider the environment according to biotopes or niches. A biotope, such as a reef, is an area where the principal habitat conditions and the living forms adapted to them are uniform. Life in a biotope is not static; some animals may wander freely from one to another. It follows from this that as living conditions of the environment become more specialized, so also will the inhabitants. It is generally thought that the more rigorous the conditions, the fewer the number of different species and the greater the numbers of individuals of those species. This situation is, for example, typical of marsh environments. The relationship between some organisms can be categorized as either symbiotic, commensalistic, or parasitic. In a symbiotic relationship, benefits exist for each interacting organism (if neither organism is dependent on the interaction it is said to be mutualistic). When one organism benefits but there is no harm to the other a commensalistic association exists—an example is one animal living and being protected in the shell of a larger and different organism. In a parasitic relationship one organism benefits at the expense of another.

When speaking of the inhabitants of a biotope, one should apply the concept of a biological community. A community is composed of organisms, occurring together, that appear to be dependent on each other or perhaps on the common environment. The community may just represent the animals living on the leaf of a particular plant or it may include a more extensive relationship, such as that between plankton and the animals that consume them—a dependency primarily based on the need for food. The food chain in the ocean is a complex system involving several levels, starting with phytoplankton (the producers), then different kinds of herbivores (plant eaters), carnivores, and finally scavengers and bottom feeders. Before considering the general population of the ocean, a brief discussion of an especially interesting biological community is appropriate.

CORAL REEFS AND ATOLLS

Coral reefs and atolls are among the more interesting biological environments in the ocean. They can be located 1,000 km (621 mi) or more from the nearest continental land mass and represent a delicate balance between the forces of the ocean and the strength of the reef.

A reef is a biological community on the sea floor that forms a solid limestone (calcium carbonate) structure strong enough to withstand the force of waves. The predominant organisms in most of these communities are corals and algae. Algae tend to grow over the coral, encrusting it, giving it strength, and forming a solid structure. This is especially important since most reefs grow at about sea level so they must be strong enough to withstand the eroding power of breaking waves.

Coral reefs require specific conditions for growth, such as a water temperature of about 20°C or 70°F, and for this reason most reefs are found in the tropics. Corals cannot tolerate low salinities and can be killed by freshwater. In some instances, the reefs must be able to grow fast enough to keep up with the rise in sea level or, in some regions, with the slow sinking of the islands upon which they are growing. The recent rise in sea level of about 130 m (426 ft) in the last 15,000 years (Figure 4-4) required that most reefs grow very fast to maintain their favorable growth position near sea level.

There are three basic types of reefs (Figure 8-13): the *fringing reef,* the *barrier reef*, and the *atoll*. The fringing reef grows out from a landmass but is attached to it. An example is the reef that borders the Florida Keys. A barrier reef is separated from the landmass by a lagoon. This type of reef, as exemplified by the Great Barrier Reef of Australia, can be a very imposing structure. An atoll (Figure 8-14) is an oval-shaped reef surrounding a lagoon; it is not associated with an obvious landmass. Atolls commonly are found rising abruptly from the deep sea and some are very large. Kwajalein, for example, in the Pacific, is about 65 km, or 40 mi, long and 30 km, or 18.6 mi, wide.

The conditions on a reef vary from the quiet of the lagoon to the breaking waves of the outer part of the reef and are reflected by the different types of coral growing in these areas (Figure 8-15). In areas of intense wave action, the coral must be strong enough to withstand the waves; more delicate forms can grow only in the quiet areas.

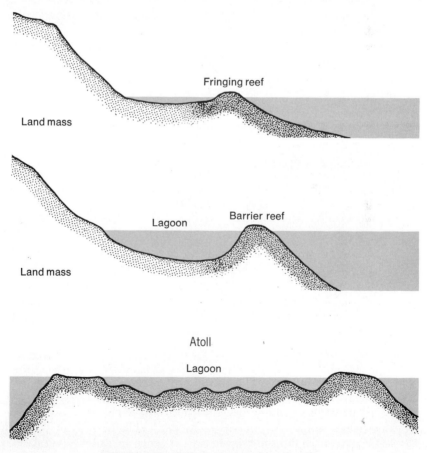

FIGURE 8-13   Different types of coral reefs.

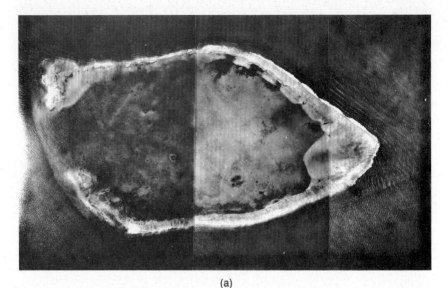

(a)

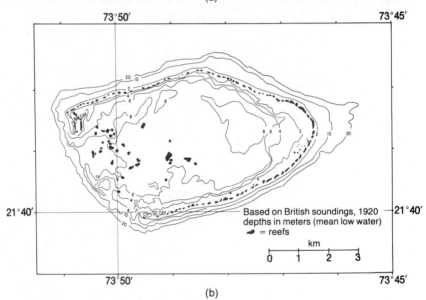

(b)

FIGURE 8-14 (a) Hogsty Reef, an atoll in the Bahamas. (Official U.S. Navy photograph.) (b) Chart of the lagoon and a shallow-water area of Hogsty Reef (depth in meters). (Photograph courtesy of J. D. Milliman.)

The finding of numerous coral atolls in the middle of the deep ocean, without any visible landmass, puzzled many early scientists. How could these shallow-water animals have established an existence in water several kilometers deep? During his voyage on the *Beagle*, Charles Darwin saw some reefs and atolls, and

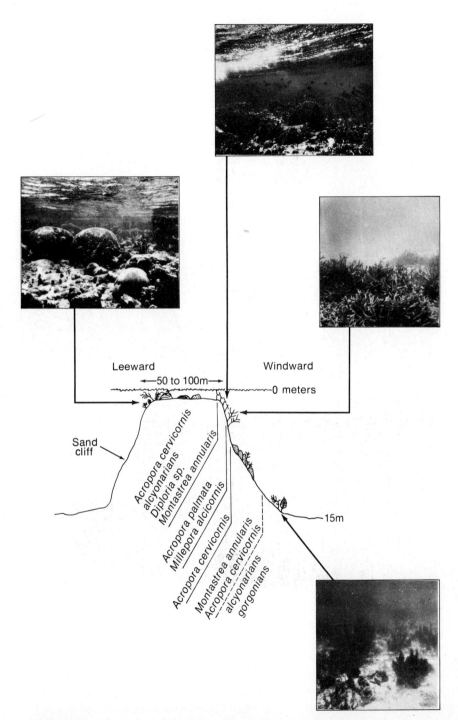

FIGURE 8-15 Illustration showing the different forms of coral that grow on the various parts of a reef. (Photograph courtesy of J. D. Milliman.)

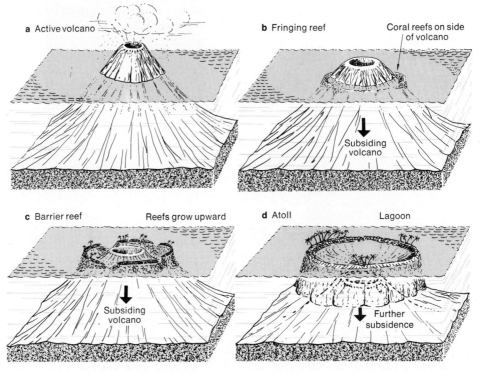

FIGURE 8-16 Darwin's theory of the formation of an atoll.

on the basis of this trip he formulated a hypothesis to explain their origin. His theory, as shown in Figure 8-16, is very simple. He suggested that a volcanic island initially provided a shallow-water base for the growth of a fringing reef. Then there was a slow subsidence of the island, during which time the reef continued to grow. Eventually the island and the reef became separated by a lagoon, like a barrier reef; further sinking then drowned the volcanic island, resulting in an atoll.

It was not until almost 100 years later that Darwin's hypothesis was found to be essentially correct. The evidence came from drillings on Eniwetok and Bikini atolls. These drillings found over 1,000 m (3,281 ft) of coral underlain by volcanic rock.

The Population of the Ocean

A chapter of a book or even an entire book is inadequate for anything but a very superficial description of the animal and plant life in the sea. The organisms of the ocean are considered here by two aspects: first, by their mode of locomotion and habitat, and second, by their taxonomic classification.

Considering mode of locomotion or movement, the organisms of the ocean can be divided into three large groups: the benthos, the nekton, and the plankton.

BENTHOS

The term *benthos* comes from the Greek word for deep or deep sea. The benthos (Figure 8-17) are those organisms that live on or below the ocean bottom. Some, such as barnacles and oysters, have planktonic larval forms that eventually settle and attach themselves to the bottom for their adult lives. Other types of benthic life (the term benthic life refers to those organisms living on the bottom and is essentially equivalent to benthos), such as worms and clams, may burrow into the bottom, while others, such as starfish and echinoids, may creep slowly over the bottom. In shallow water, some plants can form part of the benthos.

Benthic organisms may have geologic importance because they modify and change some physical and chemical properties of the sediment during their lives on the sea floor. Also, many of the fossils in the sediments are the remains of benthic organisms. Bacteria, which constitute a small part of the benthos, can be very important in the food chain in the deep sea by producing food from inorganic substances.

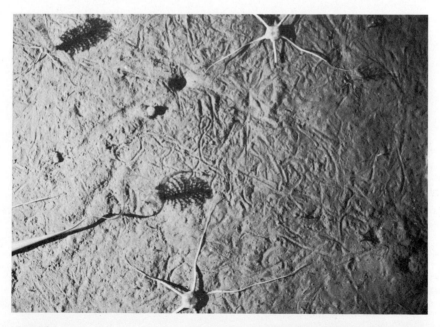

FIGURE 8-17 Bottom photograph showing some benthic forms of life at a depth of about 1,300 m (4265 ft). Note the numerous tracks on the bottom. (Photograph courtesy of Woods Hole Oceanographic Institution.)

NEKTON

Nekton (the term derives from the Greek word for swimming) include those animals that are able to swim freely, independent of current movement (Figure 8-18). This group (plants are not represented in this category) encompasses many advanced forms of animal life, such as fish, whales, and other mammals.

Nekton have the ability to search actively for food and avoid predators. These animals can also migrate extensively throughout the oceans and are found both on or near the ocean surface (Figure 8-18) or at depths on the ocean bottom (Figure 8-19).

Nekton are commercially important for man and, because of their feeding habits, strongly affect other forms of life. Since nekton feed mainly on plankton, they can limit and control the phytoplankton population.

Nekton are generally the least restricted form of animal life in the sea. Although they may inhabit different parts of the pelagic environment during their

FIGURE 8-18 Some porpoises playing on the surface of the ocean. Note the blowholes through which they breathe. (Photograph courtesy of R. K. Brigham.)

(a)

(b)

FIGURE 8-19 Some animals of the deep sea. (a) Taken at a depth of about 1,600 m (about 5,250 ft) on the continental slope off Massachusetts. The animal on the right is a fish, about 61 cm long (about 2 ft), commonly called "rattail." On the left is a small eel. Note the numerous burrows and depressions that were made by animals that live in the bottom. (b) Also a rattail, taken at a depth of about 4,000 m (about 13,100 ft) on the continental rise south of Nova Scotia. In this picture the camera has fallen over and is taking a picture while lying on its side. (Photograph courtesy of C. D. Hollister.)

lives, their distribution is still somewhat limited by temperature and pressure. The influence of these and other environmental factors is not so well understood as it is for the benthic and planktonic forms of life.

PLANKTON

Plankton (from the Greek word for wandering) are the third group of marine organisms. These organisms are usually small (Figures 8-20, 8-25, and 8-26) with very weak or limited powers of locomotion and are moved mainly by ocean

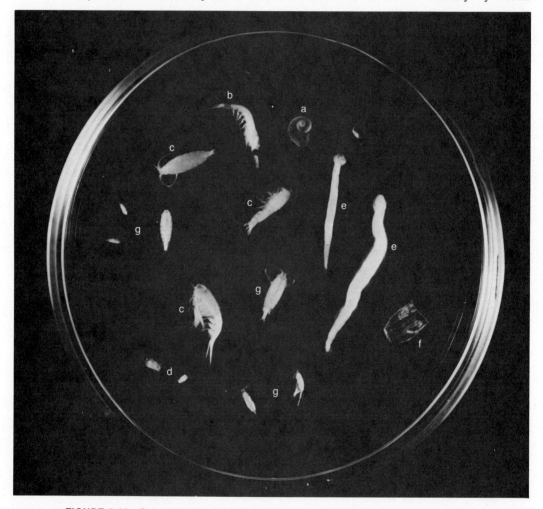

FIGURE 8-20 Selected representatives of some zooplankton: (a) pelagic gastropod, (b) euphausid, (c) amphipods, (d) ostracods, (e) chaetognaths, (f) bell-shaped trachomedusa, (g) copepods. Diameter of dish is 10 cm (about 4 in.). (Photograph courtesy of George D. Grice.)

currents. Plankton can be either animals (zooplankton) or plants (phytoplankton). Most are microscopic but the group includes some large floating forms, such as jellyfish and sargassum weed. Plankton comprise the largest group, by numbers, of organisms in the ocean. Many animal forms in the ocean have a plankton stage, usually at birth, during which they float freely in the ocean.

Phytoplankton are probably the most important individual form of life in the sea. These organisms, which by photosynthesis convert water and carbon dioxide into organic material, are the base of the oceanic food chain.

The distribution and growth pattern of phytoplankton (as plants they must live in the euphotic zone) show pronounced vertical and seasonal variations. The vertical distribution is due to the depth that the light can penetrate, a depth that may be only 1 m (3.28 ft) in nearshore turbid waters to as much as 200 m (656 ft) in the open ocean (Figure 8-21). Seasonal changes are due to complex relationships among light, nutrient supply, temperature, and herbivores (zooplankton that eat plants), among other factors. The growth rate of phytoplankton can be very high, sometimes as much as six cell divisions per day (each cell division may produce a new organism).

Zooplankton include representatives from almost every group of marine organism. These plankton may be one of two types: *holoplankton* (permanent plankton), which spend their entire life as plankton, or *meroplankton* (temporary plankton), which spend only a portion of their life (usually the larval stage) as plankton. Meroplankton become nektonic or benthic forms at other life stages. Zooplankton are very important in the economy of the sea because they eat and thereby concentrate phytoplankton, making phytoplankton-derived food or energy more easily available for the higher forms of life (when they

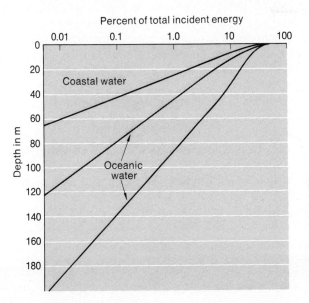

FIGURE 8-21 Percentage of total incident energy reaching different depths of the ocean for different types of water. Note that the degree of light penetration varies with water type and that for some water types only 5 percent of the light penetrates to a depth of 20 m (about 65 ft). (After Jerlov, 1951.)

eat the zooplankton in turn). Zooplankton that feed on phytoplankton are frequently called *grazers* or *herbivores* because of their position in the food chain.

Some plankton are especially valuable for oceanographers because certain species are characteristic of particular bodies of water. These species, called *indicator species*, can be used to point out the origin and movement of the water body. Many areas of the sea floor are covered with the shells, or tests, of planktonic organisms. Geologically, perhaps the most important of these organisms is the protozoan Foraminifera (Figure 5-41), whose tests provide valuable information about the history and climate of the ocean at the time these organisms lived.

No taxonomic classification has yet been established for the animals and plants of the world that is without some ambiguity. In general, the major or uppermost category is called a kingdom. Each kingdom is further subdivided into phylum and eventually to species. The order of classification is as follows:

Kingdom
　Phylum
　　Class
　　　Order
　　　　Family
　　　　　Genus
　　　　　　Species

The common and more traditional classification divides all organisms into just two kingdoms: the plant kingdom and the animal kingdom. But there are some plants and bacteria that could just as well fit into the animal kingdom. Another more modern system is one that consists of four kingdoms: Monera, Protista, Metaphyta, and Metazoa. The Monera kingdom includes primitive one-celled organisms, without a membrane separating the cell material from the cell nucleus. Its two phyla are bacteria and blue-green algae. Protista includes organisms that have nuclear membranes and some ability to move. It includes all the one-celled organisms not included in Monera and some plants, including green, brown, and red algae, dinoflagellates, and diatoms. Important marine animals are foraminifera and radiolarians. The Metaphyta kingdom includes the common land plants that have roots, stems, and leaves. These types of plants are relatively rare in the ocean and are found only in certain coastal areas. Metazoa includes all the multicellular animals. The traditional classification of the plant and animal kingdom is used in this text because of its basic simplicity and fairly common acceptance.

PLANTS OF THE SEA

Marine plants are very different from land plants. Seawater can be compared with soil water in that it carries the nutrients necessary for plant life. On land, plants had to develop extensive root systems to obtain water and food, and leaves to obtain oxygen and carbon dioxide. In the ocean, however, the plants are completely surrounded by nutrient-carrying water so no real root systems are necessary except for attachment of some plants in shallow water. Large portions of the ocean floor are unfavorable for plant growth, however, because of insufficient light for photosynthesis. Some plants in shallow water can be attached to the bottom and receive sufficient light but only a small portion of the sea floor—less than about 2 percent—is shallow and solid enough for attached plants.

Because only a small portion of the ocean is available for attached plants, most plants are planktonic. These microscopic, usually single-celled, plants produce the bulk of organic material in the ocean and occur in almost unbelievably large quantities. All the ocean's animals either directly or indirectly (by eating animals that feed on plants) feed on these plants; yet, one should remember that the phytoplankton are restricted—because of their dependence on light—to the upper layers of the ocean, while the consumers of the organic matter produced by plants can occur throughout the ocean.

The plant kingdom can be divided into four major divisions:

1. *Spermatophyta:* seed and flowering plants.
2. *Pteridophyta:* ferns.
3. *Bryophyta:* mosses.
4. *Thallophyta:* algae and fungi.

Only spermatophytes and thallophytes are found in the ocean. Spermatophytes are considered to be the highest type of plant. Apparently these plants did not originate in the sea but entered it from the land or freshwater. Only a few marine species of spermatophytes are found. Perhaps the most important form is *Zostera* or, as it commonly is called, *eelgrass*. This plant (Figure 8-22) possesses true roots attached to a stem, or *rhizome*. It can reproduce by seeds or by sending up new leaves from its rhizome. Eelgrass grows in water usually less than 5 m or

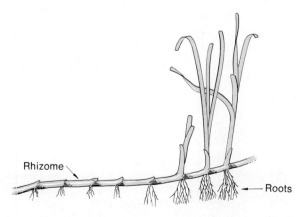

Rhizome

Roots

FIGURE 8-22 *Zostera*, a nearshore Spermatophyta.

about 16 ft deep, where wave action is not very strong. When the plant dies and is broken, its pieces may be carried out to sea and become an important source of food in some areas. Where it produces a thick growth, *Zostera* provides protection for many forms of animal life. An entire biological community, ranging from small diatoms to nearshore animals, is dependent on the eelgrass environment. In the early 1930s a disease killed most of the eelgrass on the Atlantic coast. This caused considerable damage to the scallop population because of the latter's dependence on eelgrass for refuge. Fortunately for the scallop fisheries, the eelgrass recovered and has repopulated most of its original habitat.

The more common thallophytes have no true roots, stems, or leaves. The most important members of this division are algae (Figure 8-23), which are commonly divided into five groups based mainly on their color:

1. Blue-green algae (Myxophyceae)
2. Green algae (Chlorophyceae)
3. Brown algae (Phaeophyceae)
4. Red algae (Rhodophyceae)
5. Yellow-green algae (Xanthophyceae)

The first four are usually attached plants, and the yellow-green algae are mainly planktonic.

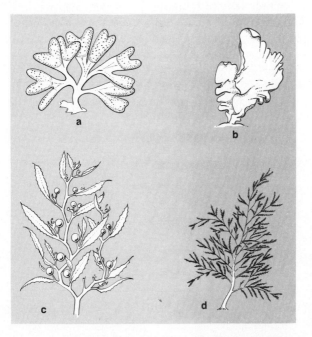

FIGURE 8-23 Some common forms of marine algae: (a) *Fucus*, a brown alga; (b) *Ulva*, a green alga; (c) *Sargassum*, a brown alga; (d) *Polysiphonia*, a red alga. (From H. U. Sverdrup, Martin W. Johnson, and Richard H. Fleming, *The Oceans: Their Physics, Chemistry, and General Biology*, copyright 1942. Reprinted by permission of Prentice-Hall, Inc., Englewood Cliffs, N.J.)

Blue-Green Algae

Blue-green algae are most abundant in rivers and lakes and are not too common in the ocean except near some rivers and tropical regions. These organisms are generally very small and poorly developed; some are floating forms. Even though they are called blue-green algae, some may be other colors due to accessory pigments within the plant. A dramatic example of this is the floating form *Trichodesmium*, which has a red color. It is this plant that often gives the Red Sea the color that earned this body of water its name.

Green Algae

Green algae are found mainly in freshwater, in the region where freshwater and saltwater are mixed, and in the shallower parts of the littoral system. They rarely occur below a depth of 10 m (32 ft) and are thus restricted to the well-lighted part of the ocean. A common form is the alga *Ulva*, or sea lettuce [Figure 8-23(b)]. Green algae can sometimes impart a distinct green color to the water; they can also form an algal slime on boats and other submerged objects.

Brown Algae

Brown algae are found mainly in the marine environment. These algae, which include kelp and *Sargassum*, are the largest of the marine algae; one, *Nereocystis*, may be as much as 61 m (or about 200 ft) long. This plant forms kelp beds that commonly grow near some coasts. These algae are sometimes harvested from boats, by cutting off the top layers of the kelp. The harvest is valuable as a food and a source of iodine, potash, and iodine-based drugs.

Brown algae are considered the most advanced of the thallophytes. The large forms that make up the kelp beds are attached to the bottom by a *holdfast*, a branched structure that attaches to a rock (Figure 8-24). From the holdfast extends a long tube (*stipe*) that ends in a large circular ball. Both the ball and the stipe are hollow and filled with gas. The gas gives the plant buoyancy, permitting it to float. At the end of the bulb are long, thin fronds that have the appearance of leaves.

There are other forms of brown algae; some are small, delicate branching plants, others, big and broad like *Fucus* (Figure 8-23). One brown alga, *Sargassum*, is found far from land on the open ocean. *Sargassum* actually grows in tropical areas but when torn loose by waves it will float and drift with the currents. It also grows and multiplies while it is drifting. Large quantities of these algae have been accumulated by currents in an area of the North Atlantic that is called the Sargasso Sea. Before the plant dies and sinks, it forms a unique environment that provides shelter for many different kinds of animals.

With some exceptions, for example, *Sargassum*, brown algae tend to be best developed in cooler parts of the ocean, especially in nearshore, rocky-bottomed areas.

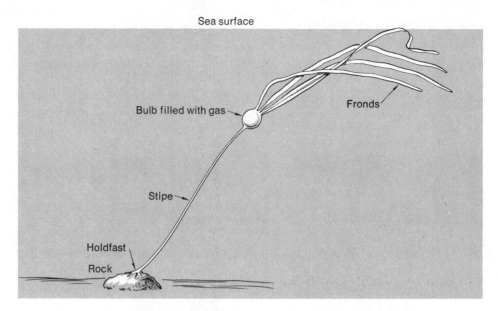

Sea surface

Bulb filled with gas

Fronds

Stipe

Holdfast

Rock

FIGURE 8-24   General structure of a large brown alga.

Red Algae

Red algae are among the prettiest organisms in the sea. They are red colored (although some are purple, brown, or green) because of the abundance of red pigments. Like other plants, red algae contain green chlorophyll pigment but the green color is masked by other pigments. Red algae extend farther out to sea than the other forms. Thus proceeding out from shore into deeper water one would successively observe green, brown, and then red algae, with some overlap. Red algae are widely distributed throughout the ocean but they tend to be more abundant in warmer waters. Some forms, such as *Lithothamnion*, an alga that lives with coral, have the ability to precipitate calcium carbonate.

Red algae can be an important producer of organic matter, especially on the outer areas of the continental shelf. Some species are commercially valuable because they are a source of agar, which is used, among other things, as a thickening agent for ice cream and other products.

Yellow-Green Algae

The yellow-green algae are a group of several different types of organisms. Whatever their classification, the yellow-green algae are a very important group of plants in the ocean. They are mainly planktonic organisms containing chlorophyll and are capable of photosynthesis. Because they can float, they are found in the surface waters of all the oceans and are not restricted to the nearshore areas, as are the other species of algae.

Perhaps the most important of this group is the diatom, a microscopic organism found both in saltwater and freshwater. Diatoms are single-celled organisms that can combine with other individuals to form long chains or groups (Figure 8-25).

In some areas of the ocean, the production and growth of diatoms is especially high. Diatoms can multiply in a geometric manner (that is, 2 become 4, 4 become 8, 8 become 16, and so on) and in this manner their numbers can become very large in a short period of time (days). Diatoms have siliceous shells and if enough of them accumulate on the ocean bottom after death, they can form a deposit called *diatomaceous ooze*. There are large areas of the ocean (Figure 5-42) that are covered with this type of deposit.

FIGURE 8-25 Electron microscope photographs of some common diatoms. The diatoms are about 20 to 80 μ (microns) (about 0.0008 to 0.0031 in.) in diameter except for the lower left, which is actually a colony of diatoms. (Photographs courtesy Dr. Susumu Honjo.)

Diatoms have the ability to form resting spores when the environmental conditions become unfavorable. They can remain in this configuration for long periods of time; when conditions improve, they start reproducing again.

Another important member of the yellow-green group is the dinoflagellate (Figure 8-26). Some of these organisms are animallike (they eat food rather than photosynthesizing it), and others are definitely plants. Dinoflagellates have flagella, or "tails," that they can use for locomotion and to search for nutrients or better environmental conditions. Many of these organisms are luminescent; when excited they impart a glowing color to the water.

(a)

(b)

(c)

(d)

FIGURE 8-26   (a) and (b) Electron microscope photographs of some dinoflagellates and (c) and (d) coccolithophoridae. The white bar is 10 μ long (about 0.0004 in.). [Photographs courtesy of Dr. David Wall (a) and Dr. Susumu Honjo (b, c, d).]

Dinoflagellates tend to be more abundant in warmer waters; diatoms, in colder waters. A large sudden growth, or bloom, of dinoflagellates will discolor the water, causing a "red tide." Humans can be poisoned if they eat mussels or clams that have fed on certain species of dinoflagellates and diatoms during a red tide period. The tests, or shells, of dinoflagellates are easily destroyed after death; thus these organisms do not form extensive deposits on the sea floor.

Another group of yellow-green algae consists of the smaller Coccolithophoridae organisms that are about 5 to 10 μ in diameter. These organisms were first recognized by geologic oceanographers who noted their calcareous shells in deep-sea sediments. They were not collected in plankton nets because they were small enough to pass through the holes of the nets. The contribution by the Coccolithophoridae and other small yellow-green algae to the economy of the sea is not adequately understood at present.

In summary, the plants of the sea are very different from those of the land. In the ocean there are few areas where attached plants can grow and receive enough light for photosynthesis. Therefore, the organic producers in the ocean must be microscopic floaters that can utilize the light available in the upper 200 m (656 ft) of the ocean. Even though the plant life is microscopic, its numbers are sufficient to feed the remaining population of the ocean.

BACTERIA

Bacteria, although poorly understood, are important organisms in the ocean for the production of organic material. The varieties of bacteria appear almost infinite, as they are found in almost every environment. Bacteria are known to grow in hot, salty springs where temperatures approach 100°C (212°F) as well as in near-frozen brines having temperatures of −2°C (about 28°F). Bacteria can live in oxygen-free or oxygen-saturated conditions whether these exist in surface waters or at the greatest depth of the ocean. There are two types of bacteria: the *heterotrophs* and the *autotrophs*. The heterotrophs use organic material obtained from other organisms for their nutrition. This group is probably the more common one. Autotrophic bacteria use inorganic material and carbon dioxide to form organic material and in this respect are similar to plants. The amount of organic material produced by bacteria in this manner is relatively very small when compared with that produced by plants, but this small quantity may be important in the deep sea where other sources of food are not available.

ANIMALS OF THE SEA

The number of phyla in the animal kingdom varies with the classifier. In any case, all phyla have marine representatives, whereas some do not have a terrestrial or land representative. The ocean is the area where the animal kingdom has had its best development and land is where plants have had their best development.

There is one fundamental difference between plants and animals. Plants have chlorophyll and can produce their own food; animals do not have chlorophyll and must eat plants or other animals to obtain their necessary food.

The important marine animal phyla are described in the order of one interpretation of their increasing biological complexity. For a more detailed discussion of individual animals of the sea, consult a good biology text.

Protozoa

The phylum Protozoa consists mostly of single-celled microscopic organisms, including the orders Foraminifera and Radiolaria (Figures 5-40 and 5-41). Shells of these animals cover large areas of the ocean bottom (Figure 5-42). Most of these organisms live in the surface waters of the ocean and when they die, their shells settle to the bottom. Radiolaria and Foraminifera can be used to date and correlate sediments because certain species are characteristic of different geologic time periods. The shells of most Foraminifera are composed of calcium carbonate; the Radiolaria shells are siliceous (composed of silica and oxygen). Some Foraminifera are benthic and live on the bottom.

Another group of Protozoa is the dinoflagellates (see Figure 8-26), some of which, as previously discussed, are more typical of plants. The "animal" dinoflagellates are also important; these small organisms reproduce very rapidly and as voracious feeders, they can consume a large quantity of phytoplankton.

Also included in this phylum are two animals that may be more familiar to you: *Amoeba* and *Paramecium*; however, neither of these animals is particularly important in the ocean.

Porifera

Porifera, or sponges, are multicellular benthic animals that occur in many different forms. They are classified on the basis of composition of their internal skeletons, which may be calcareous, siliceous, or spongin material—the component of commercial sponges.

Coelenterata

The phylum Coelenterata contains animals that are more complex because they have developed tissue and also a high degree of polymorphism (individual species may occur in a variety of forms). Three classes of this phylum are worthy of mention. The class Hydrozoa includes some jellyfish, such as *Physalia*, more commonly known as the Portuguese man-of-war. Large jellyfish, sometimes as much as 2 m (6.4 ft) in diameter, belong to the class Scyphozoa. The third class, Anthozoa, contains most corals, sea anemones (Figure 8-27), and alcyonarians (Figure 8-28). Corals are important because their calcareous skeletons can form the core of large coral reefs (Figure 8-15).

PLATE 1 A sea spider (*Pycnogonida*) photographed from *Alvin*. (Photograph courtesy of Woods Hole Oceanographic Institution.)

PLATE 2 Millions of these small (about 5- to 7.5-cm, or 2- to 3-in., long) lantern fish were observed in a deep-scattering layer from *Alvin*.

PLATE 3 A crustacean (*Pandalus*). (Photograph courtesy of P. A. Shave, Marine Biological Laboratory.)

PLATE 4 A ctenophora (*Pleurobrachia*). (Photograph courtesy of P. J. Oldham, Marine Biological Laboratory.)

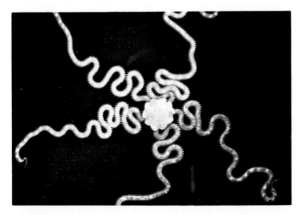

PLATE 5 An echinoderm (*Amphioplus*). (Photograph courtesy of P. J. Oldham, Marine Biological Laboratory.)

PLATE 6 A crustacean (*Byblis*). The specimen has been stained. (Photograph courtesy of P. J. Oldham, Marine Biological Laboratory.)

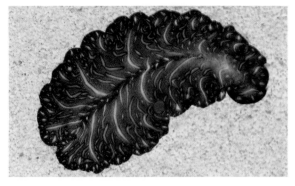

PLATE 7   This is a turbellarian (worm) from the Great Barrier Reef (Heron Island). Turbellarians avoid daylight and crawl about on the undersurface of corals. They are extremely fragile, fragmenting immediately when handled. (Photograph courtesy of Ederic Slater.)

PLATE 8   A large jellyfish common to the eastern seaboard of the United States. (Photograph courtesy of George Lower.)

PLATE 9   This annelid worm lives in a tube secreted by its own body. The appendages are used as paddles to keep the water circulating through its tube, thus bringing in oxygen and small animals upon which it feeds. It is luminescent, which seems strange since there is no opportunity for any other animal to appreciate its beauty. (Photograph courtesy of George Lower.)

PLATE 10   A hermit crab removed from its shell. Its habit of living in an abandoned snail shell has resulted in the loss of the heavy exoskeleton characteristic of other crabs. (Photograph courtesy of Ederic Slater.)

PLATE 11   This is a small shark embryo removed from the uterus of the mother. The long umbilical cord is attached to the half-empty yolk sac, which supplies nourishment to the embryo during its early life. (Photograph courtesy of George Lower.)

PLATE 12   This colorful annelid (worm) lives in a tube among the corals of the Great Barrier Reef. The feathery heads extend when feeding and are quickly retracted into the tube when disturbed. (Photograph courtesy of Ederic Slater.)

PLATE 13   An acorn barnacle (*Balanus*). (Photograph courtesy of P. J. Oldham, Marine Biological Laboratory.)

PLATE 14   An anemone (*Metridium*). (Photograph courtesy of P. J. Oldham, Marine Biological Laboratory.)

PLATE 15   A nudibranch. (Photograph courtesy of P. J. Oldham, Marine Biological Laboratory.)

PLATE 16   A coral (Coelenterate). (Photograph courtesy of P. J. Oldham, Marine Biological Laboratory.)

PLATE 17 A bryozoan (*Bugula*). (Photograph courtesy of P. J. Oldham, Marine Biological Laboratory.)

PLATE 18 A squid (*Loligo*). (Photograph courtesy of P. J. Oldham, Marine Biological Laboratory.)

PLATE 19 A tiny shrimp that lives in the mantle folds of an oyster where it is nourished by the organisms brought to it as the oyster feeds. (Photograph courtesy of Ederic Slater.)

PLATE 20 A sea vase (Tunicate). (Photograph courtesy of P. J. Oldham, Marine Biological Laboratory.)

PLATE 21 A basket star. The five principal arms are divided into a great many smaller branches. This is a bottom view. (Photograph courtesy of George Lower.)

PLATE 22 A sea urchin from the Great Barrier Reef. (Photograph courtesy of Ederic Slater.)

PLATE 23 A nudibranch from the Great Barrier Reef. This one is called "Spanish Dancer" because of its agility when swimming. (Photograph courtesy of Ederic Slater.)

PLATE 24 The *Hero*, an Antarctic oceanographic research vessel that is operated for the National Science Foundation. (Photograph courtesy of William R. Curtsinger.)

PLATE 25 A satellite photograph taken over the Pacific Ocean from a height of about 35,780 km or 22,240 statute miles. Note the low pressure area over southwestern United States and Mexico and a large storm area west of South America. (Photograph courtesy of NASA.)

PLATE 26 A satellite photograph taken over the Atlantic Ocean from a height of about 35,780 km or 22,240 statute miles. A low pressure storm area can be observed just west of the Mediterranean Sea. (Photograph courtesy of NASA.)

PLATE 27 The tanker *Torrey Canyon* lying off the coast of England. Note the oil (dark area) leaking from the vessel.

PLATE 28 Pair of Black Skimmers, nesting. (Photograph courtesy of Karin Engstrom.)

PLATE 29 *Artemisia* (a plant) and bird tracks. (Photograph courtesy of Karin Engstrom.)

FIGURE 8-27 Some anemones photographed in the Gulf of Maine at a depth of 194 m (636 ft).

FIGURE 8-28 An alcyonarian photographed at a depth of about 3,000 m (9,800 ft) on the continental rise off New York. (Photograph courtesy of Dr. David W. Folger.)

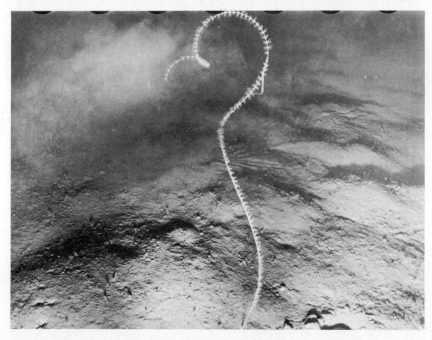

The phyla Platyhelminthes, Nemathelminthes, and Trochelminthes include the unsegmented varieties of worms. Platyhelminthes are flatworms, and include *Planaria*, the "cross-eyed" worm commonly studied in high school or college biology, and the tapeworm found in humans, which is only one of the many parasitic genera in this phylum. There is a wide range of size for adult individuals in this phylum, from as small as 1 cm (0.4 in.) or less to over 20 m (65 ft). Some of the larger forms have been called sea monsters; however, even though they are long, they are very thin and not really up to the standards of a "Loch Ness sea monster."

The Nemathelminthes are the round, or thread, worms. Most forms are parasites. One land variety, hookworm, or *Trichinella*, is a parasite dangerous to humans. Trochelminthes are the so-called wheel worms. Although most examples of this phylum are freshwater inhabitants, some occur as marine plankton.

Chaetognatha

Chaetognatha is a relatively small phylum for which few genera are known. Chaetognaths are very voracious and can quickly consume an entire spawn of small fish. They are small (a centimeter or less in length), transparent creatures that resemble worms and are commonly called arrow, or glass, worms (Figure 8-20).

Annelida

Annelida are the earthworms, leeches, and sandworms. They have segmented, elongated bodies. Most marine annelids are benthic, with many living in burrows. The burrowing forms are important because of their incessant churning and mixing of the upper 10 to 20 cm (4 to 8 in.) of bottom sediment.

Arthropoda

Arthropoda is the largest phylum in the sea, both in numbers and total mass. Arthropods have an external skeleton and numerous jointed appendages. This phylum includes three very important classes: Insecta, Arachnoidea, and Crustacea. Insects, which form a very large and important class of land animals, are almost totally absent from the sea. Only *Halobates*, a water strider, spends its entire life in the sea, living on the water surface.

Arachnoidea includes horseshoe crabs and sea spiders, or pycnogonids. The latter are not true spiders. The arachnids, like insects, form a relatively minor group in the ocean.

The most important class of animals is Crustacea. Crustaceans are divided into several subclasses. One subclass, Cirripedia, includes barnacles. Adult barnacles generally have hard shells and live as attached forms on the ocean bottom, on fixed objects (pier pilings, for example), or on other animals. They have a

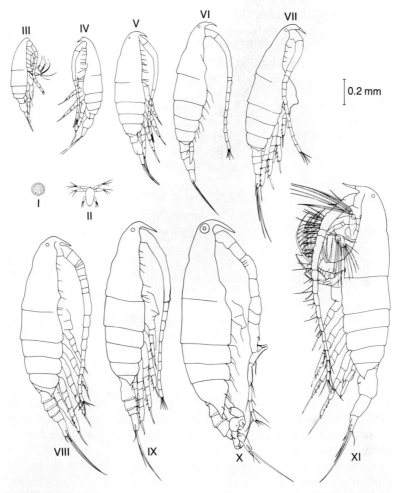

FIGURE 8-29 Most of the growth stages of the copepod *Labidocera aestiva*. All the stages are drawn to the same scale. Some parts of the head appendages are not shown until stage 11, which is an adult stage. (Photograph courtesy of Dr. George D. Grice.)

pelagic larval stage that accounts for their occurrence on objects such as whales or ships. Their growth on ships causes a fouling problem, and they must be removed periodically from most vessels.

The most numerous of the crustaceans are members of the order Copepoda (Figure 8-29). Copepods are important because they eat phytoplankton and concentrate it in their bodies for other larger animals to eat. In this manner they form an essential link in the food chain of the sea. Most copepods are pelagic zooplankton (Figure 8-14).

Probably the most commonly known order of crustaceans is Decopoda; it

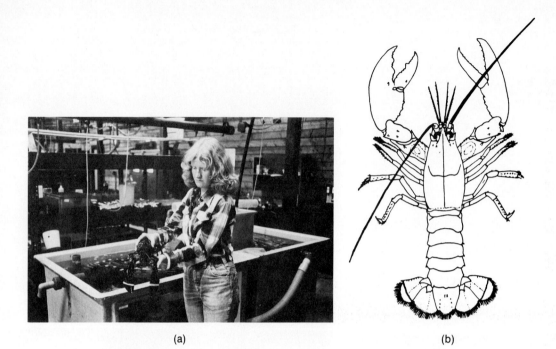

(a)   (b)

FIGURE 8-30 (a) Marine scientist holding a lobster that is being raised in captivity. Note that its claws are secured to prevent it from damaging other lobsters (and scientists). (Photograph courtesy of Phyllis Laking.) (b) The so-called American lobster, *Homarus americanus*.

includes shrimps, crabs, and lobsters. Most of these animals are benthic, and only a few are pelagic forms. This group of animals is rather important to those (Figure 8-30) who fish and trap for them; however, they are less important in the overall picture of the ocean.

Another order, Euphausiacea, also are a common zooplankton. These animals are somewhat larger than copepods and are more advanced in their development. Euphausids, along with some copepods, are capable of extensive vertical migrations, sometimes traveling several hundreds of meters a day. Euphausids (krill is an example) are the favorite food of some whales (Figure 8-31).

The order Amphipoda consists mainly of benthic organisms with some pelagic forms. Recently some extremely large amphipods were photographed feeding at a bait can set at a depth of over 5,000 m (about 16,400 ft) in the North Pacific Ocean (Figure 8-32).

Mollusca

Mollusks have a soft body covered in most genera by a hard shell. This shell may have one or two parts or eight segments. Three classes of Mollusca are noteworthy: Gastropoda, Pelecypoda, and Cephalopoda.

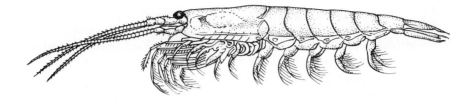

FIGURE 8-31 The common Antarctica krill, *Euphausia superba*.

FIGURE 8-32 Photographs of large amphipods from a depth of 5,304 m (17,402 ft) in the North Pacific. Pictures were taken with a unique system that consists of a weighted can of bait with a camera suspended above it. The camera takes pictures at selected time intervals, it is hoped, of animals attracted to the bait. Eventually the camera is released from the bait can and floats back to the surface where it is retrieved. The largest amphipod measured in this lowering was about 28 cm (about 11 in.) or about twice the length of any previously recorded amphipod. A full account of this experiment is given in R. R. Hessler, J. D. Isaacs, and E. L. Mills, *Science,* **175** (1972), pp. 636–37. (Copyright 1972, by the American Association for the Advancement of Science.) In the lower left-hand picture two fish are also feeding from the bait can (which is a 5-gal can). (Photographs courtesy of Dr. Robert R. Hessler of Scripps Institution of Oceanography.)

Gastropods are terrestrial and benthic-living snails, slugs, and floating forms such as pteropods. These animals have a "foot" that is used in movement on the bottom. In most forms the foot is attached to a hard spiral convolute shell, which can be absent from some of the planktonic species. Pteropods, one of these planktonic forms, may settle on the bottom and form extensive sea-floor deposits after death.

Pelecypods are clams, oysters, scallops, mussels, and the like. These are the typical bivalved mollusks. Most marine animals of this order live either attached to or burrowed in the bottom. Some forms burrow into wood and can cause extensive damage. Two genera, *Terrdo* and *Bankia*, have pelagic larval forms that can bore holes up to 30 cm (about 1 ft) long into wood. Pelecypods constitute an important source of food to man. The settlements of early man can often be identified by larger piles of empty shells, mainly pelecypods.

The cephalopods include squid, octopus (Figure 8-33), and nautiloids. In these animals the solid foot typical of the other classes is divided into arms or tentacles. The squid is a very common animal in the ocean and an important source of food to many other creatures. The giant squid, which may be as long as 18 m (about 59 ft), is the largest living invertebrate animal known. Invertebrates are animals without backbones and include all the phyla mentioned in this chapter except the last one, Chordata.

Echinodermata

Echinodermata is an exclusively marine phylum that includes sea stars, sea urchins, and sea cucumbers. These animals have a five-sided pseudosymmetry and an internal skeleton. Most of the forms are benthic. Four classes are worthy of note here.

The class Holothuroidea includes sea cucumbers (Figure 8-34), which can live in all depths of the ocean. Sea cucumbers have one very peculiar ability: They can eviscerate. In times of danger, they can discharge their internal organs, leaving a meal for their tormentors, and later regenerate another set of organs.

(a)

FIGURE 8-33 A pugnacious octopus that when annoyed by the submersible *Alvin* decided to attack its mechanical arm. (Photographs courtesy of Woods Hole Oceanographic Institution.)

(b)

(c)

(d)

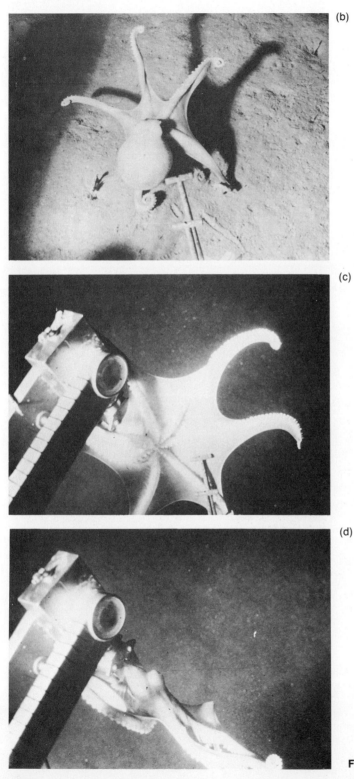

FIGURE 8-33 *(cont.)*

(a)

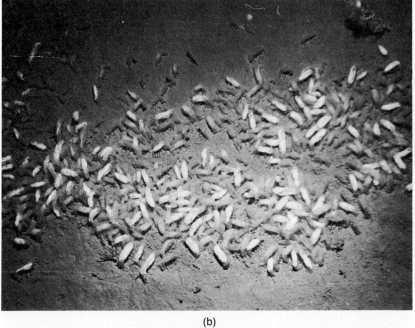

(b)

FIGURE 8-34 (a) Picture of a sea cucumber photographed at a depth of 2,615 m (8,580 ft), about 185 km (100 nautical mi) south of Block Island (near the tip of Long Island). (Photograph courtesy of Dr. Bruce C. Heezen, Lamont–Doherty Geological Observatory.) (b) A large group of holothurians photographed on the continental slope off Virginia at a depth of 1,615 m (5,298 ft). The concentration may be due to a large piece of food that they have found.

Sea stars are a member of the class Asteroidea and are well represented the world over. They are frequently seen in the littoral regions but also occur in the deeper parts of the ocean. Starfish, as they are commonly called, feed on oysters and clams and can cause considerable damage to these valuable food resources. In the past, when fishermen caught starfish they would cut them in half, thinking they had killed the animals, and throw them back into the ocean. Unfortunately for the fishermen (and clams), starfish can regenerate their lost arms, so by cutting them in half, the fishermen were actually doubling their numbers.

More recently a dramatic increase in the population of the starfish *Acanthaster* (Crown of Thorns) has been noted on some coral reefs in the Pacific. This creature feeds on coral and has been responsible for large-scale destruction of reefs in some places, such as Guam. There as much as 90 percent of the coral in some areas was destroyed in a $2\frac{1}{2}$-year period. The reasons for the explosive growth of this form of starfish are unknown. In some areas the rapid growth has quickly decreased, again for unknown reasons, to more normal conditions.

The class Ophiuroidea includes the brittle stars, animals that resemble starfish but that have a more distinct central area. These animals live in the deepest parts of the ocean. The class Echinoidea includes sea urchins and sand dollars.

Chordata

Chordata is a large and very important phylum. Chordates all possess a *notochord*, a series of elements that forms an axial-supporting structure for the animal's skeleton. This phylum has been divided into four subphyla: Hemichordata, Cephalochordata, Tunicata, and Vertebrata. The Hemichordates are mainly certain wormlike organisms, such as acorn worms (Figure 8-35). Cephalochordates are also a worm group. Both of these subphyla are of minor importance in the ocean.

Tunicates are marine filter feeders, animals that feed by filtering out organisms and detritus from the water (Figure 8-36). Some are attached to the bottom while others are planktonic forms. Tunicates can feed on similar items that small fish and other animals eat, thus depriving them of their food.

Vertebrates include all animals with vertebrae and are the most highly developed form of life. Vertebrata includes six important marine classes: Cyclostomata, Elasmobranchii, Pisces, Reptilia, Aves, and Mammalia.

The group Cyclostomata comprises primitive fish such as lampreys and hagfish. They have neither articulated jaws nor scales, and many are parasites or scavengers. The lamprey attaches itself to the side of another fish, makes a hole through the body wall, and then sucks out the fish's body fluids, killing it.

Elasmobranchii are the cartilaginous fish such as sharks (Figure 8-37), skates (Figure 8-38), rays, and chimaera. They have scales that do not overlap, as they do in bony fish. Their mouths are not located at the head (anterior) but somewhat in front of and under it. Sharks are the largest of these fish, some being over 15 m (49 ft) long. At one time, sharks used to be fished for extensively because of the nutritional value of their liver. Nowadays, sharks are commonly eaten in

FIGURE 8-35 An abyssal acorn worm photographed at a depth of 4,735 m or 15,535 ft in the Kermadec Trench. The worm's body ends shortly past the first bend, and the rest is its track. Note another set of tracks in the upper left-hand corner of the photograph. [From D. W. Bourne and B. C. Heezen, *Science,* Fig. 2, **150** (October, 1965), pp. 60–63; copyright 1965 by the American Association for the Advancement of Science.]

FIGURE 8-36 Extraordinary picture of a colony of salps (a tunicate) photographed while floating by *Alvin*. These long chains or colonies may be some of those unexplained sea monsters seen floating in the sea. A small jellyfish is also seen in the middle of the figure. (Photograph courtesy of Woods Hole Oceanographic Institution.)

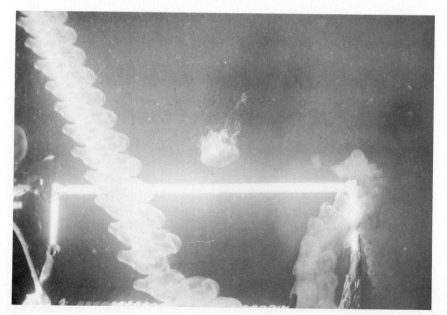

FIGURE 8-37 A large Lemon shark. Note the attached remora, or sucker fish. (Photograph courtesy of Marineland of Florida.)

FIGURE 8-38 A small skate traveling along near the ocean bottom. (Photograph courtesy of Woods Hole Oceanographic Institution.)

some Asian countries, and they are a gourmet food in parts of Europe and North America.

The class Pisces, or fish, is characterized by overlapping scales, an anterior mouth, and a bony skeleton. Of all the animals in the sea, Pisces are the most important marine animal for humans (Figure 8-39). About 70 million tons of fish are caught and consumed by people each year (see Chapter 12). There are over 25,000 different species of fish. Most fish are pelagic, living in all depths of the ocean; however, there are many species that prefer staying near the bottom

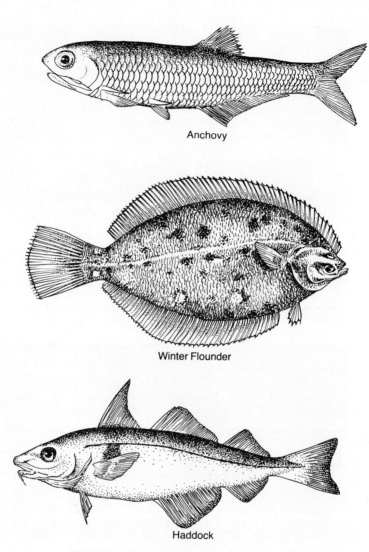

Anchovy

Winter Flounder

Haddock

FIGURE 8-39 Some commonly harvested fish.

(Figures 8-10, 8-11, 8-32, and 8-40), while others carry out most of their activities near the surface. Some fish (called *anadromous* fish) are born in fresh water, then spend most of their time in the ocean, and return to fresh water to spawn. Examples are salmon, striped bass, sturgeon, and smelt. A *catadromous* fish is one that does the opposite—it lives in fresh water but spawns in the ocean; eels are an example.

In the ocean the commonest members of the class Reptilia are turtles and snakes. None of these are very important to the overall economy of the ocean. As air breathers, they must live in surface waters.

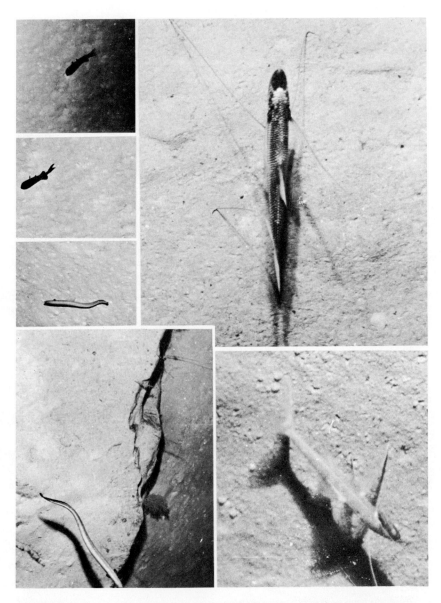

FIGURE 8-40 Some bottom-living fish photographed near Bermuda. The one in the upper left is from a depth of about 1,300 m (4,265 ft). (Photograph courtesy of C. D. Hollister.)

The class Aves, or birds, along with Mammalia, are warm-blooded animals and thus have the ability to maintain a certain body temperature. Birds do not actually live in the ocean, but many are dependent on the ocean for food and return to land only to breed. Guano, the accumulated fecal waste product of birds

Baleen whales

Blue whale
(sulphur-bottom whale)

Fin-back whale
(common rorqual)

Toothed whales

Common porpoise

Bottle-nosed dolphin

White whale (beluga)

Narwhal

False killer whale

Cuvier's beaked whale

Pilot whale
(blackfish)

Bottlenose whale

Killer whale

Sperm whale

0 1 2 3 4 5
meters

Greenland right whale

(bowhead whale)

Sei whale

Humpback whale

California gray whale

Pygmy right whale

FIGURE 8-41   Relative size of various whales. (From Dale E. Ingmanson and William J. Wallace, *Oceanography, An Introduction* (2nd ed.). Copyright 1979 by Wadsworth Publishing Company, Inc. Printed by permission of Wadsworth Publishing Company, Belmont, Calif., 94002.)

and other animals, occurs on some nesting areas and forms valuable fertilizer deposits.

The class Mammalia is the most advanced group of organisms living in the sea. These animals are warm blooded and air breathing. There are three orders with marine genera worthy of note here: Carnivora, Pinnipedia, and Cetacea. The carnivores include the sea otter and polar bear, and the pinnipedes include seals, sea lions, and walruses. Cetacea are the dolphins, porpoises, and whales. Whales are the largest creatures known on earth (Figure 8-41); the blue whale can be over 30 m (about 100 ft) long and weigh about 150,000 kg or 150 tons. Some whales may become extinct because of extensive hunting of them as a source of food and for their oils and fats. Many whales are filter feeders whose diet consists mainly of plankton; others, such as the sperm whale, have teeth and feed on organisms like squid.

The whale is a unique animal in the ocean, being both the largest animal and a mammal, and one of its most magnificent creatures. They have been immortalized both in books (*Moby Dick*, for example) and song (especially touching are the "songs" that the whales themselves make). Nevertheless, these creatures have been hunted for over 1,000 years for their meat and oil.

Many have questioned whether it is ethical to kill whales anymore, especially since so many alternative sources of food and oil exist. Likewise, can the present whale population stand continued hunting without some species' becoming extinct? Since many whales inhabit the deep-sea portions of the ocean, international cooperation is necessary to stop or significantly reduce their harvest.

The 10 principal types of whales are listed in Table 8-3. The largest is the blue whale, which is the largest animal that has ever lived. Note that most feed almost exclusively on plankton, especially krill (small shrimplike crustaceans usually no longer than 5 cm (about 2 in.) (Figure 8-31).

The International Whaling Commission, established in 1948, meets on a regular basis to establish quotas for catch of the various species by different countries. The principal oceanic whaling nations at this time are Japan and the U.S.S.R.; several other countries have coastal whaling industries (Table 8-4). Regulations require a three-fourths vote, but if a country objects, the regulation is not binding on that country. At present, however, regulations are generally accepted. In recent years there has been increased emphasis, with some success, on reducing hunting pressure on whales. It has been hard to document completely the effect of whaling on the whale population, but there appears to be a continual decline in the size of the whales caught. According to Allen (1980) the average weight of whales caught in 1932 was 66 tons, and had decreased to 20 tons by 1978. Some of this decrease is due to the catching of different species of whales, but some is also due to the increased killing of younger ones. At present the quantity of whale meat is only about 0.2 percent of the world's total food catch.

TABLE 8-3 Principal Characteristics of the Large Commercial Whales.

Type of whale	Distribution	Breeding grounds	Breeding behavior	Usual size (ft)[a]	Food
Sperm Whale	Worldwide — Breeding herds in tropical/temperate	Oceanic	Polygynous	M 35–60 F 30–38	Squid, fish
Right Whale	Worldwide — Cool temperate	Coastal	Mixed breeding herds	40–60	Copepods, other plankton
Bowhead Whale	Arctic — Close to edge of ice	—	Mixed breeding herds	40–60	Krill
Gray Whale	North Pacific — Large N–S migrations along coasts	Coastal	Mixed breeding herds	35–46	Benthic invertebrates
Humpback Whale	Worldwide — Large N–S migrations along coasts	Coastal	Mixed breeding herds	35–50	Krill, fish
Blue Whale	Worldwide — Large N–S migrations	Oceanic	Mixed breeding herds	70–100	Krill
Fin Whale	Worldwide — Large N–S migrations	Oceanic	Mixed breeding herds	58–85	Krill, other plankton, fish
Sei Whale	Worldwide — Large N–S migrations	Oceanic	Mixed breeding herds	45–57	Copepods, other plankton, fish
Bryde's Whale	Worldwide — Tropical/warm temperate	Oceanic	Mixed breeding herds	40–50	Krill
Minke Whale	Worldwide — N–S migrations	Oceanic	Mixed breeding herds	23–33	Krill

SOURCE: Data from Allen, 1980.
[a] The usual size given is the range covered by most of the commercial catches. The maximum sizes given have generally been taken from Mackintosh, 1965.

TABLE 8-4 Numbers of Whales Taken in a Recent Year by Member Nations of the International Whaling Commission, and by Nonmember Nations

*MEMBERS			*NONMEMBERS		
Whaling in 1978	**Numbers taken**[a]	**Not whaling**	**Whaling in 1978**[a]	**Numbers taken**	**Not whaling**
Australia	624	Argentina	Chile	62	About 120
Brazil	1,030	Canada	Peru	1,511	countries
Denmark	182	France	Portugal	238	
Iceland	580	Mexico	S. Korea	574	
Japan	8,999	Netherlands	Spain	224	
Norway	1,772	New Zealand	Various[b]	275	
U.S.S.R.	12,139[c]	Panama			
U.S.A.	29[d]	S. Africa			
		U.K.			
Total	25,355 (89.8%)			2,884 (10.2%)	

Source: Data from Allen, 1980.
[a] The figures refer to the 1976–77 and 1977 seasons.
[b] Independent catcher–factory ships registered in various countries.
[c] Includes aboriginal catch of 186 gray whales.
[d] Aboriginal catch of bowheads.

The Organisms and the Ocean

PLANTS AND THE OCEAN

Plants are the key to life in the oceans. These organisms are the primary producers of organic material, and almost all other forms of life are dependent on them for food. This process of organic production is called *photosynthesis*; it is a process common to all plants whether they are floating in the ocean, attached to rocks in the shallow parts of the ocean, or rooted on land.

Photosynthesis is dependent on the plants' being able to receive light. The photosynthetic reaction is an endothermic reaction; that is, it requires energy.

$$CO_2 + H_2O + nutrients + solar\ energy \longrightarrow organic\ matter + O_2$$

Plants contain chlorophyll, a green pigment that allows them to utilize energy from sunlight. This energy, combined with carbon dioxide, water, nutrients, and vitamins, produces organic matter and oxygen (see equation above). The reverse of this reaction, in which organic matter is consumed, is called *respiration*. This process uses oxygen and ultimately returns the nutrients to the water. These two processes together make up the organic cycle in the ocean (Figure 8-42).

Light is essential for plants; it determines the depth to which they can live in the ocean. As seen from Figure 8-21, generally little light is present below a depth of about 200 m (656 ft); thus for plants to survive they must live above this

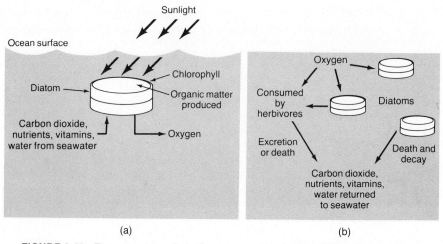

FIGURE 8-42   The organic cycle in the ocean—photosynthesis (a) and respiration (b).

depth. Plants that are attached to the bottom thus are restricted to only a small shallow portion of the ocean. Most plants in the ocean, therefore, are planktonic and float.

For plants to float, however, is not so simple as it sounds. The density of living protoplasm is generally greater than that of seawater; shells are even denser. Plants have evolved several adaptations to enhance their floating ability, the most common of which is to increase their surface area. The frictional resistance of the plant to the water is increased as the surface area to volume ratio is increased. A large surface area, besides helping the plant to float, has another advantage: It brings the plant in contact with a larger amount of nutrients. Other adaptations for flotation include special shapes of some shells that retard sinking, thin shells, and secreted oils or fats that lower the bulk density of the plant. Some phytoplankton occur in colonies of long chains or ribbons (Figures 8-25 and 8-26). This formation of colonies apparently helps to prevent sinking.

Once a plant sinks to a depth where there is insufficient light for photosynthesis, the plant is in danger of dying unless carried back up by currents. There is a certain point, called the *compensation depth*, at which the oxygen produced by photosynthesis equals the amount the plant needs for itself. This is not the lowest depth where photosynthesis can occur, but below the compensation depth the plant cannot really be considered a producer of organic material since it is consuming more than it is producing. The depth of the compensation zone is a function of the intensity of the light, which is affected by factors such as suspended material in the water and turbulence. In coastal water the compensation depth is generally around 20 to 30 m (about 65 to 100 ft). In the open ocean it can be as deep as 100 m (328 ft) or more.

The amount of photosynthesis can be closely related to available light (Figure 8-43). In the upper few meters of the ocean there can be an inhibition to photo-

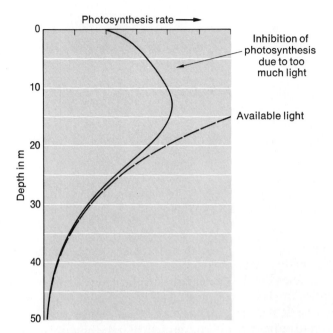

Photosynthesis rate ⟶

Inhibition of
photosynthesis
due to too
much light

Available light

Depth in m

FIGURE 8-43 Photosynthesis as related to available light.

synthesis—apparently due to too much light. Another requirement of the photosynthesis equation is carbon dioxide, which, because of its usual abundance in
the sea, is rarely a limiting factor.

The supply of nutrients can be a critical point for photosynthesis. Some of
the elements known to be essential for plant growth, such as potassium, sulfur,
and magnesium, are usually found in sufficient quantities. Other important nutrients, like nitrogen, phosphorus, and iron, are present in smaller amounts; in
some instances the low amount may be a limiting factor for photosynthesis.

A detailed discussion of the nutrient distribution in the ocean was presented
in Chapter 7. One point that should be emphasized is that the nutrient distribution
in the ocean is not uniform (Figure 7-21). There are seasonal changes (Figure 8-
44) that relate to nutrient use by phytoplankton during their periods of extensive
growth and reproduction (Figure 7-22). There are also geographical differences
that cause variations in concentration and growth of phytoplankton. For example,
in the Arctic the yearly production of organic matter by phytoplankton is relatively
low because of the ice cover, which limits the penetration of light. In the Antarctic
there is a strong water circulation pattern that supplies large quantities of nutrients
to the surface waters; the supply is so large that it is not depleted despite areas
of high productivity.

Nutrient distribution also shows vertical changes. Generally their concentrations are low in the surface waters, where they have been utilized by the
phytoplankton, and relatively high in deeper waters because of their subsequent
decay or consumption (Figure 7-21). This difference is also a seasonal phenomenon, the contrast generally being greater during periods of rapid growth.

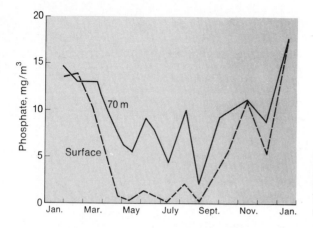

FIGURE 8-44 Seasonal changes in phosphate concentration as measured in the surface waters and at a depth of 70 m (about 230 ft) in the English Channel. (After Atkins, 1926.)

In some instances phytoplankton production decreases even though the nutrients are present in apparently sufficient quantities. In these instances, growth of the plants is limited by perhaps another essential element. Many elements are found in very small quantities in plants, but their significance is not fully understood; vitamins must also be important. Elements such as copper, zinc, and manganese have been shown to be beneficial to the growth of some phytoplankton. Silicon is important to many plants, especially the diatoms, which require it for shell production. In areas where the supply of silicon is low, diatoms tend to have thinner shells, apparently reflecting the scarcity of the element.

The above factors are important to the growth of plants; there are other factors, somewhat less obvious, that affect the metabolism of plants and the density of their distribution.

Salinity is important in that it influences the osmotic pressure between the plant and the environment. In this respect salinity can limit the type of plant that can grow in a particular environment.

Temperature can affect the metabolism and distribution of plants. For example, a twofold or threefold increase in metabolism may occur with a 10°C (about 18°F) rise in temperature. Marine organisms tend to be more sensitive to overheating than overcooling. Certain species of plants are common to polar areas; others, to temperate or equatorial regions. Temperature also affects water viscosity, thus causing some plants to sink.

Light and, to a lesser but important degree, temperature influence the seasonal changes in the phytoplankton population. In the temperate and high latitudes, phytoplankton blooms usually occur sometime in the spring and, after quickly reaching a peak, the population rapidly decreases. This decrease is usually accompanied by a decrease in nutrient concentration. In many areas a similar, though usually smaller, increase in phytoplankton growth occurs in the autumn, followed by a decrease extending over the winter, in turn followed by the spring bloom (Figure 8-45). The exact time of high growth varies according to local conditions. The spring bloom may be caused by an increase in the incident radiation and the development of a stable upper water layer. When this upper layer becomes shallow enough for the phytoplankton to be kept within the well-lighted region, a bloom can occur. If mixing by winds extends too deep, the phytoplankton, which are at the mercy of the currents, are carried below the compensation depth and have insufficient light to bloom. With the beginning of spring there is an increase in water temperature in the upper layers that causes a stratification to the water, which suppresses vertical mixing. In regions of ice and high river runoff, a sharp salinity gradient, due to overlying fresher water, suppresses vertical mixing in a similar manner. This means that the plants are confined to the upper sunlit area and thus have a chance to bloom.

A high concentration of nutrients is necessary for rapid growth. The nutrients in the surface waters that were depleted by the spring and autumn blooms are replenished over the winter by mixing of the water, especially during storms. During the spring and summer months the surface waters are heated, producing a thermocline (Figure 9-7), or temperature stratification. This stratification restricts the transport of the nutrient-rich bottom waters into the surface-water layers where they could be utilized in photosynthesis. Thus the thermocline can act as a brake holding back the growth of the phytoplankton. During autumn and winter, storms and cooling surface waters tend to break up the thermocline and permit mixing of the water layers. In this manner a supply of nutrients is carried to the surface, which sometimes results in an autumn bloom. This bloom is generally smaller than the spring one because less light is available.

In some areas of the ocean, enough nutrient-rich bottom water is moved to the surface by various types of vertical water motion to support a large phytoplankton population. These motions include upwelling, divergence, turbulence,

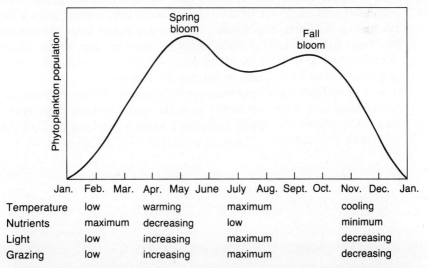

FIGURE 8-45 Diagrammatic representation of yearly changes in the phytoplankton growth and environmental factors such as temperature, nutrients, light, and grazing.

and convection (see pages 313–316). Thus the regions of vertical water motion may also be areas of high organic production.

Areas of upwelling can occur along a coast where the winds blow away the surface water that, in turn, is replaced by nutrient-rich deeper water (Figures 9-15 and 9-16). Turbulence is especially important in shallow areas characterized by strong tidal action, such as the Bay of Fundy. Convection occurs in areas having strong seasonal temperature changes. In winter the water is cooled, becoming denser, and may eventually sink, to be replaced by nutrient-rich deeper water.

In conclusion, the growth processes of marine plants are similar to those of land plants: Both must have enough light and nutrients and an otherwise hospitable environment. The difference between marine and land plants is the result of the smaller concentration of nutrients in the ocean and the fact that plants in the ocean have to float in the surface waters to avoid being carried to the depths where the available light is insufficient to sustain plant life. The growth of these floating plants is of paramount importance in the ocean since all forms of animal life are dependent on them for food.

Animals and the Ocean

The study of animals (and to a lesser degree plants) in the ocean is complicated by the difficulties of sampling them adequately and their greater mobility.

Marine organisms can occur in one of three types of distribution: even, random, or clumped (Figure 8-46). In an even distribution, each organism is an equal distance away from its neighbor, which is not a common occurrence. A rare example is ophiuroids, or brittle stars, which sometimes live in such a proximity that each one of their arms just avoids contact with a neighbor. In a random distribution, no obvious relationship exists between the distribution of any two individuals. This distribution is also very rare in the ocean. The most common type of distribution of organisms in the ocean is the clumped one, wherein animals or plants are found in patches or clumps.

The difficulties of sampling a plankton population are further complicated since some may have the ability to make small vertical movements. With luck one might sample a clumped population using a plankton net, but the probability of finding the same population again and obtaining another sample is small. The

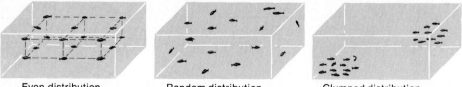

Even distribution   Random distribution   Clumped distribution

FIGURE 8-46   Types of distribution of animals in the sea. Most of the organisms tend to be clumped.

difficulty is even more pronounced with free-swimming nekton. Experiments have shown that when two similar nets are towed through the water at the same depth, but a few meters apart, the quantity of organisms caught in one net may vary greatly from that in the other net.

Organisms that live in the ocean must adapt to the conditions of their environment. Those living in the littoral region must withstand the large temperature and salinity changes that occur within this area. Shelled animals are streamlined and strengthened to withstand the crash of waves; barnacles and similar forms are strongly attached to the bottom or to rocks.

Despite the more arduous physical conditions in the nearshore region, there are ecological advantages, such as abundant food, oxygen, and light. Growth of floating and attached plants is usually high in this environment, therefore there are large quantities of organic matter for food. This growth is sufficient to feed the benthic population and others, as usually demonstrated by a thriving benthic community in the littoral region.

Moving out toward the deep sea, one can observe different zonations of the marine organisms. The reasons for animals inhabiting one region of the ocean rather than another are generally not fully known; however, factors such as temperature, feeding habits, and light obviously must be important.

Temperature is important, for reasons other than those previously mentioned, because it establishes faunal boundaries in the ocean. These boundaries exist in horizontal directions—there are animals typical of equatorial, temperate, or polar regions—and vertical, with depth. Some nekton, however, can move freely from one area into another. The migration pattern of bluefin tuna shows that they can range over most of the ocean. Other animals are more restricted and spend their entire lives within a narrow temperature range.

Temperature also has an effect on the development of some animal forms. Fish tend to develop more rapidly and reach sexual maturity more quickly in warmer regions. In colder areas the fish take longer to develop but generally grow larger.

Salinity is generally not an important factor in the deep ocean because there are only small changes there. Nevertheless, oceanic animals generally cannot tolerate large salinity changes and will succumb if carried by currents into an area of significantly different salinity.

Perhaps the most important factor in the distribution of animals in the ocean is food. It certainly is critical in the deep sea where much of the food has to settle through the overlying waters. It is thought that the food supply in the deep sea is very small. This probably is true in areas very far from land and in areas underlying those with low organic production in the surface waters. In other localities the food supply may be adequate for the relatively small population that lives in the deep sea. Any material, such as plants or dead animals, that sinks below a depth of about 2,000 m (6,562 ft) stands a good chance of reaching the bottom because of the small number of pelagic organisms below this depth, and, therefore, there is the correspondingly small probability of the material's being eaten before reaching the bottom [Figure 8-34(b)]. Bacteria can also be an important source of food in the deep sea.

Almost all types of marine organisms are present in the abyssal zone; fewer are found in hadal depths. The density of life is lower in the deep parts of the ocean than in shallower areas. Animals on the deep-sea floor, as far as we know, also tend to be clumped, perhaps in response to local food supplies, such as a large dead organism. The dominant benthic animals in the deep sea are the coelenterates, echinoderms, crustaceans, and mollusks. Relatively little sampling has been carried out for abyssal and hadal pelagic forms.

An interesting question concerning the deep-sea fauna is, Where did they originate? One possibility is that they slowly moved down from the shallower depths. In seeking an answer, one has to remember that the temperature of the ocean bottom has changed throughout geologic time. There is good evidence, based on isotope studies, that the bottom water temperature may have been 7 or 8°C (12–14°F) warmer 25 million years ago. When the temperature dropped, it could have killed some earlier deep-sea forms. New animals could then have migrated down into the deep-sea area. The recent discoveries of new life forms from the oceanic spreading centers have added another confusing dimension for the evolution of deep-sea life (Figures 6-21, 6-22, and 8-1).

Light is a very important factor in the ocean. The deep sea, however, is essentially completely dark if one considers only solar light. Many deep-sea animals, and even some plants, have the ability to produce light, a phenomenon called *bioluminescence*. Apparently this incompletely understood phenomenon occurs in all parts of the ocean. It can be observed at sea or from beaches when certain dinoflagellates are present in the water. They can give an eerie blue color to a ship's wake or to breaking waves.

There may be several possible uses for bioluminescence. In the dark sea, it may help individuals of the same species to locate each other for breeding. For some deep-sea fish, light is used for feeding: These animals have evolved light organs to attract other creatures. Some squid use bioluminescence for protection. When attacked, they eject a cloud of glowing material that often confuses their antagonists. But for many animals of the sea, such as the planktonic dinoflagellates, the function of bioluminescence is not understood.

An interesting phenomenon of the ocean that appears to be related to light is the deep-scattering layer. The deep-scattering layer was first observed as echoes on depth sounders. Observers on ships traveling on water several hundreds or thousands of meters deep would notice a broad area of sound reflection on their echo sounder that rose toward the surface at dusk and sank at dawn (Figure 8-47). At first, the reflection was confused with the bottom. Some scattering layers appeared as even bands of faint echoes, others as dark, individual inverted U-shaped echoes. The apparent relationship of the reflection with the rise and fall of the sun strongly suggested a biological origin.

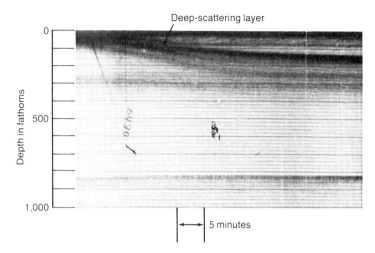

Deep-scattering layer

Depth in fathoms

0

500

1,000

← → 5 minutes

FIGURE 8-47 Deep-scattering layer observed in the eastern Pacific. The descent of the layer (see on the upper left of the record) coincides with sunrise. Note that once the sun rises, the layer stays at a depth of about 150 fathoms (about 270 m). Another layer at about 300 fathoms (about 540 m) has remained stationary during the night and sunrise. (Photograph courtesy of R. H. Backus.)

Net tows made in the layers sometimes collected euphausids; other times, fish or squid. Apparently the organically caused, sound-reflecting, deep-scattering layer occurs in many areas of the ocean; it is not always caused by the same type of organism. One type of scattering layer, called *Alexander's Acres* (Figure 8-48) has been observed by scientists in *Alvin*, and appears to be related to individual schools of small lantern fish.

Another question concerning deep-scattering layers is, Why do these animals migrate up and down, sometimes as fast as 5 m (16 ft) per minute? Since the layers usually move up when the sun goes down and down when the sun rises, a reasonable explanation is that the movements are in response to light. Sometimes when a bright moon is present, the layers will descend slightly. The answer is probably not quite so simple—other factors, such as internal physiological rhythms and feeding, may also be important.

The color of many marine organisms is directly related to the presence of light in their environment. Fish in shallow water sometimes have protective coloring, being dark on top and whitish underneath. This type of coloring tends to make them obscure against the background. For example, a predator looking at them from above would see their dark top against a likewise dark ocean; similarly, looking from below, it would see a light-colored fish against a similar light background. Other organisms can change color or are mottled to make themselves inconspicuous in their surroundings.

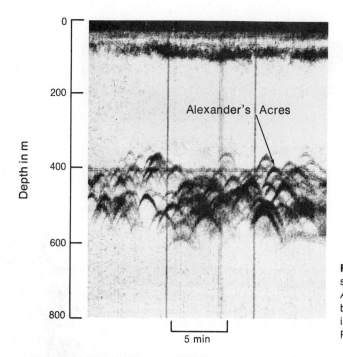

FIGURE 8-48 Echo-sounder record showing the deep scattering layer called *Alexander's Acres.* Note the difference between this layer and the one shown in figure 8-47. (Photograph courtesy of R. H. Backus.)

Organic Production

The production of organic matter by plants (mainly phytoplankton) is clearly the most important biological process in the ocean; without it the basic characteristics and abundance of marine life would be impossible. Organic production can vary considerably with time, location, and other factors. In the polar regions, for example, production is very high in the summer months because the sun shines almost 24 hours a day. In the winter the days are essentially dark and production is very low. Production in the temperate regions was previously discussed. In the tropics, where the light reaching the ocean is fairly constant, production proceeds at a relatively constant rate throughout the year.

When a marine biologist discusses organic production, he or she is referring to the amount of organic matter produced in a unit area or volume (square meter or cubic meter) during a unit time (day or year, for example) and the amount synthesized from inorganic salts by plants. The gross production is the total amount produced, some of which will be utilized by the plant itself during its respiration process. The standing crop is the actual number of organisms. This number is not an absolute measure of productivity but may be related to it. One reason for this is that the production can be high but the standing crop can be kept low because of consumption of the plants by herbivores.

Organic production can be measured by several techniques, mostly based on the photosynthetic equation (see page 273). Other methods include measuring the flow of nutrients going through the organic cycle or the amount of chlorophyll in the water. The latter is an indication of the number of plants present.

One widely used technique of measuring organic production involves ^{14}C. A known amount of carbon dioxide containing ^{14}C is added to seawater containing a known amount of normal carbon dioxide and phytoplankton. The amount of carbon fixed by the plant into organic matter can be calculated by measuring the amount of ^{14}C in the plants at the end of the experiment. An estimate or measurement of the ^{14}C lost by the plant during respiration must be made.

The standing crop can be measured by filtering seawater and collecting the plankton. This method leaves much to be desired, as many small forms, such as coccolithophores (Figure 8-26), can flow through even the finest nets. Thus only when the phytoplankton population is composed of larger forms will filtering give an accurate estimate of the standing crop. Another method requires measurement of chlorophyll, which must come from the plant's cells. Some problems with this method are that the amount of chlorophyll relative to the amount of carbon in the plant can vary and that the quantity of chlorophyll present varies with different species.

Another way of measuring organic production is to monitor the change in the total quantity of nutrients within a particular area. The loss of nutrients, such as nitrogen or phosphorus, can be assumed to be due to their being added to organic matter during photosynthesis. This method has difficulties because one cannot accurately estimate other losses, such as those due to inorganic precipitation or recycling of dead material. In general, the methods of measuring organic production have some drawbacks. They can, however, be used to give a fairly good approximation of relative organic production in different parts of the ocean.

Variations in organic production with latitude have been previously discussed. There are variations in organic production in inshore waters [less than 50 m (164 ft) deep], intermediate waters [100 to 200 m (328 to 656 ft) deep], and offshore waters [deeper than 1,000 m (3,281 ft)]. It has been found that the daily range of the rate of production in these areas is similar but that over a period of time, such as a year, there is a considerable difference in total production (Figure 8-49). Organic production in the nearshore waters generally exceeds that of the intermediate waters and offshore waters, and the production in intermediate waters usually exceed that of the offshore waters. The most reasonable explanation for this observation is that the nearshore waters are richer in nutrients than the offshore waters. This factor appears to be more important than the fact that light penetration is less in nearshore regions than in deep-water regions because of increased turbidity in nearshore waters due to waves and river runoff. Strong wave action in nearshore waters can stir up the bottom sediments and thus also decrease light penetration.

Clouds can locally affect productivity. With a massive cloud cover, little light will reach the sea surface. Winds also affect productivity because choppy water reflects more light than a smooth surface. Wind also mixes the water, providing a continuous supply of nutrients to the surface waters. Temperature,

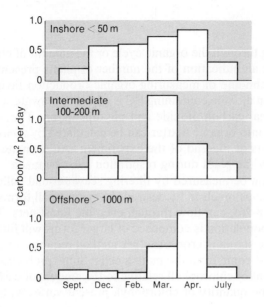

FIGURE 8-49 Comparisons of the daily gross organic production for various months at inshore, intermediate, and offshore stations. The production is measured in the surface waters of these areas. (From Ryther and Yenstch, 1958.)

through its effect on the thickness and depth of the thermocline, is also important. Nutrients will not be recycled from deeper waters if the thermocline is well defined.

The average amount of organic production in the ocean is difficult to determine accurately. Areas of high production, such as Georges Bank, can have values as great as 300 g of carbon per square meter per year (g C/m²/yr). In other areas, organic production can be one-hundredth of this amount. Several estimates place the average gross production of the ocean at about 50 g of carbon per square meter per year (Table 8-5). Areas of high productivity are shown in Figure 12-16. To determine the total mass of organic production, one must consider the thickness of the photic zone (which can be as much as 200 m thick). One estimate is that 20 billion tons (20 trillion kg) of carbon are incorporated into living plant material each year (Table 12-6). Production of organic matter on land is about 25 billion tons per year, or slightly more than that of the ocean. The production from land is more impressive when one considers that only 28 percent of the world is land and much of it is not capable of being used or actually used for agriculture, whereas organic matter can be produced over all the ocean as well as to depths of about 200 m (656 ft).

Estuaries are areas of extremely high organic productivity. This is due, in part, to the large amounts of nutrients carried by rivers into the estuaries and commonly kept there by estuary circulation systems. Many estuaries are flanked by marshes and tidal flats, which are also extremely productive areas. Organic production within some estuarine areas can, in some instances, be higher than the most productive areas on land. Estuaries are also important for their breeding areas for many fish and other forms of marine life. Unfortunately, it is this critical environment that also suffers most of the immediate effects of pollution (see Chapter 13).

TABLE 8-5 Organic Productivity Measurements
from Some Different Areas

Area	g C/m²/day	g C/m²/yr
Open ocean waters	0.05–0.15	18–55[a]
Equatorial Pacific	0.50	180[a]
Equatorial Indian	0.20–0.25	73–90[a]
Upwelling areas	0.50–1.00	180–360[a]
Sargasso Sea	0.10–0.89	72
Continental shelf off New York	0.33 (mean)	120
Fladen Ground, North Sea	—	57–82
Kuroshio Current	0.05–0.10	18–36[a]
Arctic Ocean	0.005–0.024	1
All oceans estimated mean	0.137	50

SOURCE: Data from Ryther, 1963.
[a] Seasonal cycle assumed negligible; annual production computed from daily rates.

GRAZING

In some areas an abrupt decrease in the phytoplankton population during the spring or autumn blooms does not coincide with the depletion of the nutrients. The decrease is often due to grazing (that is, the consumption of the phytoplankton by zooplankton). There is usually an inverse relationship between the numbers of zooplankton and phytoplankton in any given area. Some scientists suggest that this is because the zooplankton tend to avoid large numbers of phytoplankton. Others suggest that the plant population is kept small by the feeding activity of the zooplankton. The grazing is probably the more effective of the two. Many herbivores, such as copepods, have very large appetites for phytoplankton. Most of these animals feed by filtering water and removing phytoplankton in the process; in this manner they can rapidly reduce a local plant population.

The efficiency of zooplankton in catching phytoplankton decreases considerably as the number of available phytoplankton decrease, so some of the plants almost always survive. This is very important because it permits the phytoplankton population to start increasing again when conditions become favorable.

The production of organic material is also reduced by respiration of the phytoplankton as well as by grazing. A model such as is shown in Figure 8-50 can be used to predict the population of an area if the important factors like penetration of solar energy, nutrients, and quantity of zooplankton can be estimated.

In summary, the main production of organic matter in the ocean is by floating plants or zooplankton, with lesser amounts contributed by attached plants in shallow water and by bacteria. The key ingredients in producing organic matter by photosynthesis are water, carbon dioxide, nutrients, sunlight, and numerous other compounds and elements. Production rates are generally controlled by

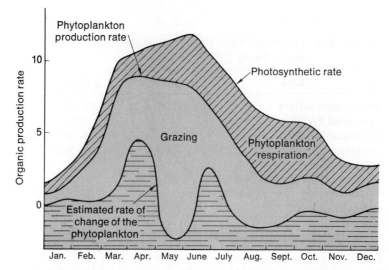

FIGURE 8-50 Estimated rates of production and consumption of carbon by plankton. The top curve is the photosynthetic rate. The middle curve is the phytoplankton production rate. It is obtained by subtracting the respiratory rate from the photosynthetic rate. The estimated rate of change of the phytoplankton is obtained by subtracting the zooplankton grazing from the phytoplankton production rate. (After Riley, 1946.)

availability of nutrients and amount of sunlight. In certain areas where the waters are well mixed, either seasonally or continuously, production can be high because of replenishment of nutrients to the near-surface layers from deep waters.

Some of the organic matter produced by plants is also consumed by them in their respiration. The remaining amount, or net production, is the main source of food, either directly or indirectly, for most of the animals in the ocean.

The Food Cycle

Up to now we have considered the different organisms and their responses to various factors of the environment. Another fundamental relationship in the sea is the food cycle. The cycle starts with the production of organic matter by phytoplankton. Organic matter is then consumed by herbivores, the zooplankton. The herbivores are then consumed by a higher form of animal (like sardines), which ultimately is consumed by a larger predator (like tuna). This highly simplified scheme is shown in Figure 8-51. By this method, some have estimated that 1,000 kg of plants will produce 1 kg of the larger predators, based on a 10 percent conversion efficiency going up from each level. But there are many other complex relationships that exist in the sea than are indicated by this simple estimate. A more complicated and realistic picture is shown in Figure 8-52. This diagrammatic representation also shows the importance of the nutrients and bacteria in the cycle.

The marine oceanic food cycle differs from the land cycle mainly because the primary producers in the ocean, plants, must be small and must float. Animals

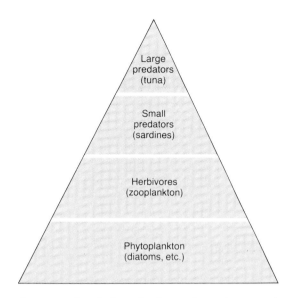

FIGURE 8-51 A simplified illustration showing the different feeding levels in the ocean.

have evolved that feed on the microscopic plants, thereby concentrating the organic matter. Because plants are so numerous, it follows that the consumers of plants must also be numerous.

To completely understand the food cycle among the plankton, nekton, and benthos, one should examine the feeding habits and interrelationships of all the individual species—a very large task.

The question of efficiency (for example, does 1,000 kg of phytoplankton produce 100 kg of zooplankton?) is also very complex. The exact efficiency is variable but each additional step in the cycle does cause a net loss of organic

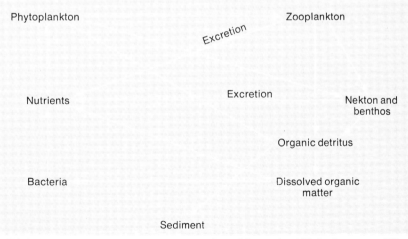

FIGURE 8-52 A diagrammatic representation of the cycle of life in the ocean. (After Raymont, 1963.)

material. Generally, therefore, there is a low degree of efficiency in the transfer of organic matter from the phytoplankton to the fish. The point of efficiency is one that must be considered if one wants to feed larger portions of the world's population from the sea. Perhaps we should either take our food from a lower part of the food cycle or find a way to make the transfer of organic matter more efficient.

Summary

Biological oceanography emphasizes the study of the animals and plants in the ocean and their interaction both with themselves and their environment. In spite of a long history of biological research into the ocean, much basic information still remains to be discovered; dramatic evidence of this is the recent discoveries of new organisms along parts of the ocean ridge system.

The biological oceanographer, because of sampling limitations and the way plants and animals tend to be distributed in the ocean (generally clumped), has considerable difficulty in ascertaining the actual amount of organisms present, or returning to an area for resampling. Often research is made under simulated oceanographic conditions in a laboratory or controlled conditions in the field.

Organisms that live in the ocean have several advantages over terrestrial ones, but they also have disadvantages. For plants, only a small portion of the sea floor is adequate for attached growth, so most plants have had to evolve as floating organisms (phytoplankton) so that they could remain in the surface waters of the ocean where they can receive the light necessary for photosynthesis. The production of organic matter by plants is a key process in the ocean, since it produces the food needed by all the other organisms for survival. The major ingredients for producing organic matter by photosynthesis are water, nutrients, carbon dioxide, sunlight and some other trace compounds and elements. In general, it is the nutrients and sunlight that are limited. Certain areas of the ocean, especially where the waters are well mixed and the nutrients are recycled to the surface waters, are especially productive. Since the plants that produce this food are so small, another type of organism called zooplankton is necessary to accumulate the organic matter (by eating the phytoplankton) into large enough particles for animals higher in the food chain.

The marine environment can be divided into two major realms: benthic (the bottom) and pelagic (the overlying water). Life is generally more abundant in the nearshore regions where the food supply is more abundant. Organisms can be classified by their mode of locomotion or habitat (benthos—those that live on or in the bottom; nekton—those that swim; and plankton—those that float) as well as by their taxonomy. Most marine organisms are fundamentally related to each other via the food cycle. The cycle is initiated by organic matter production by phytoplankton, which are then consumed by herbivores (zooplankton) which in turn are consumed by a higher form of life, such as small fish like sardines. Finally these are eaten by bigger predators, such as tuna.

Suggested Further Readings

ALLEN, K. R., *Conservation and Management of Whales*. Seattle: University of Washington Press; London: Butterworth & Co., 1980, 107 pp.

BURGESS, J. W., and E. SHAW, "Development and Ecology of Fish Schooling," *Oceanus*, **22**, no. 2 (1979), pp. 11–17.

COKER, R. D., *This Great and Wide Sea*. Chapel Hill: The University of North Carolina Press, 1947.

DALE, B., and C. M. YENTSCH, "Red Tide and Paralytic Shellfish Poisoning," *Oceanus*, **21**, no. 3 (1978), pp. 41–49.

DARWIN, C., *The Voyage of the Beagle*. New York: Harper and Row, 1959.

DUDDINGTON, C. L., *Flora of the Sea*. New York: Crowell, 1967.

GRASSLE, J. F., "Diversity and Population Dynamics of Benthic Organisms," *Oceanus*, **21**, no. 1 (1978), pp. 42–49.

HARBISON, G. R., and L. P. MADIN, "Diving—A New View of Plankton Biology," *Oceanus*, **22**, no. 2 (1979), pp. 18–28.

HARDY, A. C., *The Open Sea: Its Natural History*. Boston: Houghton Mifflin, 1965.

HASTINGS, J. W., "Bioluminescence," *Oceanus*, **19**, no. 2 (1976), pp. 17–27.

HEEZEN, B. C., and C. D. HOLLISTER, *The Face of the Deep*. New York: Oxford University Press, 1971.

HENNEMUTH, R. C., "Marine Fisheries: Food for the Future?" *Oceanus*, **22**, no. 1 (1979), pp. 2–12.

IDYLL, C. P., *Abyss: The Deep Sea and the Creatures That Live in It*. New York: Thomas Y. Crowell, 1964.

ISAACS, J. D., and R. A. SCHWARTZLOSE, "Active Animals of the Deep-Sea Floor," *Scientific American*, **233**, no. 4 (1975), pp. 84–91.

ISAACS, J. D., "The Nature of Oceanic Life," *Scientific American*, **221**, no. 3 (1969), pp. 147–62.

JANNASCH, H. W., and C. O. WIRSEN, "Microbial Life in the Deep Sea," *Scientific American*, **236**, no. 6 (1977), pp. 42–65.

MCCONNAUGHEY, B. H., *Introduction to Marine Biology* (3rd ed.). St. Louis, Mo.: C. V. Mosby, 1978.

PARSONS, T. R., M. TAKAHASHI, and B. HARGRAVE, *Biological Oceanographic Processes*. Elmsford, N.Y.: Pergamon Press, 1977, 332 pp.

RESECK, J., Jr., *Marine Biology*. Reston, Va.: Reston Publishing Co., 1979, 257 pp.

RICKETTS, E. F., and J. CALVIN, *Between Pacific Tides*. Stanford, Calif.: Stanford University Press, 1968.

STEELE, J. H., "Patterns in Plankton," *Oceanus*, **23**, no. 2 (1980), pp. 2–8.

TAIT, R. V., and R. S. DESANTO, *Elements of Marine Ecology*. New York: Springer-Verlag, 1972.

TEAL, J., and M. TEAL, *Life and Death of the Salt Marsh*. Boston: Little, Brown, 1969.

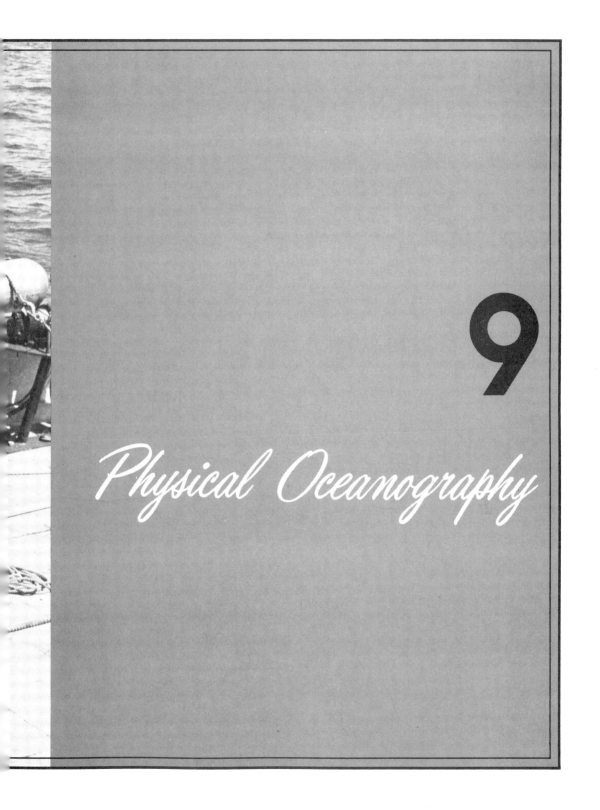

9

Physical Oceanography

PHYSICAL OCEANOGRAPHY COMBINES the study and measurements of the physical properties of the sea and their variations in time and space with a theoretical study of the processes that control the state of those physical properties. The first aspect is mainly a description of the properties and motion of the ocean. The second aspect is associated with theoretical physics.

Physical oceanography is probably that field of oceanography least dependent on the other fields, but it cannot be wholly separated from them. For example, water chemistry is important in determining seawater density, and variations in density, in turn, affect the circulation and currents of the ocean. The extent of some ocean currents was first determined by following the marine organisms living in them. Surface distribution of nutrients (and therefore the areas of biological activity) in the ocean is, in part, controlled by the physical conditions of the sea. Currents, either catastrophic ones such as turbidity currents or regular ones, distribute sediment over the ocean. The bottom topography of the ocean, which may be modified by currents, can also affect the large-scale oceanic circulation. Finally, the interaction of the ocean and the atmosphere is an extremely important influence on the earth's weather and climate.

History of Physical Oceanography

People's first interest in the sea was a practical one. They wanted to know the best way of going from one place to another. Early exploration of the sea probably started with the Phoenicians and the Greeks who established lines of commerce in the Mediterranean and Atlantic several hundred years before the birth of Christ. Much of the information obtained by early sailors was not recorded but kept as a state or family secret. It was not until the fifteenth and sixteenth centuries that people developed a realistic understanding of the dimensions and shape of the ocean. The voyages of Magellan, Diaz, and Columbus and later of Drake and Cook were of primary importance to this understanding.

The first expedition to measure the depths of the ocean was that of Sir Clark Ross from 1839 to 1843. Early attempts at mapping oceanic currents were made by Benjamin Franklin and later by Findlay and Maury (see Chapter 2). Maury examined log books of numerous ships and noted that oceanic currents are related to the wind. His published findings (1855) combined with Findlay's of 1854, and those of the later Challenger Expedition (1872–1876), provided a basis for understanding the dynamic structure of the ocean. One especially unique expedition was that by Nansen to the Arctic to study how winds and currents affect pack ice. Nansen built a specially strengthened ship, *Fram*, that was frozen into an ice flow and drifted from 1893 to 1896 almost reaching the North Pole. *Fram* was eventually freed from the ice and returned to its home port in Norway. In addition to this outstanding work, Nansen also invented the Nansen bottle (Figure 9-2).

Marine disasters have, unfortunately, been very influential in promoting oceanography. For example, the sinking of the *Titanic* in 1912 after it hit an iceberg led to the establishment of the International Ice Patrol. Much was learned about oceanic circulation by studying the movements of ice.

It was not until the German Meteor Expedition (1925–1927) that oceanographic expeditions changed from those of a worldwide and general style to those of a more localized and detailed nature. The members of the Meteor Expedition studied the South Atlantic Ocean and deduced deep oceanic circulation by numerous measurements of temperature and salinity.

During World War II, the lack of knowledge about the physical aspects of the ocean gave considerable impetus to studies of subjects like the propagation of sound through water and the prediction of wave height in coastal areas (for troop landings). Many of these efforts are still priorities in research. More recently, large-scale, often international, efforts have become an effective way of studying the often subtle aspects of physical oceanography.

Present Objectives

Research in recent years has emphasized the theoretical causes for the observed distribution of various seawater properties, circulation patterns, the origin of waves, and other aspects. The development of new sensing devices permits accurate and inplace measurements of temperature, salinity, currents, and other variables (Figure 7-2). Satellites have also become an important tool for physical oceanographers (Figures 14-13 and 14-14). Advances in data processing, especially use of modern computers, allow meaningful results to be obtained quickly, in some instances even before the vessel has left the study area.

Physical oceanographers are especially interested in the interrelationships of the atmosphere and the ocean. These interrelationships are important controls in both atmospheric and oceanic circulation. Understanding and prediction of weather, both oceanic and terrestrial, await more thorough study of the air–sea interactions (see Chapter 10). Such studies can be aided by the use of unattended oceanic buoys or satellites that can transmit information directly to the laboratory.

The ability to predict specific characteristics of the ocean is very important. At present, only tides, surface waves, and tsunamis (large ocean waves produced by submarine earthquakes) can be determined in advance. Predictability of the characteristics of the upper layers of the ocean would be especially valuable for study of sound propagation and marine biological problems.

Recent years have seen increased interest in predicting oceanic circulation. An understanding of circulation requires knowledge of how the physical properties of seawater are changed by heating and cooling, how the atmosphere interacts with the ocean, and how the earth's rotation affects the ocean. To obtain this knowledge requires a combination of mathematical studies and extensive and elaborate field measurements.

Problems such as large-scale oceanic circulation can best be solved by synoptic measurements (many measurements at the same time over a large area). Synoptic observations in the past required numerous expensive surface ships but recently developed fixed buoys and platforms used in conjunction with airplanes, submarines, satellites, and surface vessels make the procedure now more efficient. Use of man-made contaminants, such as radioactive isotopes like the tritium produced from atmospheric atomic bomb blasts, can provide valuable data about oceanic circulation. One such new program is called Transient Tracers in the Ocean (TTO) and has already provided some interesting data on water motion (Figure 7-24). These large programs require the cooperation and participation of scientists from many countries. One of these programs, called MODE (Mid-Ocean Dynamics Experiment) involved half a dozen research vessels, scientists from 15 institutions, and sophisticated technology. The main objective of this program was to study the motion of eddies, or large circular flows of water that extend from the surface to the ocean depths. These eddies, which are about 200 km (about 124 mi) in width and last about 2 months, are similar to weather systems in the atmosphere. The success of MODE led to a larger experiment called PO-LYMODE. It emphasized more sophisticated studies of oceanic eddies over periods of up to 18 months. Several different types of eddies have been found including small ones (50 km or 31 mi in diameter) that did not reach the ocean surface; in the Pacific, eddies up to 1,000 km (621 mi) in diameter were identified. These masses of water have been compared to underwater storms with "water winds" of several 10's of centimeters per second. They can affect large areas for periods of several months. In doing so, they are carrying water, essentially undiluted, over large horizontal distances. One of the key questions is how these eddies interact with the large-scale oceanic circulation patterns.

Perhaps somewhat related to oceanic eddies are the studies being made of Gulf Stream rings. These are large loops or meanders that pinch off from the Gulf Stream and form a circular ring of water having strong currents within them (Figure 9-1). The rings are often about 150 to 300 km (93 to 186 mi) in width and may extend to depths of 3,500 m (about 11,500 ft), almost reaching the bottom.

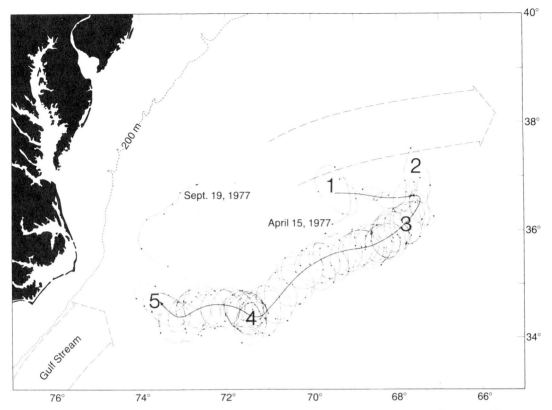

FIGURE 9-1 Trajectory of a satellite-tracked, free-drifting buoy looping around the center of a Gulf Stream ring during the period April 15 to September 15, 1977. The movement of the ring can be shown by the shifting center of the loops. The number 2 refers to position during the April 26–29 period; 3, to May 15–17; 4, to July 31–August 2; and 5, to September 13–15. During April the ring became connected with the Gulf Stream and moved rapidly (0.5 kn) eastward. In May it separated from the Gulf Stream and began its southward drift (0.1 kn). In September the ring rapidly and completely coalesced with the Gulf Stream and was lost. (Photograph based on the work of Dr. Phil Richardson, Woods Hole Oceanographic Institution.)

Often 10 or more of these rings may exist, and several have been tracked for months. Some rings may exist for 2 to 3 years. Similar types of rings have been found in the Kuroshio Current region, the Pacific analog of the Atlantic Gulf Stream, and along the Antarctic Circumpolar Current. Like eddies, these rings can move large masses of water, transporting organisms or pollutants to other areas in the process. The interfacial area between a ring and the surrounding seawater has disturbed or distorted physical characteristics and could be a locality where submarines might be harder to detect. (Submarines are often detected by sound velocity measurements that would be harder to make in such regions because of the varied physical conditions.) As in the case of eddies, little is known about how these rings affect some of the larger aspects of the ocean. This question will be among the more important ones considered in the coming years.

Two other areas of active research are those of coastal upwelling and climate (see Chapter 10). Both of these have direct influence on the quality of life on earth and offer the possibility of being controlled or modified by human activity.

Instruments of the Physical Oceanographer

The physical oceanographer is interested in measuring the important physical properties of seawater, such as its temperature, salinity, and density. These measurements can then be used to deduce evaporation, heat exchange, currents and water movements, and other physical processes occurring in the ocean.

There are three common ways of measuring the temperature of ocean water. The most frequently employed is a very accurate recording thermometer. This thermometer is usually attached to a Nansen bottle (Figure 7-6) and lowered into the ocean (Figure 9-2). When the Nansen bottle is at the desired depth, a weight (called a messenger) is dropped down the wire and when it hits the bottle, it overturns, releasing another messenger to perform the same task to other bottles farther down the line. When the bottle overturns, it traps a sample of seawater that can be used for salinity determination or other measurements. The overturning of the bottle also breaks the mercury column inside the thermometer, preserving the temperature measurement made at that time. If this did not happen, the recorded temperature would change and rise as the bottle was lifted back to the ship and passed through the warmer surface waters. Usually two of these thermometers are used. One is exposed to the pressure of the overlying seawater; the other is protected from this pressure. The pressure of the seawater squeezes the unprotected thermometer causing an anomalous temperature reading proportional to the pressure. The difference in the temperature between the two thermometers is then a measure of the pressure, or depth, since pressure is closely related to depth. A high degree of accuracy is necessary for these temperature measurements; values should be accurate to better than 0.02°C (almost 0.03°F). Accuracy is necessary because temperature has a very important effect on density and other physical properties and also because temperatures in the deep sea are relatively constant and have such extremely small variations that only a very accurate thermometer could record them. The lowering of several Nansen bottles at one locality constitutes a *hydrographic station.*

These hydrographic stations can take considerable time and obtain measurements at only a few discrete points rather than throughout the water column. In recent years techniques have been developed that can continuously measure temperature, salinity, and other parameters. A simple device for measuring temperature is the *bathythermograph*, or BT (Figure 9-3), which can be quickly lowered from a vessel even while it is in motion. Pressure- and temperature-sensitive elements produce a continuous plot of temperature versus depth on a

FIGURE 9-2 Nansen bottle being used on a hydrographic station. Scientist's left hand is on the container that holds the two reversing thermometers. (Photograph courtesy of Woods Hole Oceanographic Institution.)

coated glass slide. The BT, which is simple to use, is not as accurate as a good thermometer but has the advantage of producing a continuous picture. An expendable version of the BT also exists that can transmit its data directly to a surface vessel or an airplane (Figure 3-16).

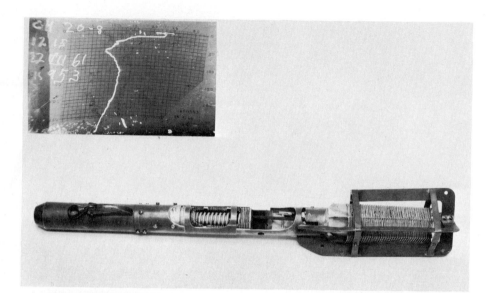

FIGURE 9-3 A bathythermograph, or BT. This device is lowered from a ship and will produce the record shown in the upper left. The horizontal scale is temperature; the vertical is depth. The notation indicates the cruise and time of measurement. Note the rapid decrease in temperature (called the thermocline) within the upper part of the ocean. (Photograph courtesy of Woods Hole Oceanographic Institution.)

Other new instruments combine the accuracy of thermometers with the speed and continuous measurement capabilities of the BT by using various sensors and by either transmitting the data directly to the ship by telemetry or storing it until the instrument is retrieved. Because they can also measure several variables and are very accurate and sensitive, these devices can be used to examine the small-scale details of the various physical properties of the ocean (Figure 3-24).

Speed and direction of currents in the ocean are measured by both direct and indirect methods. One direct method is ship drift, in which the difference between the anticipated and the actual arrival point of a ship is assumed to be due to currents [Figure 9-4(a)]. Another method of measuring currents is to put objects into the water and either note their movements or let them drift freely unobserved [Figure 9-4(b) and (c)]. Drift bottles usually have a card inside them offering a small reward if the finder returns the card and tells where and when he found it. Drift bottles have certain disadvantages. Their exact routes to the places where they are found is not known, nor is the influence of the wind on their drifts. Nevertheless, by plotting the launchings and recoveries of many of these bottles, a general notion of large-scale surface oceanic circulation can be gained.

Drogues can be used to measure currents slightly below the surface, away from the direct effects of the wind [Figure 9-4(c)]. A sophisticated type of drogue is the *Swallow Float*, a sealed aluminum tube that is dropped into the ocean. The

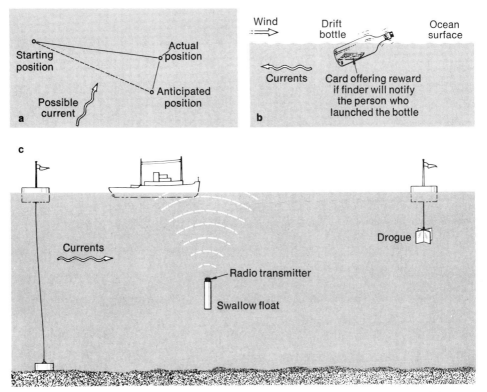

FIGURE 9-4 Some direct methods of measuring currents: (a) ship drift; (b) drift bottles; (c) drogue and Swallow Float. In the last method, the ship maintains its position relative to a buoy and observes the movement of the drogue (visually) or Swallow Float (electronically).

instrument is heavier but less compressible than saltwater and at a certain depth its density equals that of seawater. (In other words, water has a relatively greater increase in density with depth than does the Swallow Float.) If a sound device is attached to the float, its movement at a given depth can be monitored. The surface ship must, however, have some method of positioning itself, as it may also drift with the surface currents. A buoy is often used and the movements of the ship and of the Swallow Float or drogue are determined relative to the buoy. Measurements obtained from the Swallow Float show that the deep ocean, which had been thought to be relatively stagnant, can have currents moving at a speed of almost 50 cm per second.

Other types of current-measuring devices are telemetering buoys, free-fall devices (Figure 9-5), and current meters (Figure 9-6). Current-measuring devices can be suspended from fixed objects such as buoys and lightships and can monitor the currents for long periods of time. More sophisticated devices transmit the data to shore-based laboratories or computers. Current-measuring instruments can record either the average current velocity over a long period of time or the

FIGURE 9-5 Free-fall instrument that obtains profiles of current velocity and sea-water density between the sea surface and bottom. Velocity profile is inferred from measurements of weak voltages induced as the instrument is carried through the earth's magnetic field by the ocean currents encountered in its vertical traverse of the water column. (Photograph courtesy of Dr. Tom Sanford, University of Washington.)

instantaneous value. A current measured by an averaging device, such as flowing to the north at 50 cm per second may actually have been moving at other speeds part of the time or even in another direction.

There are several problems in leaving instruments unattended in the ocean. These include loss (or theft), damage due to storms, and biological growth [Figure 9-6(b)] that can reduce the capabilities of the instrument.

Indirect methods of measuring currents include the measurement of temperature, salinity, oxygen, or other properties of seawater. Because many of these properties are strongly influenced or determined by surface phenomena, their

(a)

FIGURE 9-6 (a) A current-measuring device. Water moving by the rotor on the bottom of the instrument causes it to turn; the number of turns (recorded inside the instrument) per unit time is a measure of the current. Instruments like this are commonly used in buoy systems. (Photograph courtesy of Woods Hole Oceanographic Institution.)

character or quantity at depth or along a distance from a source can be a measure of oceanic mixing and currents.

The distribution of certain organisms can also indicate current systems. The presence of floating organisms in an area far removed from their usual habitat suggests that they were transported there by currents.

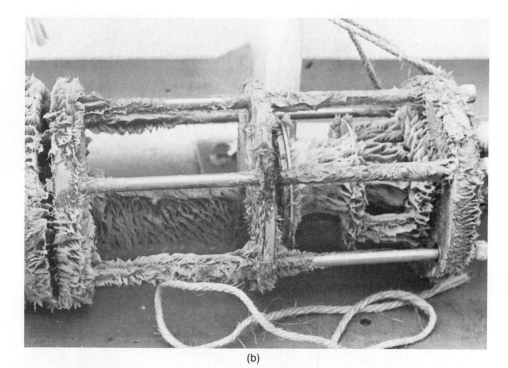

(b)

FIGURE 9-6 (b) Biological growth on a current meter left on the New England shelf at a depth of 85 m (about 279 ft) for 6 months. (Photograph courtesy of John Vermersch.)

Physical oceanographers, like other oceanographers, are using modern electronic techniques of instrumentation as well as the more conventional equipment described above. Satellites, airplanes, and large electronically laden buoys provide data about various oceanic parameters. Analysis of these data is commonly done by high-speed computers, in many cases directly on the ship (Figure 3-4).

General Characteristics of the Ocean

SALINITY

Three very important properties of seawater are its salinity, temperature, and density. Salinity is defined as the total amount of dissolved material, in parts per thousand by mass, in 1 kg (about 2.2 lb) of seawater when all the bromine and iodine have been replaced by an equivalent amount of chlorine, all the carbonate has been converted to oxide, and all the organic matter has been oxidized. It generally is impractical to analyze for every component of seawater; usually just a single element related to salinity, such as chlorine, is measured (salinity ‰ = 1.8066 × chlorinity ‰). This technique is possible because the major components of seawater have an essentially constant ratio to each other (Forchhammer's Principle, see pages 181–182). Salinity usually is determined by measuring the electrical conductivity of seawater, a method accurate to 0.002 ‰. Continuous measurement of salinity by *in situ* measuring devices (Figure 3-24) is possible but is less accurate. The conductivity of seawater is a measure of its ability to transmit an electrical current and is dependent on the concentration of ions (charged atoms) in the water. By comparing the conductivity of a sample of seawater with that of a standard, an estimation of salinity can be made.

The principle factors influencing salinity are evaporation and precipitation. Two less important factors are freezing or melting of ice (when seawater freezes the salts are left behind) and runoff from land. In areas of high evaporation, such as the Red Sea, salinity can be as high as 40 ‰. In most parts of the ocean, however, the salinity range is from 33 to 37 ‰, with a median value of about 34.7 ‰. The higher salinity values occur near the arid equatorial areas; the lower values, near the polar regions. Actually about 75 percent of the water in the ocean has a salinity range between 34.50 and 35.00 ‰, so accurate measurements are necessary to resolve small differences. Differences of a few hundredths of a part per thousand can be important for some oceanographic processes.

The distribution of salinity with depth in the ocean, which is very similar to the temperature distribution, can be expressed as three or four zones (Figures 9-7 and 9-8):

1. A well-mixed surface zone, 50 to 100 m (164 to 328 ft) thick, of generally uniform salinity.
2. A zone with a relatively large salinity change, called the *halocline*.
3. A thick zone of relatively uniform salinity, extending to the ocean bottom.
4. An occasional zone at a depth of 600 to 1,000 m (1,968 to 3,281 ft) in some areas, where there is a minimum salinity.

TEMPERATURE

Temperature is probably the most commonly measured oceanographic variable. It can be measured with mercury thermometers mounted on water sampling devices, such as Nansen bottles (Figure 9-2), or by electronic devices. The water obtained in the Nansen bottle is generally used for salinity determinations, as well as for oxygen and nutrient measurements. Sophisticated instrumentation, using temperature sensors and transmitting devices can permit continuous measurements of temperature as the device is lowered through the ocean (Figure 3-24). Other devices, such as infrared radiometers, when used from an airplane or satellite, can give instantaneous readings of the sea surface temperature.

The surface of the ocean is heated by the following:

1. Radiation from the sky and sun.
2. Conduction of heat from the atmosphere.
3. Condensation of water vapor.

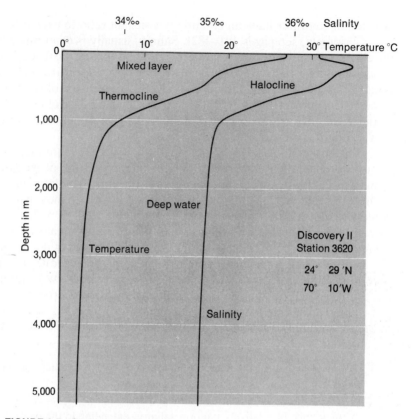

FIGURE 9-7 Typical salinity and temperature profiles. Data from *Discovery II*, Station 3620. (After Fuglister, 1960.)

The sea surface is cooled by the following:

1. Back radiation from the sea surface to the atmosphere.
2. Conduction of heat back to the atmosphere.
3. Evaporation.

Ocean currents can transfer heat from one area to another by bringing bodies of water having different temperatures in contact with each other.

The surface temperature of the ocean is closely related to latitude and time of year, since more heat per unit area is received at the equator than at the poles (Figure 9-9) and in the summer than in the winter. The usual temperature pattern with depth in the ocean (Figures 9-7 and 9-8) consists of three principal layers:

1. A warm, well-mixed surface layer, from 10 to perhaps 500 m (about 33 to 1,640 ft) thick.
2. A transition layer, below the surface layer, called the *main thermocline*, where the temperature decreases rapidly. The transition layer can be 500 to 1,000 m (about 1,640 to 3,281 ft) thick.

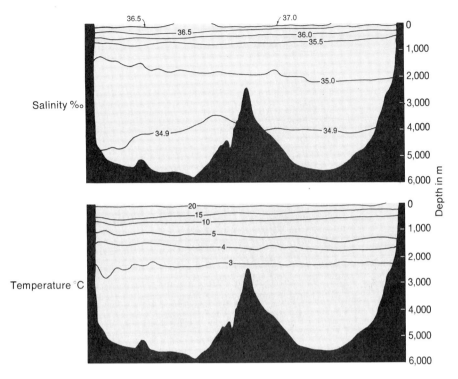

FIGURE 9-8 Salinity and temperature section across the North Atlantic. Section runs west to east at about 24° N. (After Fuglister, 1960.)

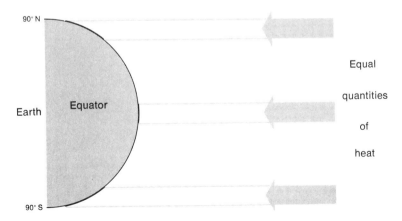

FIGURE 9-9 The difference in heat received per unit area for polar and equatorial regions.

3. A layer as much as several kilometers thick that is cold and relatively homogenous where the temperature slowly decreases toward the ocean bottom.

The main thermocline is virtually absent in polar regions since most of the ocean surface is covered with ice in winter, and solar radiation is small in summer. In the tropics, the thermocline may be close to the surface. Areas having a strong seasonal warming also may have a temporary or seasonal thermocline in the surface layer.

As we shall see in later sections, the ocean is constantly transferring heat from the equatorial regions toward the poles. On the surface this transfer is accomplished by currents, such as the Gulf Stream, which moves warm waters toward the poles. In the polar regions, the relatively denser waters of this region (see next section) slowly sink and move toward the equator forming the deeper water layers of the ocean.

Temperature and salinity values below the main thermocline usually have a very close relationship. This relationship can be used to define different water types or water masses. Temperature–salinity relationships can also indicate the source and mixing of water masses and are especially important in studies of deep-ocean circulation.

DENSITY

The third important physical property of seawater is its density. Density of seawater (its mass per unit volume) ranges from about 1.02 to 1.07 g per cubic centimeter. It is controlled by three variables whose interactions are complex: salinity, temperature, and pressure[1]. In general, density increases with increasing salinity, increasing pressure (or depth), and decreasing temperature. Thus colder, deeper, more saline water is also usually the densest water. Density of seawater can be calculated if these three variables are known precisely.

Changes in seawater density result from processes such as evaporation or heating that occur at the sea surface. We shall see in subsequent sections that the deep, or *thermohaline*, circulation of the ocean is due to density differences in the ocean.

There is a tendency, because of gravity and buoyancy forces, for denser water to sink, and less dense or lighter water to rise to the ocean's surface. This movement results in a stable density stratification (increasing density with increasing depth) to the ocean. The upper 100 m or so of the ocean are strongly influenced by wind and waves and therefore are well mixed and relatively uniform. Below this surface layer, large changes in temperature (the thermocline—see Figure 9-7) and salinity (the halocline) produce a corresponding rapid increase in

---

[1] Pressure in the ocean is expressed in decibars (a decibar is approximately 0.1 normal atmospheric pressure). A decibar is nearly equal to the weight of a column of seawater 1 m high acting on a surface of 1 cm².

density (called the *pycnocline*). Below the pycnocline are the deep, denser waters of the ocean.

If a situation develops in which more dense water overlies less dense water, the water layering is unstable and overturning will generally occur. This unstable situation could be a result of high evaporation of surface waters that increases its density. This water eventually sinks, being replaced by deeper water, until it reaches a zone of similar density.

The large density differences that produce the stable pycnocline effectively isolate the surface waters from the deep waters of the ocean. The exception is in the polar regions where the thermocline, halocline, and pycnocline are absent and therefore a strong density stratification does not exist. The absence of this stratification allows interaction between the atmosphere and eventually the deep-ocean water (see next section). The increase in oxygen concentration in deep water, for example, is generally thought to be mainly due to atmospheric exchange with surface waters in polar areas and their subsequent sinking to the deep waters of the ocean (Figures 9-17 and 9-18).

OTHER PROPERTIES

Sunlight can penetrate seawater to a depth of about 100 to 200 m (328 to 656 ft) depending on the number of particles in the water. The high value is typical in clear ocean water far from land (and sources of sediment, pollutants, and so forth), whereas in coastal waters, especially near rivers, light penetration may be limited to only a few meters. The well-lighted surface portion of the ocean is often called the euphotic zone (and has enough light for photosynthesis); the darker, deeper layers are called the aphotic zone.

Seawater has a very high heat capacity; in other words, considerable heat has to be applied before its temperature will rise. This means that seawater does not have rapid temperature changes (one of the reasons why it takes so long in the summer before the ocean becomes warm enough for swimming) and has a moderating influence on coastal climates.

Interactions between the Atmosphere and the Ocean

Interactions between the atmosphere and the ocean are complex, and in many instances it is difficult to establish cause and effect. The circulation of the ocean is dependent mainly on two atmospheric factors: wind and heating of the ocean. The ocean, which can store heat much better than the atmosphere or land, absorbs more heat per unit area at the equator than at the poles (Figure 9-9). This heat is transferred to the colder areas of the ocean by convection, or movement of the water. The heat-storing capacity of the ocean is very important in modifying and influencing continental climate. This influence can be noted on the west coasts of land in intermediate latitudes of the northern hemisphere, such as California and England, where the dominantly onshore winds transport warm air from the sea to the land.

The wind blowing on the ocean generates waves, mixes the surface waters, and removes water vapor from the sea surface. The water vapor is taken into the atmosphere by evaporation and is eventually transferred to the land as precipitation. This cycle, called the *hydrologic cycle* (Figure 7-9), is completed when the water is returned to the ocean. Near coasts the wind may cause the surface waters to move offshore and be replaced by colder, nutrient-rich, deeper waters (Figure 9-15).

The interaction of the atmosphere with the ocean produces two distinct types of circulation: a wind-driven circulation and a thermohaline, or density circulation. The wind-driven circulation is stronger than the density circulation; its major importance, however, is restricted to the upper 1,000 m. The thermohaline circulation, on the other hand, extends down to the deep sea.

WIND-DRIVEN CIRCULATION

The atmosphere, like the ocean, receives a larger quantity of solar radiation per unit area in the equatorial regions than in the polar regions. To maintain the earth's heat balance, some of this heat is transferred to the higher latitudes by the atmosphere. If the earth were not rotating, a simple circulation would probably exist between the equator and the poles (Figure 9-10). The air at the equator is heated, expands, and rises, and because of its expansion creates a low pressure area. Cooler air from the surrounding area moves toward the low pressure area at the equator and in turn is heated, expands, and rises. At the poles the air is cooled, contracts, and sinks, creating a high pressure area. The polar air moves

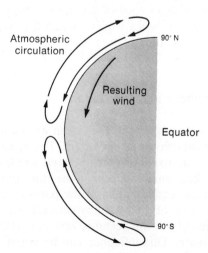

FIGURE 9-10 Possible atmospheric circulation and resulting winds for a nonrotating earth.

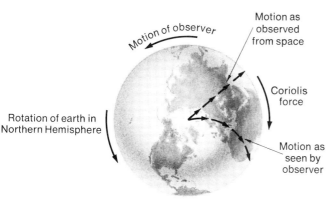

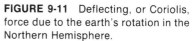

FIGURE 9-11 Deflecting, or Coriolis, force due to the earth's rotation in the Northern Hemisphere.

toward the equator (from high to low pressure), resulting in a steady wind from the north in the Northern Hemisphere.

This simple, theoretical, one-cell model of atmospheric circulation must be expanded into a more complex three-cell model for a rotating earth because the air movement has a deflection due to the earth's rotation. This phenomenon, known as the *Coriolis force* (Figure 9-11), causes the air to turn toward the right in the Northern Hemisphere and toward the left in the Southern Hemisphere. Thus instead of flowing in a straight line along the pressure gradient from high to low pressure, the wind appears to be deflected into curved paths. In the Northern Hemisphere rising warm air from the equator starts to flow north toward the pole. As it leaves the equator, it is deflected toward the right, cools and eventually descends at about 30° N latitude. Part of this air completes the *gyre*, or circular trip, and heads south toward the equator and another part continues northward toward the pole. The air traveling toward the pole is warmed and again deflected to the right, forming another cell. In the polar region the air moves downward at the pole and travels south until it is heated sufficiently to rise again at about 60° N. The resulting winds and circulation cells are shown in Figure 9-12. Areas of rising or sinking air generally have calm winds, such as the doldrums along the equator and the horse latitudes at about 30° N and S. Areas where the air is traveling along the earth's surface generally have steady and intense winds, such as the westerlies[2] and the trades.

The wind system described above exerts a stress on the ocean's surface and produces the wind-driven circulation of the ocean (Figure 9-13). The easterly trade winds form the equatorial currents common to all oceans. In the Atlantic and Pacific Oceans these currents are intersected by land and are deflected to the north and south. These deflected currents travel along the western parts of the oceans and are called the western boundary currents, which are among the largest and strongest currents in the ocean. One, the Gulf Stream (Table 9-1),

---

[2] When meteorologists refer to a west wind, they mean that it blows from the west; when oceanographers refer to a western flowing current, they mean that it goes to the west.

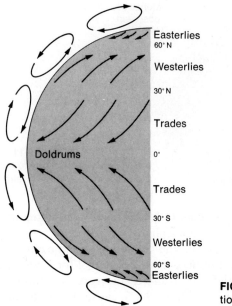

Easterlies
60° N

Westerlies

30° N

Trades

0°

Doldrums

Trades

30° S

Westerlies

60° S
Easterlies

FIGURE 9-12 Atmospheric circulation and resulting winds.

transports more than 100 times the combined outflow of all the rivers of the world. The western boundary currents are due in large measure to the variation of Coriolis force with latitude. These currents are driven across the ocean by the westerly wind and form currents that flow back into the equatorial region, thus completing the large gyre. Gyres of this type occur in the subtropic regions of the North and South Pacific, the North and South Atlantic, and the south Indian Oceans. The northern and southern gyres of the oceans are separated by an eastward-flowing countercurrent. A similar gyre, although changing direction every 6 months, is found in the north Indian Ocean. The change in direction is due to half-yearly reversals in the atmospheric circulation pattern, called *mon-*

TABLE 9-1   Velocity and Transport of Some of the Major Currents
in the Ocean

Current	Maximum velocity (cm/s)	Transport (millions of m³/s)
Gulf Stream	200–300	100
North Equatorial Pacific	20	45
Kuroshio	200	50
Equatorial Undercurrent	100–150	40
Brazil	—	10
Antarctic Circumpolar (West Wind Drift)	—	100
Peru or Humboldt	—	20

SOURCE: Data from Warren, 1966.

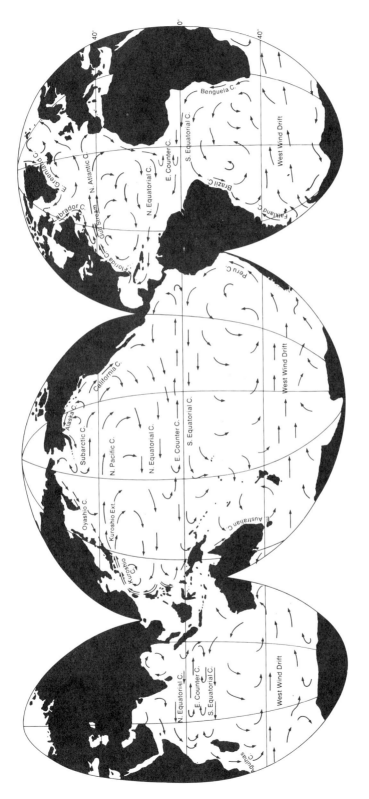

FIGURE 9-13 Major surface currents of the ocean. Winter conditions for the Indian Ocean. (After Sverdrup et al., 1942.)

soons. Smaller and weaker gyres are found in the northern subpolar regions of the Atlantiç and Pacific Oceans. No gyres, however, occur in the southern sub-polar region, probably because there are no land barriers to obstruct the flow of water and create a gyre. Therefore, the Antarctic Circumpolar, or West Wind Drift, flows completely around the world.

As previously mentioned (see pages 294–296 and Figure 9-1) parts of some of these large currents may break off as loops or meanders, called rings. These rings have almost a life of their own and may exist for several years.

The wind-driven circulation of the ocean results from differences in water pressure; these differences result mainly from changes in the slope of the sea surface due to winds. Winds blowing on the water cause the water to move and build up in the direction that the wind is blowing [Figure 9-14(a)]. This creates a pressure difference between the high and low areas (higher pressure where the water is piled up). The pressure difference generates a force that tends to push the surface water back toward the region of lower pressure. In other words, the water wants to go downhill or down the slope [Figure 9-14(b)]. Because of the Coriolis force, however, the moving water or current is deflected to the right in the Northern Hemisphere. If the pressure difference is balanced by the Coriolis force, the current is called a *geostrophic current*. Oceanographers can calculate the geostrophic current if they accurately know the horizontal and vertical distribution of temperature and salinity and can estimate the thickness of the current.

Most of the surface currents in the ocean are geostrophic but some near-surface currents may be in part nongeostrophic; these are called *undercurrents*. One of these, the Equatorial Undercurrent in the Pacific, flows from west to east at the equator 100 m (328 ft) or more below the ocean surface. This current, also called the Cromwell Current, is 300 km (186 mi) wide, only a few hundred meters thick, and has velocities as high as 3 kn (150 cm per second). Usually undercurrents flow in a direction opposite to the surface current. The origin of undercurrents and their relationships to the overlying water are not fully understood.

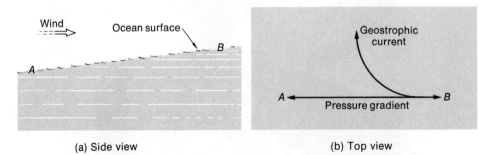

(a) Side view

(b) Top view

FIGURE 9-14   (a) Sloping sea surface produced by the wind blowing on the sea (slope is exaggerated). (b) Pressure gradient and resulting geostrophic current (for Northern Hemisphere) produced from the situation shown in (a).

UPWELLING

The action of the wind on the sea surface, besides causing horizontal movement of water, may also produce vertical motion. Upwelling, or upward motion of water, can occur when prevailing winds blow parallel to a coast. The motion of the water in many instances is offshore [Figure 9-15(a)] due to the Coriolis force, and subsurface waters are brought to the surface. If these subsurface waters are high in nutrient content, an area of high biological productivity may result. Sinking of surface water can happen by essentially the same process if the water flows toward the land [Figure 9-15(b)].

Upwelling can have a profound influence on the biological productivity of an area. It has been estimated that 50 percent of the world's fish supply comes from upwelling areas, although upwelling areas in themselves comprise only about 1 percent of the total area of the ocean. A major IDOE program called Coastal Upwelling Ecosystems Analysis (CUEA) has made an 8-year interdisciplinary study of upwelling and shown how weather systems hundreds to thousands of kilometers away can cause local currents and upwelling (Figure 9-16). Upwelling brings cool, nutrient-rich waters to the surface. This in turn permits increased growth by phytoplankton and more available food for other organisms (like fish) higher up in the food chain. Local conditions such as submarine canyons or coastal configurations can influence the strength of the upwelling process. Other processes, such as the meanders of large currents such as the Gulf Stream (Figure 9-1), may also force nutrient-rich waters toward the surface.

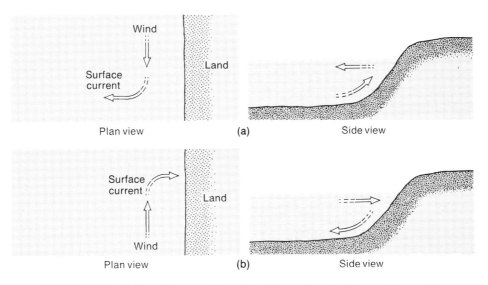

FIGURE 9-15 (a) Upwelling, or rising, of surface waters due to nearshore winds. (b) Sinking of surface waters due to nearshore winds.

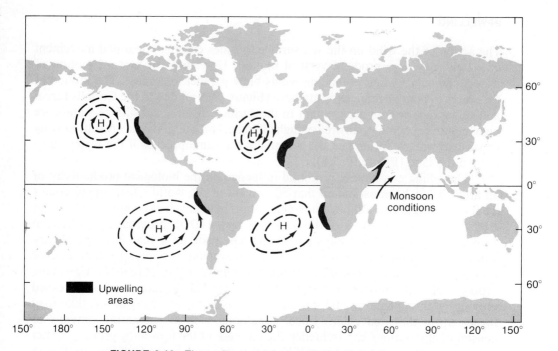

FIGURE 9-16 The major coastal upwelling areas of the world (shaded) and the weather circulation patterns that drive them. (Adapted from Hartline, 1980.)

THERMOHALINE CIRCULATION

The thermohaline circulation, generally a deep-water process, is caused primarily by variations in water density. Density differences that drive this circulation generally develop at the sea–air interface. Thus the wind-driven and thermohaline circulation systems are related.

Direct observations of the thermohaline circulation are difficult to make, mainly because the circulation rate is very slow. Most of the information concerning thermohaline circulation comes from detailed subsurface measurements of temperature, salinity, and dissolved oxygen. More recent studies using transient tracers have given new insight into these processes (see page 218).

The thermohaline circulation is a convection process whereby dense, cold water formed in high latitudes sinks and slowly flows toward the equator. In this manner much of the deep water of the ocean is formed, or assumes its characteristics. This process occurs principally in two places; in the North Atlantic and in the Antarctic.

In the North Atlantic, cold, heavy water sinks and moves south, near the bottom, across the equator (Figure 9-17). This water, called the North Atlantic Deep Water, is defined by its temperature, salinity, and oxygen and is easily distinguishable from other water masses (Figure 9-18). In the Antarctic region an

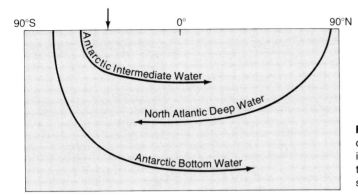

90°S   0°   90°N

Antarctic Intermediate Water

North Atlantic Deep Water

Antarctic Bottom Water

FIGURE 9-17 Diagrammatic section of the major subsurface water masses in the Atlantic Ocean. Arrow shown on top is approximate location of section shown in Figure 9-18.

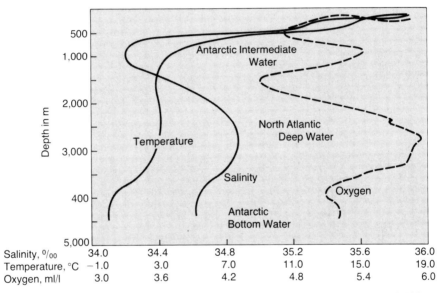

Salinity, ⁰/₀₀	34.0	34.4	34.8	35.2	35.6	36.0
Temperature, °C	−1.0	3.0	7.0	11.0	15.0	19.0
Oxygen, ml/l	3.0	3.6	4.2	4.8	5.4	6.0

FIGURE 9-18 Temperature, salinity, and oxygen measurements made at GEOSECS Station 60 in the South Atlantic (32°58′ S and 42°30′ W). The distinct variations in the measured parameters easily define the three main water masses (Figure 9-17).

Antarctic Bottom Water and an Antarctic Intermediate Water are formed. The former, one of the densest bodies of water in the ocean, travels north on the bottom across the equator. The Intermediate Water also travels north, but at a depth of about 1 km (about 0.6 mi) below the surface.

The thermohaline circulation is obviously extremely important in establishing the conditions of the deep ocean. Since the deep water has its origins in the polar or near-polar regions, it is easy to understand why these deep waters are so cold. The relatively high oxygen content of the deep waters of the ocean, relative to shallow waters, also reflects its polar origin. Without this surface

315

source of oxygen the deep waters of the ocean could be depleted of their oxygen content by the oxidation of organic matter that falls through it.

Bottom-flowing water masses can also be influenced by the topography of the sea floor. Dense waters formed in the Arctic Ocean are prevented from reaching the Atlantic by a submarine ridge. The Mid-Atlantic Ridge can also be a barrier to flow between basins of the western and eastern Atlantic.

Even though the thermohaline circulation is very slow, the bottom waters of the ocean eventually come back in contact with the surface through this circulation.

In summary, the major current systems of the ocean are due to the combined effects of the wind blowing on the ocean surface and variation in density between different parts of the ocean. The density variations are primarily the result of differences in heat received and the effects of dilution and evaporation. The directions of the currents are influenced by the rotation of the earth (Coriolis force) and the presence of and shape of the continents and ocean floor.

Wind-Generated Waves

ORIGIN OF WIND-GENERATED WAVES

Another effect of the wind's blowing on the surface of the ocean is the formation of waves (there are several different kinds of waves found in the ocean). Wind-generated waves form on the sea surface by the transfer of energy from the air to the water; the method for this transfer is not completely understood but it appears to be related to two basic mechanisms.

1. Wind is deflected as it blows over the wave profile, causing pressure differences that can supply energy to the waves.
2. Moving pressure fluctuations may react with water by resonance to form waves during turbulent wind conditions.

In discussing waves it is convenient to idealize the waveform (Figure 9-19). In this simplified picture the wavelength, L, is defined as the horizontal distance between two crests (or two other similar points on the waveform) measured parallel to the direction of travel of the wave. The period T is the time for the passage of successive wave crests past a fixed point. The wave height H is measured from the wave crest to the wave trough. The depth of water is h. The velocity of the wave, or the rate of propagation of the waveform, is the wavelength divided by the period, or $C = L/T$.

In deep water it is the waveform that is advancing; the water itself moves forward only a very small amount. Surely everyone has observed a piece of paper or a boat floating on water and going up and down with the passing waves. The

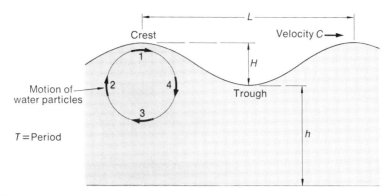

FIGURE 9-19 Simplified picture of important wave characteristics and the motion of water particles in a waveform. Circle showing motion of water particles is exaggerated in size.

waveform is moving along the surface of the water horizontally but the paper or boat is just going up and down with essentially no horizontal movement (as shown by the different numbers in Figure 9-19). Therefore, when talking about wave motion, one must distinguish between the motion of the waveform and the motion of the water particles.

When the wave is in deep water, the motion of the individual water particles at the surface is circular. However, the diameter of the circle decreases quickly with depth; for example, at a depth of one-fourth of the wavelength, the diameter is reduced to about one-fifth of its original size and the water motion is more of a back-and-forth rather than a circular motion. The velocity of the water particles also decreases rapidly with depth. Waves having a period of 10 seconds or less generally produce negligible motion below 100 m (328 ft).

The height and period of wind-generated waves are functions of three factors: (1) velocity of the wind; (2) duration, or time that the wind blows; (3) fetch, or the distance of water over which the wind blows. The absolute effect of these factors has not been resolved but some relationships are known. Wave height and wavelength generally increase to a definite maximum with increasing wind velocity and duration. The fetch is important in determining wavelength. Wavelengths of only a few meters are common in lakes where the fetch is relatively small, whereas wavelengths of several hundreds of meters are typical for oceanic waves because the fetch is large. In shallow waters, such as lakes, the water depth can become another important factor influencing the waves. The maximum height of wind-generated waves is not known, especially because of the difficulty of measuring wave height during storm periods, but estimates have reached about 25 m (82 ft).

The effects of different wind speeds and their effect on the sea surface are summarized in the *Beaufort Wind Scale* (Table 9-2). Accomplished mariners can easily estimate the Beaufort number by observing the condition of the ocean surface.

TABLE 9-2   The Beaufort Wind Scale

Beaufort number	General description of wind	Condition of sea	Wind speed (miles/hour)	Wind speed (km/hr)	Height of wave (ft)	(m)
0	Calm	Sea smooth as mirror	Less than 1	Less than 1	0	0
1	Light air	Small wavelet-like scales; no foam crests	1–3	2–5	0.5	0.15
2	Light breeze	Waves short; crests begin to break	4–7	6–11	1	0.3
3	Gentle breeze	Foam has glassy appearance; not yet white	8–12	12–20	2	0.6
4	Moderate breeze	Waves now longer; many white areas	13–18	21–29	5	1.6
5	Fresh breeze	Waves pronounced and long; white foam crests	19–24	30–39	10	3.1
6	Strong breeze	Larger waves form; white foam crests all over	25–31	40–50	15	4.7
7	Moderate gale	Sea heaps up; wind blows foam in streaks	32–38	51–61	20	6.2
8	Fresh gale	Height of waves and crests increasing	39–46	62–74	25	7.8
9	Strong gale	Foam is blown in dense streaks	47–54	75–87	30	9.3
10	Whole gale	High waves with long overhanging crests; large foam patches	55–63	88–101	35	10.8
11	Storm	High waves; ships in sight hidden in troughs	64–75	102–120	—	—
12	Hurricane	Sea covered with streaky foam; air filled with spray	Above 75	Above 120	—	—

DIFFERENT TYPES OF WIND-GENERATED WAVES

Wind-generated waves can be divided into three stages: sea, swell, and surf [Figure 9-20(a), (b), and (c)]. Waves in the area directly affected by the wind are called *sea* [Figure 9-20(a)]. Sea waves are irregular, with no systematic pattern,

FIGURE 9-20   (opposite page) Wind-generated waves. (a) Sea waves in the generating, or storm, area. (b) Swell. Note how the wave pattern has become relatively uniform. (c) Surf—breaking waves. (Photographs courtesy of Woods Hole Oceanographic Institution.)

(a)

(b)

(c)

319

and have waves of different periods and heights traveling in various directions. As these waves leave the area where they were generated (the storm region) and where they were under the direct immediate influence of the wind, the longer waves, because of their higher velocity (velocity $C = L/T$), outdistance the shorter and slower waves. The waves eventually assume a more uniform pattern, since waves of similar dimensions tend to travel together because of their similar speed. The waves in this regular pattern are called *swell* [Figure 9-20(b)]. As swell travels still farther from the generating area, they remain constant in length but decrease in height. A wave pattern is capable of traveling across an entire ocean.

The third stage of wind-generated wave is *surf*, which occurs near shore when a wave shoals and breaks [Figure 9-20(c)]. Breaking waves differ from sea and swell waves in that the water particles are no longer traveling in an orbital motion but now are moving toward the beach. This results in a large amount of energy that is directed toward the beach.

When waves arrive in shallow water, all their characteristics, except their period, change. Wavelength and velocity decrease with decreasing depth; this change is small until the depth of the water (h) equals one-half the wavelength of the wave. At this depth the wave is said to "feel" bottom, and the wave quickly increases in height. A wave breaks when the particle velocity at the crest of the wave exceeds the velocity of the wave. On a gently sloping beach this usually occurs when the ratio H/h is between 0.8 and 0.6.

If waves enter shallow water at an angle to the beach or encounter irregular changes in the nearshore bottom topography, their direction of travel changes. This change, called *refraction*, occurs when one part of the wave first reaches shallow water. This part is slowed down, causing the entire wave to turn toward the shallow water. Thus the wave crests tend to parallel bottom contours (Figures 9-21 and 9-22). Refraction depends on bottom topography, wavelength, and direction of approach. On irregular coasts, refraction causes a concentration of wave energy, or *convergence*, in topographically high areas, such as submerged ridges or elevated points. Wave energy diverges over submarine canyons or in bays. Clever fishermen moor their boats over local depressions in the bottom topography while waiting out a storm; the wave energy diverges at these areas and the danger of the boats' capsizing is reduced.

After waves break, water is carried into the surf zone and is transported toward the beach, which acts as a barrier. Some water returns seaward along the bottom of the surf zone. With the usual angular approach of the waves to the beach, water is also transported along the beach (Figure 9-23). This longshore current increases until it can overcome the incoming waves, at which time the water flows seaward in a *rip current*. Rip currents, because of their seaward flow, can be dangerous to swimmers.[3] The positions of rip currents are dependent on submarine topography, slope of the beach, and height and period of the waves.

---

[3] If caught in a rip current, let the current carry you seaward; do not try to swim against it to reach shore. As it diminishes in strength, swim parallel to the beach until the effect of the rip current disappears; then start to swim ashore.

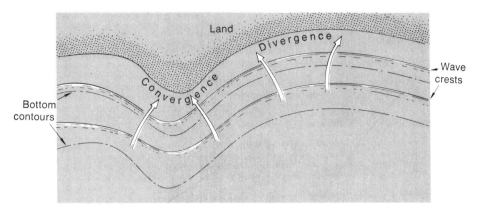

FIGURE 9-21 Diagrammatic illustration of wave refraction in the nearshore region.

FIGURE 9-22 Aerial photograph showing wave refraction. (Photograph courtesy of the U.S. Air Force, Cambridge Research Laboratories.)

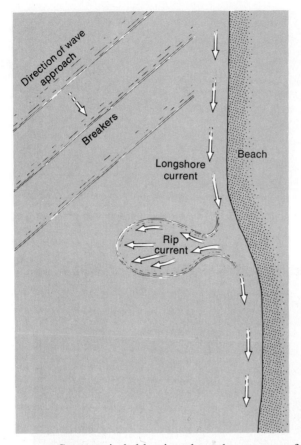

FIGURE 9-23 Longshore transport and rip currents in the nearshore region.

Strong winds blowing along the ocean surface produce a series of convection cells called *Langmuir cells*, named after the person that first described them. These cells have alternating left and right patterns, whose long axes parallel the wind direction (Figure 9-24). Often part of the cell is visible as material floating on the sea surface, which frequently accumulates over the convergence part of the cell. Langmuir cells can be important in the transport of nutrients and gases, as well as in mixing of the surface waters.

Internal Waves

An internal wave can occur between fluids of different densities. Such waves generally are not seen directly but can be detected by systematic and closely spaced temperature observations. Such observations indicate that internal waves are a common phenomenon in the ocean. They usually travel more slowly than surface waves, but they may have a greater height (remember, not at the air–sea surface). Internal waves also have been detected to break like surface waves.

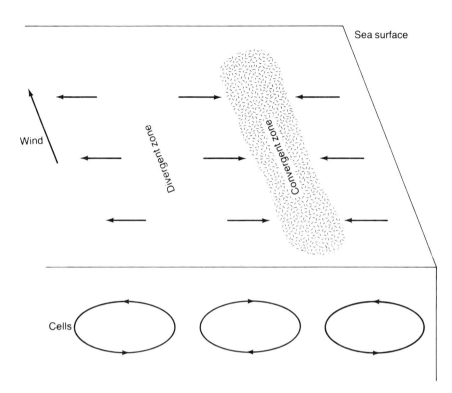

FIGURE 9-24 Diagrammatic view of Langmuir cells. Note the alternating left and right hand circulation forming divergent and convergent regions.

Their presence is sometimes hinted at by slow-moving surface slicks composed of plankton, fine-grained sediment, or surface-water contamination (Figure 9-25) that accumulate over the trough of the wave.

Any condition causing waters of different density to come in contact with each other can cause internal waves. Examples are outflow of freshwater from rivers and mixing of different water types. Tidal movements probably can cause some long-period internal waves.

Catastrophic Waves

STORM SURGES

Catastrophic waves are the result of unusual conditions, such as intense storms over or near the ocean, or submarine slumping. Catastrophic waves may often cause damage and loss of life. Strong winds, usually associated with hurricanes, can pile water up on a coast causing an exceptionally high sea level. High water levels or storm surges can be dangerous, especially if they coincide with times

FIGURE 9-25 Sea-surface expression of internal waves. The surface slicks, composed of fine-grained sediment, are related to the troughs of the internal waves. (Photograph courtesy of E. C. LaFond, U.S. Naval Undersea Center.)

of high tides in low coastal regions. In the Gulf Coast area of the United States, storm surges have been known to raise the water level as much as 7 m (about 23 ft). In 1900, over 6,000 people were drowned during such a storm in Galveston, Texas. Storm surges differ from other waves in having a gradual rise of the water level rather than a quick, rhythmic rise and fall.

LANDSLIDE SURGES

The movement of large quantities of rock or ice into the ocean due to earthquakes or glacial movements can generate immense waves. An exceptionally large wave occurred in Lituya Bay, Alaska, in 1958. It was estimated that 30,000,000 m³ (about 40 million cubic yards) of rock fell from a height of about 1,000 m (3,281 ft) into the bay, causing a wave that rose up over 500 m (1,640 ft) onto the mountainside on the other side of the bay (Figure 9-26). Over 15,000 people were drowned by a similar wave on the Japanese island of Kyushu in 1792.

FIGURE 9-26 (opposite page) (a) Aerial view of Lituya Bay, Alaska, taken in 1954, before the giant wave. (b) Aerial view of Lituya Bay, Alaska, taken in 1958, showing the wave damage. (c) Aerial view of wave damage on the north shore of Lituya Bay. View is about 3 km (about 2 mi) from the entrance. Width of the zone of destruction is about 600 m (about 1,968 ft) at the right margin of photograph. Note trees with limbs and bark removed. (All photographs courtesy of U.S. Geological Survey.)

(a)

(b)

(c)

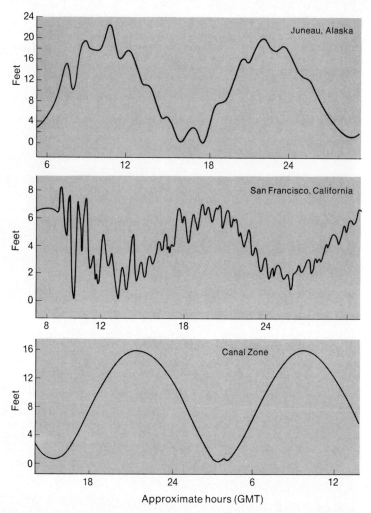

FIGURE 9-27 Tidal gauge records of the 1964 Alaskan tsunami, as felt in different cities. Note that the tsunami arrived later at the cities farther from the earthquake center. These records do not show the individual waves, but the broad scale, up-and-down movement of the water. (After Spaeth and Berkman, 1967.)

TSUNAMIS

Tsunamis are commonly called *tidal waves*; however they actually have nothing to do with the tides. Their origins can be traced to submarine movements caused by earthquakes, slumping, or volcanic eruptions. In deep water, tsunamis may have wavelengths as long as 700 km (435 mi), travel at speeds over 350 kn, and yet have wave heights, or amplitudes, of only a few centimeters. When tsunamis reach shallow water and break against the coast, they may be higher than any wind-generated wave. The destructive effect of tsunamis is strongly controlled by submarine topography, as breaking waves generally are small near projecting points of land bordered by deep water and are high near submarine ridges.

Earthquakes in Chile in 1960 and in Alaska in 1964 (Figure 9-27) produced

FIGURE 9-28 The tsunami warning system in the Pacific Ocean. Tide and seismographic stations are indicated. The lines show the travel times of a tsunami to reach Hawaii.

large tsunamis, which caused many deaths and much property damage. An international early warning system allows time for preventive actions to be taken after an earthquake but before a tsunami approaches. Sensitive seismographs, now located at stations around the Pacific, record the shock waves from the earthquake; observers can then quickly determine the position of the earthquake and calculate when the resulting tsunami may arrive (Figure 9-28). Persons in areas that could be hit by the wave can then be forewarned.

Most tsunamis in the Pacific are caused by submarine movements along the "ring of fire," an area of crustal instability that somewhat discontinuously encircles the Pacific (see page 129). Tsunamis generated in these areas will travel outward and reach most other areas of the Pacific (see Table 9-3). Fortunately, the west coast of the United States has received little damage from tsunamis for several reasons.

TABLE 9-3 Events Associated with the Prince William Sound, Alaska,
Earthquake of 1964

Time	Event
0336	Earthquake strikes the northern shore of Prince William Sound, Alaska.
0344	Seismic sea-wave warning alarm rings in Tsunami Warning Center, Honolulu, Hawaii.
0435	Kodiak, Alaska, experiences tsunami wave 3–4 m (10–13 ft) above mean sea level.
0502	First warning issued.
0555	Kodiak confirms existence of tsunami.
0700	Tsunami reaches Tofino, British Columbia, Canada.
0708	Kodiak reports waves 11 m (36 ft) at 0540; 12 m (39 ft) at 0630; 10 m (33 ft) seas diminishing.
0739	1 m (3 ft) wave arrives in Crescent City, California—some evacuees return thereafter.
0750	Four persons drowned in DePoe Bay, Oregon.
0900	Tsunami reaches Hilo, Hawaiian Islands.
0920	4 m (13 ft) wave (probably the fourth) sweeps into Crescent City—causes great damage.
1020	Tsunami reaches east coast Hokkaido, Japan.
1038	Tsunami reaches northeast coast Honshu, Japan.
1355	Tsunami reaches Kwajalein, Marshall Islands.
1910	Tsunami reaches La Punta, Peru.

SOURCE: From an Intergovernmental Oceanographic Commission publication, *Tsunami Warning System in the Pacific.*

1. The west coast is relatively stable in regard to earthquakes, at least in comparison with other areas of the Pacific.
2. Tidal waves produced in the Aleutian and South American areas approach this coast diagonally and hence are less destructive.
3. The relatively large shelf on the west coast of the United States causes the waves to lose considerable energy before reaching shore.
4. There are many high-cliffed or hard-rock coastal regions.

STATIONARY WAVES

A wave type common to many enclosed bodies of water like bays and lakes is the *stationary wave*, also called *standing wave*, or *seiche*. In a stationary wave, the waveform does not move forward but the water surface moves up and down. The motion is similar to that of soup in a bowl that has been tilted and then put down on a flat surface (Figure 9-29). The water surface remains stationary at certain locations, called *nodes*, while the rest of the surface moves up and down.

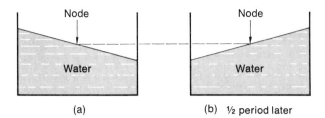

FIGURE 9-29 A simple stationary wave, or seiche.

Stationary waves can be generated by storms, rapid changes in atmospheric conditions, or sudden disturbances to the water surface. Once the wave is generated, the lake or bay will have an oscillation that is controlled by the length and depth of the basin. Stationary waves have been responsible for much property damage and loss of life.

Ocean Tides

Tides are the daily or twice daily rhythmic rise and fall of sea level. Tidal movements were observed, measured, and recorded by early people, who noted their relationship to the moon. The explanation of and attempt at tidal prediction was one of people's earliest scientific ventures. Many of the theories and techniques of tidal prediction were developed in the eighteenth and nineteenth centuries. Techniques for tidal prediction have recently been improved, mainly through the use of high-speed computers.

Tides are waves that have a period of about 12 hours and 25 minutes and a wavelength of about half the circumference of the earth (about 20,030 km or 12,450 mi).

The tidal range (maximum height at high tide minus minimum height at low tide) averages between 1 and 3 m (3.3 to 3.9 ft) but can be as high as 20 m (about 65 ft) in some areas, such as the Bay of Fundy. Exceptionally high tides are generally due to the geographic position and geometry of an area.

Most areas have either one high and one low tide each day (*diurnal* tides) or, more commonly, two highs and two lows a day (*semidiurnal* tides, Figure 9-30). Individual high and low pairs of semidiurnal tides are usually of different heights. Because the time interval between high tides is about 12 hours and 25 minutes (half a lunar day), high tide occurs about 50 minutes later every day. This simple fact shows that tides are primarily influenced by the moon. If they were mainly controlled by the sun, they would occur at the same time every day, the solar day being 24 hours long.

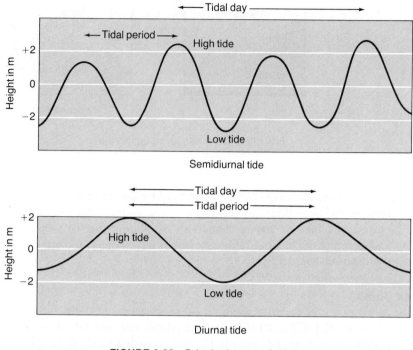

FIGURE 9-30 Principal types of tides.

The up-and-down motion of the tides has a potential for energy generation that has been applied in only a few areas. One is along the Rance River in France (Figure 14-1); this use of the ocean is discussed further in Chapter 14.

CAUSES OF THE TIDE

Tides are caused by the gravitational attraction of the sun and moon on the earth. This attraction affects water, solid earth, and the atmosphere, but the results on the last two cannot be observed by the unaided eye. The gravitational attraction between the earth and moon (Figure 9-31) is strongest on the side of the earth that is facing the moon. This attraction causes the water on the near side of the earth, N, to be pulled toward the moon. The gravitational attraction of the moon is at a minimum at the point farthest away, F, on the opposite side of the earth and this, combined with centrifugal forces, causes the water to "bulge out." The two bulges stay essentially aligned with the moon as the earth rotates relative to the moon. Because it takes the earth 24 hours, 50 minutes to rotate relative to the moon, a place on earth experiences two tidal highs and lows within this time period. The magnitude of the high tides at one place is generally different because the moon is inclined to the earth's equatorial plane [Figure 9-31(a) and (b)].

The sun also exerts a large tidal influence on the ocean, even though the

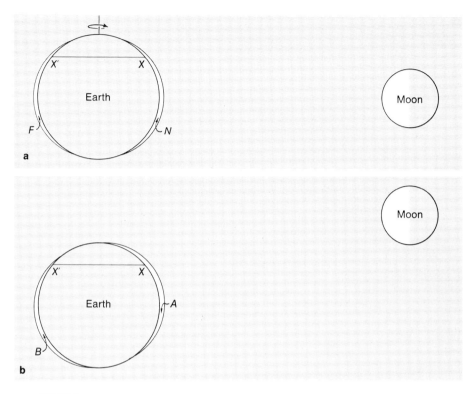

FIGURE 9-31 Illustration of the gravitational attraction between the earth and moon (dimensions are exaggerated). (a) The moon is parallel to the earth's equatorial plane. The points X and X' indicate the change in position of a point on the earth's surface after half a lunar day (12 hours and 25 minutes). This position will produce equal tides at point X or X'. (b) The moon is inclined to the earth's equatorial plane. This position will produce unequal tides at point X or X'.

tidal bulge produced by the sun is only 46 percent of that produced by the moon. The smaller effect of the much larger sun is due to its greater distance from the earth: Gravitational forces vary inversely with the square of the distance between two bodies and directly with their mass.

The effect of the sun becomes especially important when the sun and moon are lined up with the earth; the combined gravitational attraction of the two bodies produces a very strong tide, called the *spring tide*. Spring tides occur roughly every 14 days, at new and full moon (Figure 9-32). Relatively weak tides, called *neap tides*, occur when the sun and the moon are at right angles to each other, also about every 14 days, at half moon. The tidal range is higher than average during spring tides and less than average during neap tides. Tidal currents in the open ocean are relatively weak. Near land, however, they can attain speeds of several kilometers per hour. Tidal currents in shallow water and estuaries can be geologically very important. They can move large amounts of sediment that may

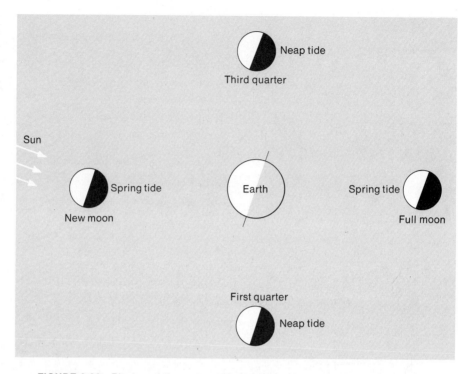

FIGURE 9-32 Phases of the moon and their associated spring and neap tides.

block harbors and eventually have to be removed by dredging. In some estuaries, during times of high tide, a large wave forms and travels upstream. This wave, called a *tidal bore*, can be as high as 3 m (9.9 ft) or more and have speeds of over 15 km per hour (about 9.3 mi per hour).

TIDAL FRICTION

Research has shown that the earth does not rotate smoothly under the tidal bulges but that there is a frictional force between the earth and the water. In theory this friction should slow the earth's rotation. Although the slowing rate is very small, the time between sunrise and sunset has increased by 0.001 second in the last 100 years. In spite of this small rate, it becomes significant over long periods of geologic time. A decrease in the earth's rotation also results in an increase in the speed of the moon's revolution, causing the moon to move slowly away from the earth. In the geologic past, if the moon had been closer to the earth, the length of the day would have been shorter and the number of days in a year would have been greater. There is evidence from growth rings of ancient corals that 400 million years ago the number of days in a year was close to 400.

Because the moon was closer to the earth in the past, tides would have been much stronger than at present. These tidal conditions could have created biological and geologic conditions different from those common today. Some scientists have suggested that the increased tides due to the proximity of the moon may have provided the impetus for the evolution of hard-shelled organisms. Soft-shelled organisms, living in shallow-water conditions, apparently would have had difficulty existing in these rigorous environments. Geologically, vast inland seas, flushed once or twice daily by the high tides, would have covered many of the low-lying areas of the world. It has been noted that some of the recent eruptions of Mount St. Helens have coincided with periods of high tide, suggesting to some that tidal forces may help the volcano erupt.

Underwater Sound

Underwater sound is an important tool for the oceanographer and is used to measure the depths of the ocean, as well as to examine the character and thickness of the earth's crust. Biological oceanographers can use sound to detect and study organisms (Figures 8-47, 8-48, and 12-18). Military uses of sound, such as for submarine detection and locating and positioning objects on the sea floor, have also encouraged the study of underwater sound.

SOUND VELOCITY

The velocity of sound in the ocean depends on temperature, salinity, and pressure (depth). Sound velocity in seawater ranges from 1,400 to 1,570 m (4,593 to 5,151 ft) per second; it increases with increasing salinity, temperature, and depth. The increase is 1.3 m (about 4.2 ft) per second for each 0.001 part increase in salinity, about 4.5 m (14.8 ft) per second for each degree Celsius (1.8°F) increase in temperature, and 1.7 m (5.6 ft) per second for each 100 m increase in depth. Although small, these changes affect estimates of water depth as determined by sound velocity. Corrections for changes in salinity, temperature, and pressure should be applied for estimates of water velocity; general correction factors have been developed for most areas of the ocean. More accurate estimates of sound velocity are possible with a sound velocimeter, a device that is lowered into the ocean to measure sound velocity directly, eliminating the need for corrections.

The vertical changes of sound velocity in the ocean can be divided into three zones [Figure 9-33(a)].

1. The surface zone of the ocean, 100 to 150 m (328 to 492 ft) in thickness, where the waters are well mixed and the sound velocity increases with depth due to the pressure (depth) effect.

2. A zone where the sound velocity decreases because of rapid temperature decreases (thermocline). Minimum values are found at about 600 m (1,968 ft) in the Pacific and 1,200 m (3,937 ft) in the Atlantic.

3. A zone where the sound velocity increases with increasing pressure and the temperature is relatively constant.

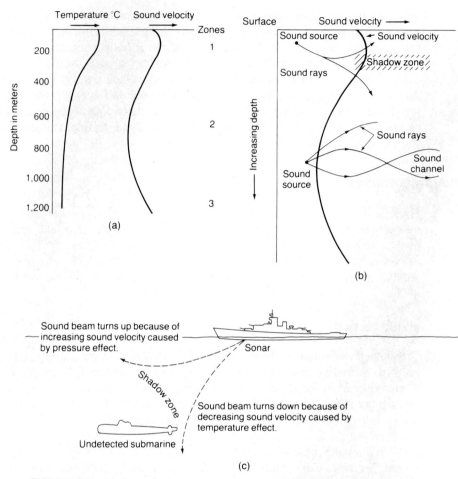

FIGURE 9-33 (a) Sound velocity profiles in the ocean. Arrow indicates direction of increasing temperature or sound velocity. Different zones are discussed in the text. (b) Shadow zone and sound channel. (c) Shadow zone formed by the refraction of sound due to changes in the velocity of sound with depth in the ocean. A submarine in the shadow zone would be very difficult to detect.

Sound waves, like ocean waves, can be refracted and hence will turn toward areas of lower sound velocity.

Refraction combined with vertical variation of sound velocity in the ocean can produce shadow zones and sound channels. A *shadow zone* is an area where relatively little sound will penetrate. It occurs in the upper parts of the ocean when a positive velocity gradient (increasing sound velocity) overlies a negative velocity gradient (decreasing sound velocity) and the sound is in the positive gradient [Figure 9-33(b)]. The sound is refracted upward in the positive gradient area and downward in the negative gradient area (in both instances toward areas of lower sound velocity), producing the shadow zone. It would be very difficult to detect a submarine in a shadow zone [Figure 9-33(c)].

A *sound channel* can occur in the area where the velocity of sound reaches a minimum value [Figure 9-33(b)]. Sound traveling in this minimum value zone is refracted upward or downward to the area of lower velocity and thus back into the minimum value zone. There is little energy loss in this zone due to vertical spreading, and sound can be transmitted for thousands of kilometers. This area, called the SOFAR (sound fixing and ranging) channel, has had sound transmitted through it over a distance of 25,000 km (about 15,500 mi). The SOFAR channel has a practical aspect as it can be used by ships in distress. An explosive charge detonated in this channel by a vessel will be detected at coastal stations and used to determine the position of the vessel by calculating the different arrival times of the sound at different stations.

As sound travels through water, it decreases in energy due to spreading, absorption, and scattering. The sound loss due to spreading is proportional to the square of the distance traveled. Sound can also be absorbed by the water and converted to heat. Absorption is proportional to the square of the sound frequency: the higher the frequency, the greater the absorption. Sound can be scattered by particles, marine organisms, gas bubbles, and the ocean bottom itself (sound is also reflected from the ocean bottom). The scattering and reflecting of sound from vertically migrating marine organisms (Figures 8-47 and 8-48) cause the phenomenon of the deep-scattering layer (discussed in Chapter 8).

Light in Seawater

Light from the sun and sky penetrates only the upper layers of the ocean. The transparency of seawater, or the depth of penetration of the light, is dependent on the amount of material that absorbs and scatters light in the water. Major absorbing and scattering materials are dissolved organic material, especially a yellow substance formed from the decomposition of organic matter, and such organic detritus as fragments of plankton.

Most of the light entering the ocean is absorbed within the upper 100 m (328 ft). The absorption of light varies with the different wavelengths of light. Blue light penetrates seawater more deeply than does red light. The blue color common to clear open ocean areas is due to the scattering and reflection of light from particles smaller than the wavelength of blue light. Seawater, which has large amounts of suspended material, including microscopic organisms, will reflect or absorb light of wavelengths different from those reflected or absorbed by relatively clear blue water and will appear more green, yellow, or brown depending on the size, color, and concentration of suspended material.

The depth to which light penetrates determines the thickness of the euphotic zone where photosynthetic production of organic matter by plants takes place. Because plants are the major source of food for organisms in the ocean, the thickness of the euphotic zone is extremely important. Planktonic plants generally do not grow where less than about 1 percent of the available light penetrates; thus most productivity occurs in near-surface waters.

Summary

Physical oceanography is a combination of the study of the physical properties of the sea and its changes with a theoretical evaluation of the processes that control these properties. As such, it probably is the most independent field of oceanography. Recent studies have shown some unique aspects to oceanic circulation including the presence of large eddies and loops or meanders of currents that spin off from currents like the Gulf Stream.

There are three basic properties of seawater that influence many oceanic processes: salinity, temperature, and density. These properties have distinctive patterns with depth and can be used to distinguish between different water masses or bodies of water. Most of the basic properties of the ocean are the result of processes occurring at the air–sea interface. Stratification of ocean water is due to density differences that, in turn, are related to salinity, temperature, and pressure (or depth). Underwater sound, an important tool for marine scientists, is also influenced by temperature, salinity, and pressure (depth).

The major oceanic current systems result from the combined influence of the wind blowing on the ocean and variations in seawater density. The thermohaline circulation, which is mainly a deep-water process, is caused by density variations that result from differences in heat received over the surface of the earth and the effects of dilution and evaporation. Currents are deflected by the rotation of the earth (Coriolis force) and by the presence and shape of the continents and ocean floor.

Under certain circumstances upwelling of subsurface nutrient-rich waters to the surface results in a unique situation very favorable for increased biological activity.

Winds blowing on the ocean surface can also form waves. Many of the wave characteristics are related to the velocity of the wind, duration of time that the wind blows, and the fetch, or distance of water over which the wind blows. Many aspects of waves change as they leave the generating area and reach the coastal zone. There is another category of waves, which includes tsunamis and storm surges, that can be dangerous and result in the loss of life and property.

Tides, the daily or twice daily rhythmic rising and falling of sea level, are caused by the gravitational attraction of the moon and the sun on the earth. In areas having a high tide range there is potential for using this motion of water as a source of energy.

Suggested Further Readings

BAKER, D. J., "Models of Oceanic Circulation," *Scientific American,* **222,** no. 1 (1970), pp. 114–21.

GOLDREICH, P., "Tides and the Earth–Moon System," *Scientific American,* **226,** no. 4 (1972), pp. 42–57.

GREGG, M., "The Microstructure of the Ocean," *Scientific American,* **228,** no. 2 (1973), pp. 64–77.

KINSMAN, B., *Wind Waves, Their Generation and Propagation on the Ocean Surface.* Englewood Cliffs, N.J.: Prentice-Hall, Inc., 1966. (Advanced)

KNAUSS, J. A., *Introduction to Physical Oceanography.* Englewood Cliffs, N.J.: Prentice-Hall, Inc., 1978, 338 pp. (Advanced)

McDONALD, J. E., "The Coriolis Effect," *Scientific American,* **186,** no. 5 (1952), pp. 72–79.

MACLEISH, W. H. ed., "Ocean Eddies," *Oceanus,* **19,** no. 3 (1976), 88 pp.

PENMAN, H. L., "The Water Cycle," *Scientific American,* **223,** no. 3 (1970), pp. 99–108.

RICHARDSON, P., "Gulf Stream Rings," *Oceanus,* **19,** no. 3 (1976), pp. 65–68.

RYAN, P. R., "A Reader's Guide to Underwater Sound," *Oceanus,* **20,** no. 2 (1977), pp. 3–7.

STARR, V. P., "The General Circulation of the Atmosphere," *Scientific American,* **195,** no. 6 (1956), pp. 40–45.

STEWART, R. W., "The Atmosphere and the Ocean," *Scientific American,* **221,** no. 3 (1969), pp. 76–86.

SVERDRUP, H. U., M. W. JOHNSON, and R. H. FLEMING, *The Oceans: Their Physics, Chemistry, and General Biology.* Englewood Cliffs, N.J.: Prentice-Hall, Inc., 1942. (Advanced)

VON ARX, W. S., *An Introduction to Physical Oceanography.* Reading, Mass.: Addison-Wesley, 1977. (Advanced)

10

Climate and the Ocean

ONE OF THE MORE interesting challenges of the future is to be able to predict and perhaps even control climate. Weather generally can be thought of as having an effect of just a few days or so. Climate, on the other hand, extends over a longer period of time, such as months, seasons, or years. Weather obviously is easier to predict, being almost directly influenced by the present temperature, pressure, humidity, and wind conditions. Climate, however, is controlled by broader and more widespread conditions, such as radiation from the sun, seawater temperature, and various complex interactions between the ocean and the atmosphere. At present, it is possible to make a good estimate of weather for only about two or three days into the future. The ability to predict weather and climate can have great benefits for agriculture and other industrial aspects. Among the advantages of long-range climatic prediction could be the adequate stockpiling of heating fuels or the planting of certain crops for specific weather patterns in specific areas.

There have been several recent technological developments that have permitted a better scientific evaluation of weather and its causes. One is the weather balloon with specific measuring and transmitting equipment (Figure 10-1). These devices can measure variables such as temperature, humidity, pressure, and other components of the atmosphere and transmit the information to land stations. A second important piece of technology is high-speed computers that allow scientists to deal with the large quantities of data obtained (and needed) to make quick predictions of weather as well as to test various meteorological hypotheses. Perhaps the most important development is the weather satellite. These devices and their observations have become so common that the pictures they transmit are shown daily on television in most, if not all, areas of the United States and Europe. The pictures permit meteorologists to detect and track numerous kinds of weather phenomena and allow early warning for potential storms (Figure 10-2).

Recent climatic changes, such as the droughts of the late 1960s and early 1970s and the abnormal cold periods in the United States during the late 1970s, have focused attention on the importance of climate. One of the major programs

FIGURE 10-1 A large weather balloon being launched. These devices can track atmospheric movements and measure pollutants or other components of the atmosphere. (Photograph courtesy of Department of Energy.)

of the International Decade of Ocean Exploration was environmental forecasting, which attempted to reduce hazards for life and property and promote better use of marine resources by developing models of the ocean and atmosphere that could be used in forecasting. This effort led the United States to develop a National Climate Program to focus on the understanding of climate. There are several international programs, including those by the World Meteorological Organization, an agency of the United Nations, that are or have been studying the interaction of climate and the ocean. One program is POLYMODE, a joint U.S.–Soviet Union study of the dynamics of large-scale eddies (Figure 9-1) in the Atlantic. Another is the North Pacific Experiment (NORPAX) that has been looking at sea-surface temperature and its interaction with the atmosphere. A third major program is CLIMAP (Climate: Long-range Investigation Mapping and Prediction), which is examining deep-sea sediments to ascertain past variations in climate and to test some of the proposed hypotheses for the causes of the ice ages (see pages 355–357). The effects of past major changes of climate have been discussed in previous chapters, particularly the dramatic and major ice ages over the last million years or so which have resulted in the lowering and subsequent raising of sea level about 130 m (about 426 ft).

(a)

(b)

FIGURE 10-2 (a) 1980 satellite photograph taken of Hurricane Allen with the outline of the Gulf of Mexico. Note the small eye of the storm in the central part of the Gulf. Allen was the second most powerful Atlantic hurricane, with winds approaching 320 km per hour (about 200 mph). These satellite photographs permitted early evacuation of potentially affected areas. Fortunately, soon after this picture was taken the hurricane slowed in its northerly movement and eventually became considerably reduced in severity before it reached the United States. (b) Hurricane Gladys, photographed from *Apollo VII* in 1968. The whirlpool pattern is typical of hurricanes. This storm caused considerable damage to low-lying areas of Florida. (Photograph courtesy of the U.S. Naval Oceanographic Office.)

It appears that much of our new understanding of climate will come from the oceans; there appear to be numerous linkages among the atmosphere, the oceans, and climate that are fundamental, critical, and just beginning to be understood. For example, the upper (3 m or about 10 ft) of the ocean contains as much heat as the entire overlying atmosphere. Water has a much greater specific heat per unit mass than air, about four times as high; this means that the thermal capacity of the ocean is 1,000 times greater than the atmosphere. Obviously, then, it takes more energy to heat water than air. The atmosphere, on the other hand, moves at a rate about 10 times that of the ocean, therefore considerable temperature differences can exist between the two and many interactions can occur. Another important point is that the atmosphere, like the ocean, has received large amounts of pollutants, in particular carbon dioxide from the burning of fossil fuels. The U.S. National Research Council suggests that in the next two centuries the carbon dioxide content in the atmosphere could quadruple, resulting in a possible surface temperature increase of 3°C (about 5.4°F). Such a change would have rather dramatic effects on the earth's climate.

The circulation of the ocean is one of the main factors in the total heat budget of the earth. This circulation is important not only to the ocean itself but also to the atmosphere and to climate. The atmosphere and the ocean are so closely intertwined that it is hard to avoid one when talking about the other (Figure 10-3). Both atmospheric winds and ocean currents move heat from low latitude regions where there is more incoming radiation to high latitude regions where there is less incoming radiation (Figure 9-9). Changes in this pattern will result in changes in climate. It has been shown in previous pages how the atmosphere, in part, influences oceanic circulation (the wind-driven circulation—see pages 308–314) and waves (see pages 316–327). It is sometimes hard to tell if variations in oceanic phenomena come from or lead to atmospheric variations.

The interactions between the oceans and the atmosphere apparently form a worldwide control on climate. A schematic illustration of the various parts of the system is shown in Figure 10-4. There are four principle components—the atmosphere, ocean, land, and ice (on land or on the ocean)—and a change in one affects the others. The key is understanding and predicting the net effect of these components on climate. Mathematical models and long-term observations of weather and climate have led to some hypotheses that are described on the following pages.

In previous chapters we have seen how the unequal heating of earth and

FIGURE 10-3 Waterspouts (photographed off the Bahamas) provide an excellent example of the interaction between the ocean and the atmosphere. (Photograph courtesy of Captain G. Stephen Gwin.)

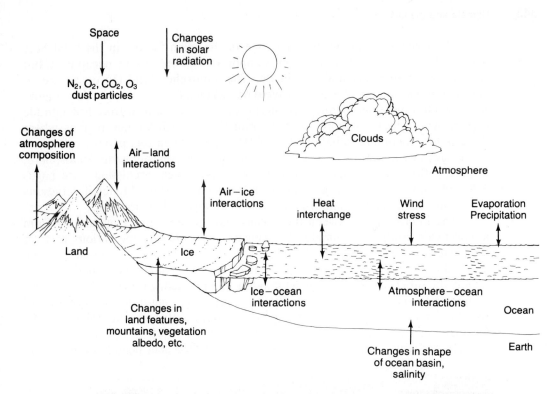

FIGURE 10-4 Components of an atmosphere–ice–ocean–earth climate system. Changes in any of the relationships can affect other parts of the system. (Adapted from U.S. Committee for the Global Atmosphere Research Program, 1975. *Understanding Climatic Change: A Program for Action*, National Academy of Science, Washington, D.C.)

ocean surfaces creates an atmospheric and oceanic circulation (Figures 9-12 and 9-13). This unequal distribution of heat is, in effect, responsible for the major currents in the oceans and the wind system over our planet. The ocean, because of its high heat capacity, is an effective buffer for both the seasonal and latitudinal changes in heat received on the earth. In the wintertime the atmosphere could cool off much more than it does, but it receives energy in the form of heat from the ocean. Likewise in the summertime, the atmosphere would get warmer if it was not being cooled by the ocean. The differences between the response of land and ocean to seasonal and latitudinal heat, in large part, cause the north and south variations in the generally westerly (from the west to the east) atmospheric circulation, especially in the middle latitudes. For example, in the wintertime, the ocean gives more heat energy to the air than does the land, resulting in a high pressure area over land whereas the opposite is true in the summertime, resulting in a lower pressure over land. This, in turn, causes meanders in the east–west circulation.

The differences in the response of land and water to heat can also affect local wind conditions. For example, rock has a lower heat capacity (easier to heat

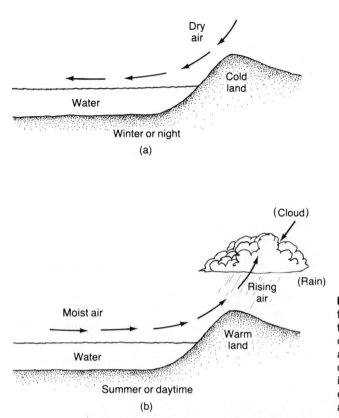

Dry
air

Cold
land

Water

Winter or night

(a)

(Cloud)

Rising
air

(Rain)

Moist air

Warm
land

Water

Summer or daytime

(b)

FIGURE 10-5 Wind patterns resulting from the difference in temperature between land and water. Where these conditions are pronounced (for example, summer conditions prevail even during the night), a monsoon pattern is said to occur. Under less rigorous conditions the winds change in a day and night manner.

or cool) than water thus in the summer the land (during the day) is often warmer than the adjacent water. Because of this, as the air over land is heated and rises, air from over the ocean moves toward the land, producing an onshore wind (Figure 10-5). At night or in the winter, the land is often cooler than the ocean and an offshore breeze may result. In the summer the air from over the ocean may be moist, leading to precipitation over the land, especially as it rises and cools. In areas where these patterns are extreme and seasonal, a monsoon condition is said to occur. In parts of India, during the summer monsoon, about 900 cm (about 354 in.) of rain have been recorded. In parts of the Indian Ocean the reversal in the monsoon conditions can be seen in a seasonal change in surface ocean currents.

Hurricanes are an especially damaging aspect of climate. The United States is presently in a relatively safe period for hurricanes, but in the earlier part of this century numerous storms with severe loss of life hit the Gulf and east coasts (Table 10-1). Over 13,000 people have been killed in the United States by hurricanes since 1900. There appears to be a periodicity to hurricanes—1940s, in the Florida region; 1950s, along the east coast; 1960s and 1970s, mainly in the Gulf of Mexico. Since the time of the last major hurricanes there has been much coastal

TABLE 10-1 Exceptionally Strong U.S. Hurricanes

1900	6,000 killed when hurricane hit Galveston Island.
1909	300 killed when storm flooded much of Louisiana coast.
1915	275 killed in Mississippi delta region by hurricane.
1919	500 killed by hurricane that hit both Key West and Corpus Christi.
1928	1,800 killed when storm hit Lake Okeechobee, Florida.
1935	400 killed when hurricane struck the Florida Keys.
1955	600 killed by hurricane that hit New England.
1957	390 killed by Hurricane Audrey, which hit Texas and Louisiana.
1965	75 killed, $1.4 billion in damage by Hurricane Betsy, which struck south Florida and Louisiana.
1969	300 killed by Hurricane Camille, which hit Mississippi.
1979	Hurricanes Frederic and David caused considerable damage.
1980	Hurricane Allen, one of the most powerful storms, caused considerable damage and loss of life in the Caribbean before expending most of its energy in the Gulf of Mexico.

development in these areas, and many of the people who live in these regions have little feel for the dangers that can be associated with hurricanes. Indeed, some of the more recent hurricane warnings have attracted as many curiosity seekers as people who have tried to avoid them.

Hurricane Frederic in 1979 was the most costly hurricane in U.S. history, resulting in $2.3 billion in damages, but fortunately, due to good early warning, only five deaths occurred (Figure 10-6). Areas most vulnerable to hurricane damage are low-lying, heavily populated coastal regions, especially those situated on barrier islands. Many experts feel that in some areas even a moderate storm could produce a major economic and human catastrophe.

One of the problems in hurricane protection is the difficulty in predicting the strength and path of one of these storms. Usually they begin in tropical regions when the ocean temperature reaches about 26.5°C or 80°F. As this occurs, the overlying moist air heats up, expands, and rises. As the air rises, the moisture in it condenses, causing clouds and thunderstorms as the air starts to move and rotate. Generally, the hurricane moves northwest or west until it encounters some westerly winds and then curves back to the east at increased speeds. Although presently there is no proven way to diffuse a hurricane, their monitoring has improved considerably through the use of satellites (Figure 10-2), reconnaissance flights, and radar.

Recent Studies—NORPAX Program

It seems fairly obvious that the physical processes of the ocean have an effect on climate, weather, and the atmosphere (Figure 10-4). The mechanisms are not completely understood, but recently some dramatic progress has been made.

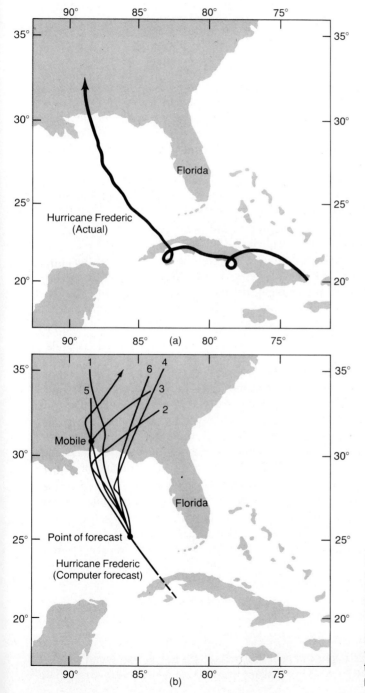

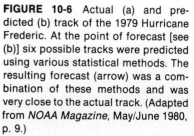

FIGURE 10-6 Actual (a) and predicted (b) track of the 1979 Hurricane Frederic. At the point of forecast [see (b)] six possible tracks were predicted using various statistical methods. The resulting forecast (arrow) was a combination of these methods and was very close to the actual track. (Adapted from *NOAA Magazine*, May/June 1980, p. 9.)

Clearly the stronger the wind, the higher the evaporation and likewise the stronger the oceanic currents. Air and water can heat or cool each other and supply moisture in either direction. The ocean is the more effective storehouse of the radiant energy that reaches the earth's surface and this energy can later be released to the atmosphere. The ocean, because of its high heat capacity and relative density stability, changes in temperature much more slowly than the atmosphere. There is a feedback between the two that is quite complex and not understood.

Probably one of the fundamental questions concerning climate is the variation of atmospheric circulation and how it is affected by the ocean, in particular by the surface temperature of the ocean. Recently it has become evident that variations of atmospheric circulation seem to be associated with subtle temperature changes in the ocean but the relationships are not often direct or clear. Although the source of heat for the earth is the sun, the net heat input is strongly influenced by clouds, ice, and snow cover as well as by less obvious items like the shape of the earth's orbit around the sun. The ocean tends to modify the temperature contrasts between the equatorial and polar regions by the surface poleward movement of water, and a returning flow at depth.

If the ocean actually influences climate, then it follows that one of the major characteristics of the ocean—its surface temperature—should be an important aspect. Research has shown that changes in ocean-surface temperature in the central Pacific seem to correlate with changes in temperature in North America. The key question then becomes, Can ocean temperatures be used to predict characteristics of the atmosphere and ultimately weather and climate? The answer is not unambiguous although some intriguing possibilities exist.

Climatic models and empirical observations suggest that variations in sea-surface temperature may be a key factor in monthly and long-term climatic patterns. Basically, it works this way, as air moves over the ocean it can have its temperature and humidity changed. Cool, relatively dry air will gain heat and moisture if it moves over warmer water. These characteristics can then affect areas downstream—rain, for example. Some studies have indicated that sea-surface temperatures can be contributing components to the quality of winter weather, wind patterns, rainfall, hurricanes, and length of the season.

There are four major factors that appear to control the variations in sea-surface temperature.

1. The amount of solar radiation absorbed in the upper layers of the ocean (great variation depending on cloud cover).
2. Horizontal and vertical movement of water (and thus heat); for example, upwelling will bring large amounts of cold water to the surface.
3. Evaporation and cooling of the surface waters (affected by the temperature of the air and wind speed).
4. Thickness of the upper, or mixed, layer of the ocean.

One of the more exciting and more informative studies of the ocean's effect on climate is that of NORPAX. This project focused on a discovery first made in the mid-1950s that there were large, so-called pools of unusually cold or warm water in the North Pacific. These pools of water are as large as 2,200 km (about 1,370 mi) wide and sometimes extend down as much as 300 m (almost 1,000 ft) below sea level. As such, they cover a significant portion of the central North Pacific and have temperatures that vary by as much as one or two degrees from the surrounding waters. According to one of the principal investigators in this project, Jerome Namias of the Scripps Institution of Oceanography, these pools of water are thought to influence the weather by affecting the flow of air over them. Thus winds blowing westward (winds are referred to by the direction they come from, not the direction they are going as in the case of ocean currents) over a warm pool could be forced to move more toward the north, which would, in turn, affect weather downstream of the wind in the United States (Figure 10-7). If there are abnormally cool and warm areas of ocean water, a similar gradient can result in the overlying air. This, in turn, can alter the behavior of atmospheric storms or even just the general circulation. These regions of temperature contrast can reach into the higher areas of the atmosphere, ultimately affecting the jet stream and causing large variations or meanders in the large airflow patterns over land. If such a pattern continues over long periods of time it can produce anomalous weather and climatic conditions. Such an effect is thought to have caused the drought conditions of 1976 in California and western Europe. Namias suggested that unusually warmer weather during the winters of the 1950s in the southeastern United States and the unusually colder winters in the 1960s and 1970s were due to these anomalous bodies of water in the North Pacific. He suggests that a pool of cool water in the central Pacific caused the prevailing westerly winds to shift toward the south in the 1960s and 1970s. Because of this, colder and drier air from Canada and the Arctic moved down into the southeast replacing the warmer, moister air that usually comes in from the Gulf of Mexico and the Atlantic. The net result was that winter temperatures were lowered by as much as 5°F in the southeastern United States. In spite of this some scientists do not share these views and feel that these pools of water have little effect on weather or climate.

More recent data seem to support the idea of Namias. In the fall of 1976 the North Pacific was anomalously cold; however, water immediately off the west coast of North America tended to be warmer. To the south, in the equatorial parts of the Pacific, temperatures were also warmer. The winter of 1976–77 produced severe cold and record snow falls in parts of the United States. The winds that usually travel eastward across the United States were deflected to the north and rivers that always froze did not freeze that year. The eastern and central parts of the United States, however, had one of the coldest winters in 177 years due to air movement in from the Arctic. According to the NORPAX scientists this weather pattern was expected because of the anomalous temperature pattern in the North Pacific at that time (see upper part of Figure 10-7). Similar types of studies have been made from data from the Gulf of Mexico and off South America with equally intriguing but not conclusive results. Continuing studies of

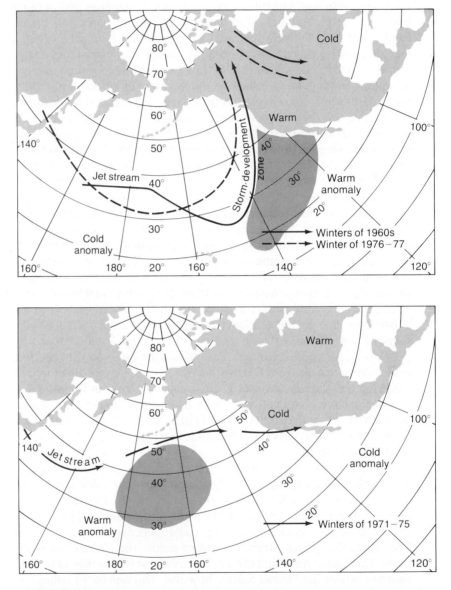

FIGURE 10-7 Diagram showing the relationship of ocean temperature anomalies and movements of the jet stream. Areas of anomalously cold and warm sea-surface temperature are thought to deflect the flow of the jet stream, which in turn strongly influences the climate in the United States. The upper figure represents more severe winter weather, whereas milder winters occur with the bottom conditions. (From *The Dynamic Ocean*, 1978 IDOE Report.)

these types may ultimately lead to improved predicting techniques for some climatic patterns.

The Carbon Dioxide Problem

The release of carbon dioxide into the atmosphere from the burning of fossil fuels (called *anthropogenic* carbon dioxide) has increased about 4 percent per year since 1860. About 50 percent of the carbon dioxide increase seems to remain in the atmosphere; the rest is thought to enter either the ocean or life forms on land. The anthropogenic carbon dioxide that gets into the ocean by gas exchange at the sea surface will work its way down into the deep sea by mixing processes. The rates and exact mechanisms of these important processes are not well known.

The amount of carbon dioxide in the atmosphere is important since it, like water vapor, is opaque to infrared radiation emitted from the earth but essentially transparent to visible incoming solar radiation (Figure 10-8). This leads to a "greenhouse effect," causing higher temperatures on the earth's surface. It is a well-accepted fact that the concentration of carbon dioxide has been increased in the atmosphere over the last two decades (Figure 10-9); many feel this could cause a rise in global temperature. The increase in carbon dioxide concentration in the atmosphere (Figure 10-9) is not surprising when one realizes that about 18 billion tons of carbon dioxide are added to the atmosphere each year from the burning of fossil fuel. There are other sources of carbon dioxide, such as the reduction of land biomass by urbanization and by burning of firewood; the exact

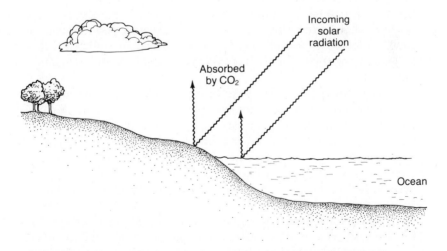

FIGURE 10-8 Illustration showing the "greenhouse effect" caused by carbon dioxide in the atmosphere. The carbon dioxide is essentially transparent to solar radiation, but absorbs the infrared radiation given off from land and water. In this manner it traps the heat leaving the earth and raises the atmospheric temperature.

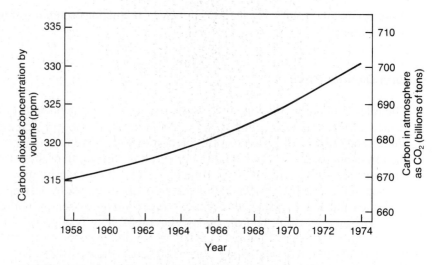

FIGURE 10-9 Average increase in atmospheric carbon dioxide concentration. (Data from observations made by Dr. C. D. Keeling and associates at the Mauna Loa Observatory in Hawaii.)

contribution of these sources is controversial, but many feel they are generally small compared to the fossil fuels. As pointed out by Brewer in 1978, the 18 billion tons in itself is not large when compared to the natural fluxes of carbon dioxide within forests and oceans. However, these fluxes were in a balanced steady state prior to industrialization and the new large input. Another confusing aspect of the new carbon dioxide input is that only about 50 percent is in the atmosphere, while the fate of the other 50 percent is somewhat controversial; many feel that some, if not most, may have gone into the ocean. The ability of the ocean to hold increasing amounts of carbon dioxide is critical in future evaluation of the greenhouse effect.

The effects of the increase of carbon dioxide in the atmosphere could be one of the more important environmental problems facing this planet. Not only must the amount of input and its fate be resolved, but perhaps more important is how these changes will affect climate. It should be appreciated that other than stopping the burning of fossil fuels, little can be done to reduce any potential effects. The problem is further complicated by having to distinguish what changes are due to increasing carbon dioxide content and which are due to other, sometimes unknown, factors. For example, the climate of the earth is presently going through a cooling trend that is thought will last another decade. This trend might be obscuring any smaller-scale heating due to carbon dioxide build up.

A 1979 National Academy of Sciences study (*Carbon Dioxide and Climate: A Scientific Assessment*) examined what would happen with a doubling of the atmospheric concentration of carbon dioxide by the first half of the twenty-first century (a rate consistent with present and projected increases). Such an increase is projected to produce a global surface warming of from 2 to 3.5°C (3.6 to 6.3°F), with greater increases in higher latitudes. This warming probably will cause geographical shifts in temperature, rainfall, evaporation, and soil moisture. On a positive side, the report suggested that although the warming will occur it could be delayed by a few decades if the intermediate waters of the ocean were to absorb some of this heat.

One obvious effect of a worldwide rise in temperature would be the melting of parts of major ice sheets and a consequential rise of sea level and flooding of low-lying areas. Such an increase could also affect many areas available for agriculture, as well as influence rainfall.

El Niño

One of the more dramatic effects of climate on ecology is the phenomenon called *El Niño* that occurs off the west coast of Peru, a highly biologically productive area. Upwelling along the coast brings up nutrient-enriched waters that lead to a growth and eventual harvest of anchovies that is so immense that, at times, it equals about 20 percent of the total food catch of the ocean. On certain occasions, however, the phenomenon called El Niño changes this pattern. El Niño is Spanish for The Child—the effect is given this name because the phenomenon occurs around Christmas time. When this phenomenon occurs warm water flows farther south along the Peru coast than usual, and eventually reaches the Chile coast. There the water acts almost like a blanket on the ocean and prevents the cold, nutrient-enriched water from upwelling. This results in a dramatic reduction in planktonic growth that directly affects the growth of anchovies as well as the sea birds that feed on the anchovies and produce an ancillary industry, harvesting of their waste products, guano. The effects of El Niño can be dramatic, reducing the fish harvest from this area by 50 percent or more. Studies of the causes of El Niño have been going on for decades to try and understand why the wind and the sea produce this phenomenon in order to be able to predict its occurrence. A breakthrough occurred in the early 1970s when Klaus Wyrtki, an oceanographer from the University of Hawaii, developed a model that appeared to explain El Niño. It was noted that there was a buildup of the trade winds prior to the development of the phenomenon. Thus Wyrtki suggested that this would cause an increase in the amounts of warm water carried westward, but when the winds relaxed the water would "slosh" to the south along the western coast of South America, producing El Niño. It was predicted, based on meteorological data, that a small El Niño would develop in 1975. Oceanographers were able to develop an expedition in time to test this prediction and indeed observed a small El Niño. It is felt that future ones can now be predicted by monitoring winds and water buildup.

Ice and Climate

Certainly one of the more dramatic features of the earth's surface is its ice cover; it is also a vivid expression of climatic differences on our planet (Figure 10-10). There have been periods of time when continental glaciers covered much larger portions of the earth, reaching, for example, as far south as New York only 15,000 years ago. Ice also covered portions of Australia, New Zealand, South America, Europe, and southern Africa. These ice ages are excellent examples of past changes in the earth's climate. The past advance of the glaciers not only meant lowered temperatures, but also lowered stands of sea level. It should be noted that glaciers and ice sheets are not only a result of climatic change but also help cause the change. Ice and especially snow-covered ice can reflect almost all incoming solar radiation. The ratio of the amount of radiation reflected to the amount of radiation received is called the *albedo* and in the case of snow-covered ice can approach 98 percent. As the temperature drops more areas become covered by snow and ice. Ice-covered areas reflect more sunlight than the uncovered ground, thus further decreasing the temperature. Obviously, this process also has some sort of feedback since if it did not, ice would eventually cover the entire ocean. The albedo effect is more important in the summer than in the winter when incoming solar radiation in polar regions is so small that the effect of the high albedo is almost negligible. The albedo of a dark-colored body such as rock

FIGURE 10-10 The U.S. Coast Guard icebreaker *Eastwind* near an exceptionally attractive iceberg in Arctic waters. (U.S. Coast Guard photograph.)

is usually very low and it thus can often achieve temperatures above that of the atmosphere (however, it should be remembered that rock will also cool off very quickly).

The variation in ice cover, especially in Antarctica, is large and should be having some, yet undefined, effect on the Antarctic climate. In areas where sea ice is forming, since salts do not get incorporated into the ice, the salinity of the remaining water will be increased in turn increasing in density and leading to the formation of bottom water.

Past Climate

Climate certainly was different in the geologic past when the geometry of the continents and ocean basins had different configurations. Over 200 million years ago a single continent called Pangaea extended almost from pole to pole and was surrounded by a major undivided ocean (Figure 4-3). Little is known about oceanic climate and circulation at that time (most marine sediments of that age have already been removed from the ocean by subduction associated with sea-floor spreading). The land climate, however, was distinguished by ice sheets at the South Pole, that occasionally extended to what is now India, South Africa, Australia, Antarctica, and South America. The then mid-latitude regions had evaporitic conditions, with salt deposits and deserts. As Pangaea eventually split apart and the forms of the different ocean basins developed, climate may have improved. With the movement of the continents different current patterns and different climates evolved (Figure 10-11). The conditions about 65 million years ago were thought to be warmer than today, with no polar ice caps. Antarctica remained attached to either Australia or South America until 35 million years ago. Thus the West Wind Drift could not be established until a clear, around-the-world passageway existed. Once this occurred it probably led to the present oceanographic circulation system whereby cold, dense waters from this region form the bottom waters common to much of the world's oceans.

One of the more dramatic examples of ocean–climate interactions occurred during the recent ice age of the last million years when the amount of ice on the earth increased dramatically and temperatures probably dropped by as much as 10°C (18°F). During this ice age there were numerous periods of advance and retreat of glaciers.

There are numerous hypotheses that have been proposed to explain the start and stop of the ice advance. Among the most popular are those involving variations in the earth's orbit, changes in the energy received from the sun, or large inputs of volcanic dust or carbon dioxide into the atmosphere. The one involving variations in the earth's orbit has been favored by several scientists and some persuasive arguments supporting it have been published. This idea, commonly called the Milankovitch hypothesis, has three basic components.

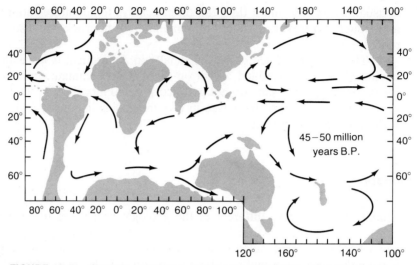

FIGURE 10-11 Oceanic circulation patterns at about 45 to 50 million years ago, based on the study of sediment cores. Compare with the present patterns shown in Figure 9-13. (Figure courtesy of Dr. Ted Moore, University of Rhode Island.)

1. Changes in the tilt of the earth's axis.
2. Changes in the precession of the earth's equinox.
3. Changes in the eccentricity of the earth's orbit.

These components (Figure 10-12) can combine, according to the hypothesis, to produce conditions that will influence the earth's climate and eventually cause glacial conditions. For example, tilt, which causes seasons, varies from about 24° to 22°—the larger the angle the more extreme the seasons. Precession of the equinox refers to the position of earth on its elliptical orbit around the sun at a particular season and has a 21,000 year cycle. Winters during the closer approach to the sun will be mild. Changes in the degree of eccentricity of the orbit will likewise influence the seasonal climate. It is the combination of these cycles that is important.

Studies of deep-sea sediments, which by nature of their location are not disturbed by advancing or retreating glaciers, using isotopic techniques, have established a general climatic curve for the past 700,00 years. By combining this information with data about the reversal of the earth's magnetic field (see Chapter 6) and making some assumptions, it has been possible to date this curve. The data show several distinct climatic cycles that are very close to those predicted by the Milankovitch hypothesis. Times of extreme seasons would not favor glacial growth because there would be considerable glacial melting during the summer and little build up of the glaciers during cold, dry winters. However, during moderate conditions, there would be less melting in the summer and rainfall (snow) would be higher in the winter, causing glacial growth.

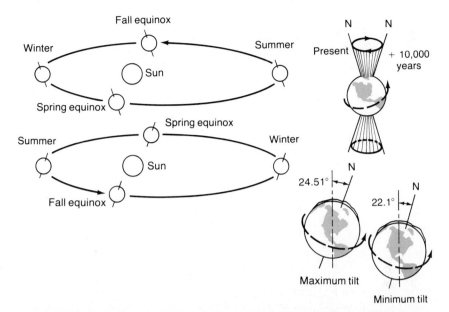

FIGURE 10-12 Variations in the motion of the earth in relation to the sun are factors that influence the earth's climate. At top left, the present orientation of the earth's axis in orbit is represented and, below, the orientation 10,000 years from now. (Relative distances are exaggerated.) At top right, diagram shows the precession (changing orientation) of the earth's axis over a period of 10,000 years and, below, the tilt of the earth's axis from maximum to minimum. (Figure courtesy of M. Kominz and Russel Kolton, University of Rhode Island.)

Other hypotheses are possible; for example, deep-sea sediments from the last 2 million years are relatively rich in volcanic ash as compared with those of the previous 20 million years. Most of the ash results from large-scale volcanic activity on land. Whether this activity has any effect on glacial growth (ash in the atmosphere reduces the amount of sunlight reaching the earth's surface) or whether it is caused by the glacial periods is unclear.

Since the end of the last glacial advance (maximum advance about 15,000 to 18,000 years ago) the climate was warming until about 6,000 years ago (Figure 10-13). An example of this effect is that semitropical plants grew in Minnesota during this period. Since then the earth has been in a cooling period. An especially cool period, sometimes called the little ice age, occurred from 1430 to 1850. Following this, until about 1940, temperatures warmed somewhat. More recently temperatures have been falling. The future is thought to include a warming phase, perhaps even higher than anticipated due to the carbon dioxide problem mentioned earlier.

One of the major programs of the IDOE was CLIMAP (Climate: Long-range Investigation, Mapping, and Prediction). This program is trying to define the past

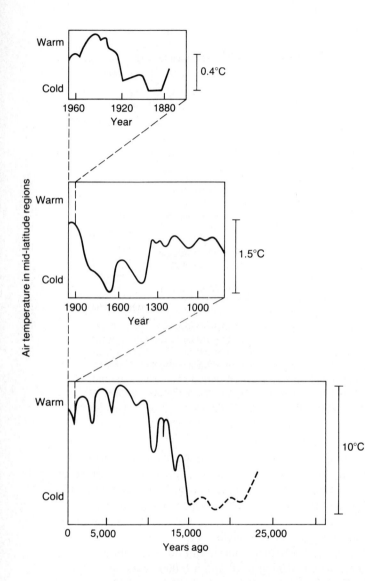

FIGURE 10-13 General trends in mid-latitude air temperature. Scales to right show the various temperature ranges. (Adapted from the *Physical Basis of Climate and Climate Modelling*, GARP Publication Series No. 16, Geneva: World Meteorological Organization, 1975.)

climatic conditions, mainly by the study of deep-sea sediments, in order to understand what caused these changes and to see if they can be predicted. By the study of fossil shells and their isotopic composition, combined with methods of dating, it has been possible to ascertain past sea-surface temperatures with an accuracy of 1 or 2°C (1.8 to 3.6°F). Combining this information with land data has permitted oceanographers to look at the recent climate of our planet.

The coming years should lead to even more integration of oceanographic and meteorological data and, if hopes are realized, a better understanding of climate. Long-term prediction and climate modification may be close at hand.

Summary

Recent studies appear to indicate a close relationship between the ocean and climate. Many of the interactions between the atmosphere and the ocean are well known, and could bear on climate and weather. As scientists learn more about climate, oceans, and the atmosphere, better predictions and perhaps even modification or control of weather or climate may be possible. The impact of weather or climate is often not appreciated—hurricanes, for example, killed more than 13,000 Americans in this century. A cold winter compared with a warm one might mean a difference of over $10 billion in fuel costs as well as several severe agricultural problems.

Many new, exciting international programs are in the process of studying ocean and climate interactions. One of the more successful of these programs is NORPAX. It has shown that variations in oceanic sea-surface temperature may affect the climatic patterns downwind. Some feel that many of the climatic extremes of North America during the past two decades may have been directly related to these temperature patterns.

One of the more challenging environmental problems is the buildup of carbon dioxide in the atmosphere. This buildup might result in a greenhouse effect with a consequent increase in global temperatures. If the hypothesis is correct, such an increase could result in melting of glaciers and a rise in sea level. One remaining problem is that not all of the carbon dioxide that enters the atmosphere is accounted for; how much enters the ocean is unclear.

Climate has varied over geologic time, having been influenced by past positions of the continents and more recently by the ice ages. Various hypotheses have been proposed to explain the ice age, but the one involving variations of the earth's orbit (the Milankovitch hypothesis) is presently in favor.

Suggested Further Readings

Carbon Dioxide and Climate: A Scientific Assessment. (Report of an Ad Hoc Study Group on Carbon Dioxide and Climate), National Research Council, National Academy of Sciences, 1979, 35 pp.

FUNK, B., "Hurricane," *National Geographic,* **158,** no. 3 (1980), pp. 346–79.

Geological Perspectives on Climatic Change. National Academy of Sciences, Washington, D.C., 1978, 46 pp.

IDOE, *The Dynamic Ocean: Its Role in Climate Forecasting*. National Science Foundation, 1978, 27 pp.

IMBRIE, J., "Geological Perspectives on our Changing Climate," *Oceanus,* 21, no. 4 (1978), pp. 65–70.

KOMINZ, M., "What Causes the Ice Ages," *Maritimes,* 23 (1979), pp. 12–15.

MOSAIC—Science in the Sea, 11, no. 4 (July/August 1980).

NEWELL, R. E., "Climate and the Ocean," *American Scientist,* **67,** no. 4 (1979), pp. 405–16.

11

The Coastal Zone

TO THE HUMAN RACE the coastal zone is probably the most important part of the ocean, since about two-thirds of the world's population live near the coast. In 1960 about 45 million U.S. citizens lived in counties that bordered the ocean; by 1970 the number had increased to over 60 million; and by 1980 to over 70 million. If the Great Lakes are included, over 50 percent of the U.S. population live within 80 km (about 50 mi) of the coast and about 130 million within 160 km (100 mi) of the shoreline.

The coastal zone is situated at the boundary of the two major environments of the earth—land and ocean (Figure 11-1). It is an area of numerous biological, chemical, physical, geologic, and meteorological interactions. The coastal zone may be defined as that part of the ocean affected by the land and that part of the land affected by the ocean. The boundary of the two environments is the shoreline, which is constantly changing in position. In the seaward direction, part, if not all, of the continental shelf may be included within the coastal zone while in the landward direction estuaries, marshes, seacliffs, the coastal plains, and other similar environments are included. Actually there really is no universally acceptable definition of the coastal zone. Some have even suggested that the entire 200-mi exclusive economic zone (see Chapter 15) be considered as part of the coastal zone.

Definitions and terminology of some coastal zone features are shown in Figure 11-2. A shore zone and shoreline can also be defined within the coastal zone; the shore zone covers that area where water and land come in direct contact, and includes the beach and surf zone. The shoreline marks the point where land and water meet; the nearshore region is that area seaward of the shoreline. The shoreline is a very dynamic area that, on a worldwide basis, has a length of over 400,000 km (about 249,000 mi). Most shoreline features are temporary since their position is affected by the height of the sea level, which is, in turn, influenced by tides, direction of wind, and strength and height of breaking waves. Thus the shoreline is constantly changing and may extend over a large area in a very short time. In addition, sea level has changed considerably over the last 30,000 years

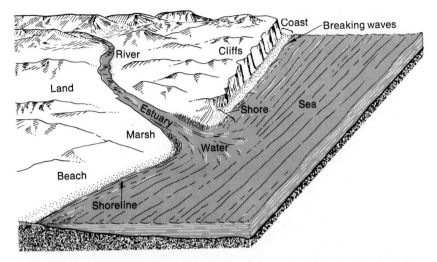

FIGURE 11-1 A diagrammatic view of some common features of the coastal zone.

because of widespread growth and melting of glaciers (Figures 4-4 and 4-5); during this time the shoreline has migrated out and back over the breadth of most parts of the continental shelf.

The extent of the coastal zone is considerable. Inland waters, such as semi-enclosed bays, estuaries, and lagoons, exceed 100,000 km^2 (over 38,000 mi^2) for the United States alone. The region seaward of the shoreline, the continental

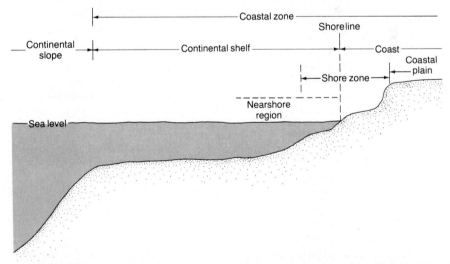

FIGURE 11-2 General features of the coastal zone. (Adapted from Inman and Nordstrom, 1971.)

shelf (discussed in Chapter 5), has a worldwide area of about 50 million km^2 (about 19 million mi^2).

It is often hard to visualize the importance of the coastal region to the other aspects of the oceans. For example, an extremely large percentage of marine organisms either begin or spend a major portion of their life in estuaries or marshes, and even more inhabit the continental shelf. As indicated in Chapter 12, over 90 percent of our fishing products come from areas that are included within the coastal zone. The coastal zone can easily be considered as the bottom step of the ocean's ecological ladder. Rivers enter the ocean through the coastal zone carrying both beneficial products such as nutrients and damaging items like pollutants and waste products from coastal industries and municipal dumping. The rivers of the world carry about 2.5 × 10^{15} g (2,500,000,000,000,000 g) of dissolved material into the ocean each year. As immense as this number is, it is only about one twenty-millionth of that already in the ocean. Much of this material will eventually reach the deeper ocean by coastal processes, such as waves, currents, and tides.

In the United States, authority for management and control of activities over the nearshore parts of the coastal zone is generally divided among several different federal, state, and local agencies. Many have overlapping and commonly conflicting responsibilities for the region. Local governments often have control over land use and waste disposal whereas state agencies may control water, pollution control, highways, and ownership of state lands. These different patterns of ownership and fragmented control can lead to complexities and incorrect decisions concerning the coastal zone. Before considering some of these problems, a discussion of some of the parts of the coastal zone and oceanographic processes in this region is appropriate.

Beaches

Beaches are the unconsolidated sediments (mainly sand or gravel) that cover most parts of the shore (Figure 11-3) and are usually directly under the influence of waves. They are generally somewhat stable but can be rapidly changed by large or storm waves. Beaches are also daily and monthly modified by tidal changes, and many also have seasonal changes. Beaches can also be strongly influenced by the work of people; the building of jetties and other nearshore facilities often causes one beach to expand at the expense of another. Erosion of beaches both natural and caused by people has led to increased recent activities toward preserving beaches.

Beaches are affected by waves in many ways. If the wave crests are not parallel to the shore as they approach the coast, refraction occurs when they reach shallow water (see pages 320–322). Refraction causes the wave crests to turn into shallow water, so the wave crests tend to parallel the depth contours. Thus wave energy converges on projecting points such as offshore bars and sea

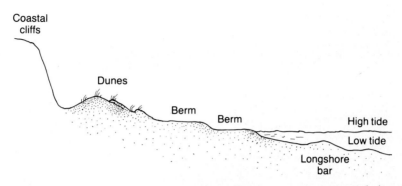

FIGURE 11-3 General characteristics of a beach. Berms are flat portions of a beach formed by wave action. The two shown in this figure are due to previous storms.

cliffs and diverges in open bay areas. Erosion by waves is therefore stronger at points of convergence, and sediment usually moves to the quieter divergent areas (Figure 11-4). If this process is allowed to go to completion a straight shoreline results.

After waves break, the forward movement of the water is toward the beach. The beach, however, is an essentially impermeable barrier. If the waves approach the beach at any angle, the discharge of water from the breaking waves is directed along or parallel to the beach, eventually forming a longshore current (Figure 9-23). This longshore current can transport sediments parallel to the beach, but eventually the water in the current builds up, so it must move seaward. When this happens a rip current (Figure 11-5) is produced; its position is dependent

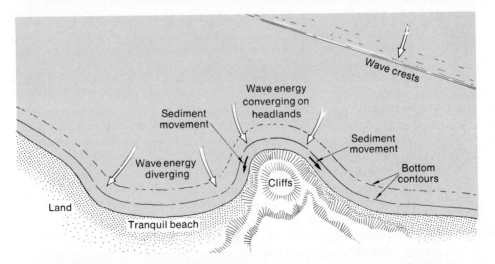

FIGURE 11-4 Convergence and divergence of wave energy due to the refraction of waves. (See also Figure 9-22.)

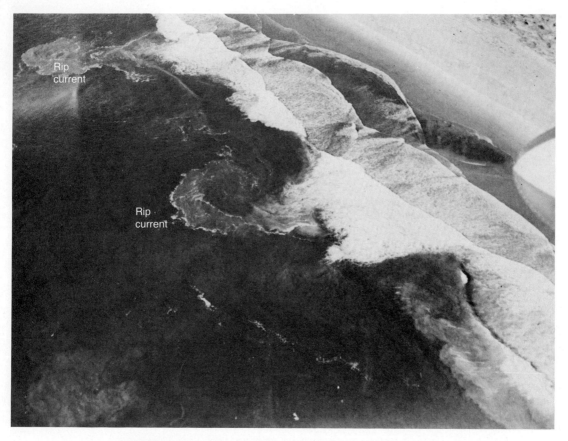

FIGURE 11-5 A series of rip currents that are distinguished by light, sediment-laden water beyond the breaking waves. (Photograph courtesy of Dr. D. L. Inman, Scripps Institution of Oceanography.)

upon the bottom topography and the height and period of the waves. Rip currents can carry sediments seaward, but the sediments can again be moved landward with the next sequence of approaching waves. Thus waves approaching a coast essentially keep the bottom sediments in somewhat of a transient state: eroding in one place, depositing in another, but eventually smoothing out the coastline—unless people interfere. If the sediments are not returned to the beach by the incoming waves, the beach is gradually eroded. This process can result in loss of homes or other structures if they are built too close to the shoreline (Figure 11-8). Some beaches lose their sediment because they are situated near submarine canyons (Figure 11-6). Sand carried by longshore currents can be intercepted by these submarine canyons and eventually carried—probably by slumping or turbidity currents—into the deep sea. Sand being moved by longshore currents can also be trapped by jetties or groins built along the coast. This usually results in deposition of sand on the upstream side of the jetty (Figure 11-7).

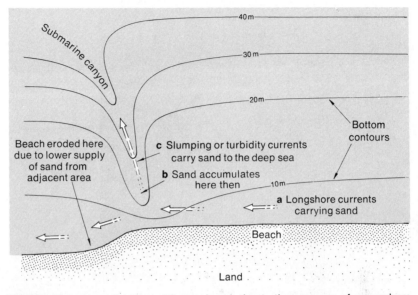

FIGURE 11-6 Loss of sediment from a beach due to the presence of a nearshore submarine canyon.

Longshore currents cause a net along-shore movement of the sediment and eventually will straighten out a shoreline. This process can take thousands of years and often causes considerable local erosion. Waterfront property owners often construct jetties or groins that interfere with the longshore current and cause the deposition of the sediment carried by the current (see Figures 11-7 and 11-8). Often erosion occurs downstream of these barriers, a process that leads to the construction of another groin or jetty, and so on.

The usual direction of movement of sand is toward the beach. During times of relatively long-period waves, like those that generally occur in the summer months, sand is picked up by waves from shallow depths and carried onto the beach. The sand is then removed, however, by the backrush of water from the wave running down the beach and is carried seaward. But since the backrush is smaller than the incoming waves, the sand is not carried as far seaward as its original position. The next wave results again in a net motion toward the beach. During times of high waves of short period (like those that generally occur in winter months), the short period keeps the sand in suspension and prevents it from settling (Figure 11-9). Therefore, much of the sand washed off the beach by the backrush of water does not settle until it is carried into deeper water outside the action of subsequent waves. This situation generally results in the loss of beach sand during the winter (Figure 11-10).

The circulation of water in the nearshore zone results in an exchange of the water in this region with the more offshore areas. In this manner nutrients, pollutants, and other material carried by the rivers are eventually distributed offshore.

(a)

(b)

FIGURE 11-7 (a) Modification of a beach by a jetty. The jetty has trapped sand moving along the beach (toward the foreground) and thus has built up the beach. (b) This photograph shows the area on the other side of the jetty, where the beaches are considerably narrower because the sand that would have been carried to them has been intercepted by the jetty. (Photographs courtesy of Jack Silver.)

FIGURE 11-8 A southern California house built too close to the shoreline and in the process of being undercut.

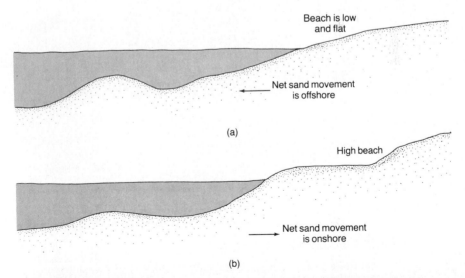

FIGURE 11-9 Typical seasonal changes along a beach. (a) Typical winter situation; (b) typical summer situation.

(a)

(b)

FIGURE 11-10 Winter (a) and summer (b) changes at La Jolla, California. About 60 cm or 2 ft of sand has been added in the summer. (Photographs courtesy of Dr. David G. Aubrey.)

Most beach sand comes from the sea floor, but its source is the land, from which it is carried to the ocean by rivers, wind, glaciers, and cliff erosion. In most areas of the world very small quantities of sediment are presently being supplied to the ocean because the recent rise in sea level has caused most rivers to deposit their sediments directly into their estuaries rather than into the ocean. The rising sea level has, in effect, caused a migration of beaches across the continental shelf to their present position.

Barrier beaches or islands, which are narrow, long ridges that sit above high tide and are separated from, but parallel to, the main coast, are an especially important aspect of the coastal zone. Examples of such beaches are Fire Island, New York; Atlantic City, New Jersey; Cape Hatteras, North Carolina; Miami Beach, Florida; and Padre Island, Texas. Because of their position, the beaches often form an effective buffer that can protect the mainland from the major effects of offshore storms. Unfortunately, their location has also made them prime sites for development. Barrier islands have been among the most urbanized regions of the United States and this use has continued, although at a reduced rate, in recent years. In some regions, such as Miami Beach, large sums of money are being spent in a usually unsuccessful attempt to restore the beaches to their original condition.

Estuaries, Lagoons, and Marshes

Estuaries, lagoons, and marshes are common to many coastal areas (Figures 11-1 and 11-11). An estuary is a body of water partially enclosed by land that has a connection to the ocean as well as to a river—in other words, an area where saltwater and freshwater mix. Within the estuary the seawater is diluted by and mixed with the freshwater.

Somewhat similar to estuaries are lagoons, which are broad, shallow areas partially restricted from the ocean by offshore barrier beaches. If the opening or access to the ocean gets closed, lagoons can eventually become freshwater lakes. Because of the recent rise of sea level, most rivers of the world now flow into estuaries. Estuaries, in general, are found where continental shelves and coastal regions are narrow and have high relief, and lagoons generally occur where the shelves and coastal regions are wide and smooth. This distribution is reasonable because most estuaries are drowned river valleys or fjords (deep V-shaped valleys cut by glaciers), and lagoons are formed mainly by the buildup of offshore bars in low-relief areas.

Many estuaries or coastal areas are bordered by a narrow strip of vegetation or wetland area containing either salt marshes or mangrove swamps. Mangroves are generally found only between 30° N and 30° S, or in the equatorial regions, whereas marshes occur in the more temperate climates (Figure 11-12). This distribution, it appears, is controlled by the apparent inability of mangrove seedlings to survive freezing temperatures.

Typical estuaries along the U.S. coast are Chesapeake Bay, Puget Sound, and Delaware Bay. One of the best examples of a lagoon is Laguna Madre along the south Texas coast. Here a long barrier island, Padre Island, isolates much of the Texas coast from the Gulf of Mexico.

FIGURE 11-11 A complex coastal area. Movement of sand from right to left has deflected the path of the main river to the left and built a series of bars off the coast. Behind the bars are marshlands. In the left foreground is another river meandering over an area of small ponds. The entire area (Cape Cod, Massachusetts) was covered by glaciers 15,000 years ago.

Estuaries and lagoons, like beaches and other parts of the coastal region, are really temporary features. With time, both will be destroyed either by the buildup of marshes or by the cutback of the coastline by marine erosion. In the former case, freshwater marshes form at the inland areas and saltwater marshes or bars tend to close the seaward part of the estuary. Estuaries and fjords that are cut into solid rock are less likely to be affected by these processes. In many instances rivers carrying large quantities of sediment will fill in their estuary. In addition, large quantities of sediment can be carried into the estuary from the offshore areas. Thus if the estuary is an important navigation channel, like Chesapeake Bay, it may need almost continual dredging to keep it open.

If sea level were to lower, most estuaries would rapidly disappear, to be replaced by rivers cutting into their deposits. If sea level were to remain constant or rise just a few meters, the processes of destruction would not be stopped. Only a considerable rise in sea level would maintain or rejuvenate present estuaries.

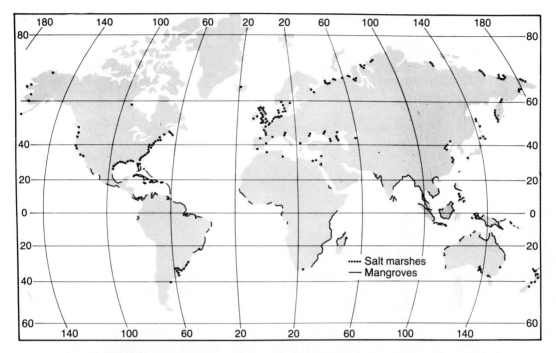

FIGURE 11-12 World distribution of well-established salt marshes and mangroves. Since the data are mainly from the literature, all such areas are not represented. (From Valiela and Vince, 1976.)

Thus it also follows that estuaries were not common in the geologic past except during periods of rising sea level or lowering of the land.

Most estuaries have a two-way movement of water. The freshwater from the river flows seaward along the surface while the saltier, and thus denser, ocean water flows landward along the bottom. The amount of mixing between the two different waters is determined by factors such as wind, tidal range, shape of the estuary, and relative inflow of river and ocean water. In areas where the mixing is low, a salt-wedge type of estuary develops (Figure 11-13). Here the saltwater and freshwater remain almost completely separated from each other, whereas in other areas, because of the above factors, the waters are more mixed (Figure 11-13).

Salt marshes are generally located in intertidal areas, along the banks of tidal rivers, or behind barrier beaches. They are extremely productive areas and support a large marine population including fish, birds, shellfish, and plants. Marshes also form a protective barrier against storms and high seas to the land behind it. The organic production of marshes, increased by nutrients supplied by rivers, can be as high as 5 to 10 tons per year of organic matter per acre compared with 1 ton per year per acre for a wheat field or less than 0.5 ton per year for the

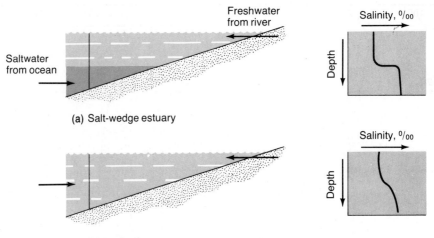

FIGURE 11-13 Cross section and salinity–depth profiles across a salt-wedge and well-mixed estuary. Note that the vertical line on the cross section indicates the position of the salinity–depth profile shown on the right.

open ocean or for a desert. This large production of organic matter results in food that can be consumed by the numerous organisms that inhabit the area. Most marshes have only a few species of life, but they are usually quite abundant. A plant zonation, common in most marshes, is related to exposure to the sea (Table 11-1). The plant *Zostera* (Figure 8-22) is mostly covered by water; proceeding toward higher ground the sequence of plants usually is *Spartina*, *Salicornia*, and *Distichlis*.

If one takes a core sample from a marsh, it may be possible to determine how the marsh is forming. For example, if *Spartina* was found to be growing over *Salicornia*, it would indicate a relative rise in sea level in the marsh area, since *Spartina* will grow closer to sea level than *Salicornia*.

The accumulation of large quantities of plants such as *Salicornia* and *Spartina* in the marsh can form a deposit or soil called *peat*. Because peat forms at or near sea level, it can also indicate the past position of sea level. Many peat

TABLE 11-1 Typical Plant Zonation
in a Saltwater Marsh

Plant	Exposure
Zostera	Always submerged
Spartina	Submerged twice daily
Salicornia	Submerged once daily
Distichlis	Submerged a few times a month

deposits have been left on the shallow portions of the continental shelf because of the recent rise in sea level.

Deltas

Deltas are large accumulations of sediment that are deposited at the mouth of a river. The word delta (Δ) was first applied by Herodotus in the fifth century B.C. to the triangular area off the tributaries of the Nile River (Figure 11-14). The amount of sediment carried by some rivers is immense. The Mississippi River carries as much as 300 million tons of sediment per year, while the Nile used to carry as much as 140 million tons of sediment per year (before the building of the Aswan Dam). The present Nile Delta covers an area of over 100,000 km² (over 38,000 mi²). All rivers transport sediment to the ocean or into whatever other body of water they flow. Usually some or all of the sediment is deposited when the river enters a slower flowing or standing body of water, such as an estuary, ocean, or lake. In the ocean this sediment may be moved away from the river mouth by longshore currents, ocean waves, or tidal currents. Occasionally some large rivers carry more sediments than these marine forces can move (or conversely, the marine forces are too small to move the sediment) and in these instances the sediment accumulates at the river mouth forming a delta. As the delta builds up, the river usually forms distributaries, or channels across the delta, and continues its flow toward the ocean [Figure 11-14(b)]. Coarse-grained sediment is deposited along the distributary, forming the channel; whereas finer-grained deposits are deposited between the channels (commonly called flood-plain deposits). Eventually, the distributary builds itself so far seaward that it can no longer be maintained by the river flow. When this happens, the river changes course and forms a newer and shorter distributary. In this manner the delta continuously builds seaward. After a distributary is abandoned, the reworking effects of waves and currents remove the fine-grained deposits and carry them seaward, leaving a deposit of coarse-grained sediments at the edge of the delta [Figure 11-14(b)].

Deltas are generally very flat areas that are extremely fertile because the sediment deposited by the river often is rich in nutrients. Thus there is a tendency for large settlements to occur on deltas, and indeed many early civilizations settled and started in such areas. Unfortunately, the low elevation and relief of deltas also makes them extremely vulnerable to flooding by storms. Also, deltas usually are areas having high oil or gas potential, or both (see Chapter 12, page 405).

Coastal Problems

Erosion is one of the more important problems of the coastal zone. A recent U.S. Army Corps of Engineers study showed that about 42 percent of the U.S. shore-

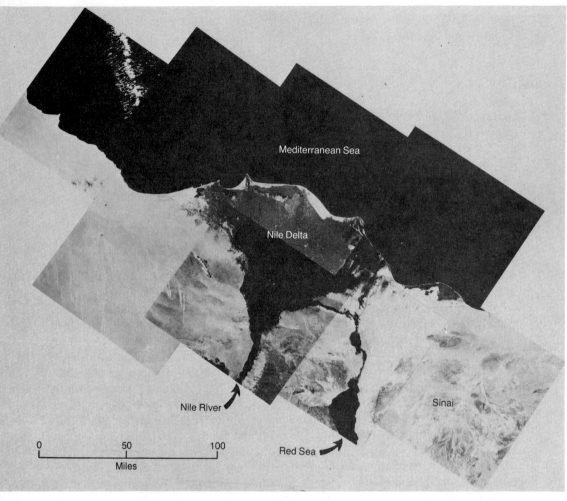

(a)

FIGURE 11-14 (a) Composite of several satellite photographs showing the Nile Delta.

line, excluding the Alaskan coast, is eroding (Table 11-2). The problem is most critical in the North Atlantic area where population is densest and where up to 85 percent of the shoreline is privately owned. Part of the coastal erosion problem around the world is related to the general slow rise in sea level due to glacial melting and to subsidence of coastal areas in some localities. The important factors affecting coastal erosion are, according to Inman and Brush (1973), degree of exposure to waves and currents, supply of sediment and runoff to the coast, shape of the coast and adjacent continental shelf, tidal range and its intensity,

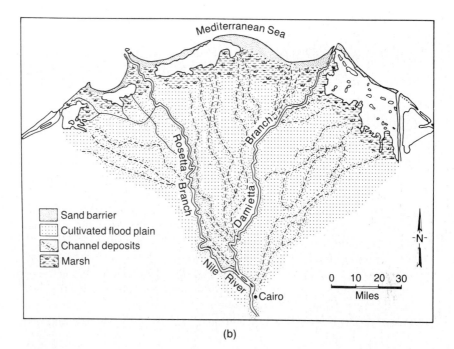

(b)

FIGURE 11-14 (b) Main geomorphic features of the Nile Delta. (Adapted from Wright & Coleman, 1973.)

TABLE 11-2 Erosion of U.S. Shoreline

Region	Total shoreline (mi)	Significant erosion (mi)	Shoreline being eroded (%)
North Atlantic	8,620	7,460	87
South Atlantic–Gulf	14,620	2,820	19
Lower Mississippi	1,940	1,580	81
Texas Gulf	2,500	360	14
Great Lakes	3,680	1,260	34
California	1,810	1,550	86
North Pacific	2,840	260	9
Alaska	47,300	5,100	11
Hawaii	930	110	12
Total for nation	84,240	20,500	24

SOURCE: National Shoreline Study of the U.S. Army Corps of Engineers.

and coastal climate. Storms or hurricanes can have an especially destructive effect on beaches. People's activities and construction, however, have probably had a more profound recent influence on coastal areas (Figure 11-7).

BEACHES

Erosion is generally especially pronounced on beaches and coastal cliffs.. Changes tend to be less where the coast is rocky or where the beach sediments are mainly gravel. Especially dramatic effects can occur along gently sloping sandy beaches where the sediments can be easily moved by waves, currents, and especially storms. Beaches are generally eroded in winter by short-period storm waves and built up in the summer by relatively calm seas that move sediment landward (Figures 11-9 and 11-10).

The alternating erosion and deposition of beaches is a natural phenomenon that produces minimal changes over short periods of time, if not interfered with. It is just this interference that causes increased erosional problems. Basically there are three ways in which people can affect coastal and shoreline areas: land recovery by dredging and filling, damming rivers, and the like; construction of jetties or other coastal structures; and the development or destruction of coastal dune areas. Coastal dunes act as a preventive barrier to inland erosion, but when they are removed or breached, increased erosion can occur. Dunes are especially sensitive to abuse and can be easily damaged by vehicular traffic, paths, or roads. Any of these activities break down the plant structure that anchors the dune, and once a gap exists in the dune pattern, storms can penetrate and erode them.

ESTUARIES AND MARSHES

Several other coastal environments are unique and deserve special care. These include estuaries and wetland regions such as salt marshes and mangrove swamps (Figure 11-15). Sedimentation rates in estuaries can be very high because of the material carried in by the rivers. Most estuaries have resulted from the recent rise of sea level (about 130 m or 426 ft in the last 15,000 years) that has drowned the river and created the estuary. Environmentally and esthetically, estuaries are very appealing areas and 7 of the world's 10 largest cities are in such regions.

Marshes and estuaries are extremely fragile environments and, unfortunately, are also attractive areas for real-estate development. Many marsh and estuary areas of the United States have been filled in, and then developed by having houses, factories, or harbor facilities built on them. By 1975, over 25 percent of the estuaries and marshes in the United States had been severely modified by such building activity. In Connecticut, for example, over 50 percent of the original marshland has been destroyed and of the 14,000 or so acres (1 acre equals 4,047 m^2) remaining in 1969, about 200 acres are being filled in each year (Figure 11-16). As cities grow, in some localities, the marshes are slowly built over and developed (Figure 11-17). In recent years public action against such development has slowed the destruction of these valuable environments.

Marshes commonly contain many food products, such as clams, oysters, scallops, and fish, that are important to the local economy. They can also have

FIGURE 11-15 Air view of a salt marsh on Cape Cod, Massachusetts. This entire marsh is drained by a single major channel. Although no major freshwater stream enters this marsh, generally the main channel is a river or estuary. (Photograph courtesy of Dr. Ivan Valiela.)

FIGURE 11-16 Industrial development in a salt marsh area. In the foreground is a waste disposal area that is slowly covering the marsh. (Photograph courtesy of Dr. Ivan Valiela.)

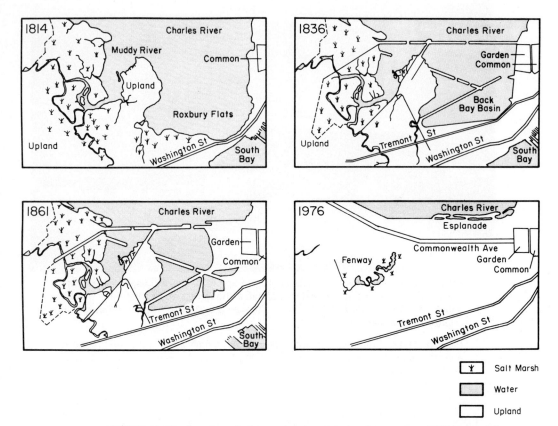

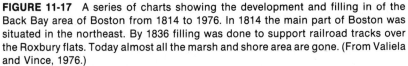

FIGURE 11-17 A series of charts showing the development and filling in of the Back Bay area of Boston from 1814 to 1976. In 1814 the main part of Boston was situated in the northeast. By 1836 filling was done to support railroad tracks over the Roxbury flats. Today almost all the marsh and shore area are gone. (From Valiela and Vince, 1976.)

other important values; for example, marshes can remove some pollutants from the water, in particular, nitrogen and some metals. Experiments have shown that there are actually beneficial effects that can result from adding sewage to coastal marshes. In addition to removing pollutants, the nutrients in the sewage increase the biological productivity of the marsh.

PEOPLE'S ACTIVITY IN THE COASTAL ZONE

Many of the problems of the nearshore region come from the building of coastal structures that interfere with the circulation of water and sediment and thus cause increased deposition in some locations and erosion elsewhere. These coastal structures, described below, especially can interfere with longshore transportation and prevent replenishment of beaches (Figures 11-7 and 11-18), which generally results in erosion on the downstream side of the structure and accretion on the upstream side. The effect can be compounded by a series of structures—an effect that generally occurs since the owner of the downstream property is "forced" into taking some sort of action to prevent the now-increased erosion of his beach. It perhaps is hard to visualize the amount of sand that moves along a beach, but many ocean beaches can have as much as 0.5 to 1 million m³ (0.65 to 1.3 million yards³) of sand transported across them in an average year.

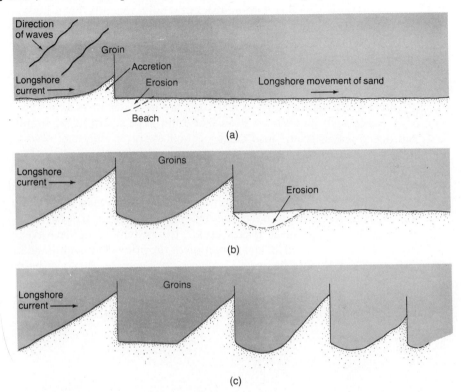

FIGURE 11-18 Events that can happen with the installation of groins or jetties into a longshore current system. Once one groin is added others become needed. (Adapted from Inman and Brush, 1973.)

The most common types of coastal structures or mechanisms to prevent erosion are the following:

1. *Groins or jetties*—structures built perpendicular to the beach into the water that interrupt the longshore movement of water and sand.
2. *Seawalls*—rigid structures built parallel to beaches to withstand and reflect oncoming waves.
3. *Breakwater*—offshore, massive structures situated parallel but off the shore that absorb the breaking of the wave before it reaches the shore. It generally produces

a quiet zone behind it that serves as an accumulation place for sand (and erosion elsewhere).

4. *Revetments*—layers or blankets of strong, nonerodable material placed on the shore to prevent erosion.

5. *Artificial fill*—material added (often yearly) to replenish the beach—an alternative to trying to prevent erosion. Often a combination of techniques is used.

Some new and innovative structures have recently been built to diminish and dampen the effects of waves and other forms of energy that cause coastal erosion. These include floating breakwaters that can adjust their direction with changing currents or winds. Such devices, if big enough, can provide shelter for even large vessels, although none have yet been built. Perhaps more realistic are groups of tires tied together that can act as a simple, but cheap, breakwater (Figure 11-19). Another idea came from a group of scientists at Scripps Institution of Oceanography headed by Professor John Isaacs. The idea, simply stated, is that ocean waves would lose much of their energy if they met rows of spherical balls attached to the sea floor but floating close to the surface (Figure 11-20). Preliminary tests show that the system is effective enough to remove up to 50 percent of the wave energy. The floats behave somewhat like upside down pendulums that have a back-and-forth movement faster than that of the incoming waves. Thus the floats move in the opposite direction of the waves and dissipate much of the energy, or as Isaacs says, "the system beats the waves to death." Such a system costs only a small fraction of the price of conventional breakwaters and offers much more flexibility because, for example, they can be moved.

Anything done in an attempt to protect or restore the shoreline should be done with considerable thought and, if possible, research. The conditions initially observed might not represent the average or even typical situation.

FIGURE 11-19 Groups of scrap tires tied together to form a cheap, but efficient, floating breakwater. (Photograph courtesy of University of Rhode Island.)

FIGURE 11-20 A tethered-float break-water designed to dissipate the energy of wind or boat-generated waves in nearshore water. (Official U.S. Navy photograph.)

Use of the Coastal Zone

Many of the problems concerning the use of the coastal zone could be solved if they were independent problems, but, unfortunately, they are not. Many aspects of coastal zone use have a negative or positive effect on another use. Consider, for example, two sets of conflicting possible uses.

1. Developing offshore or nearshore deep-draft oil terminals versus protection and conservation of the shore zone for recreation and the preservation of its natural resources.
2. Development of waterfront homes (Figure 11-8) or larger developments that destroy, damage, or change the beachfront area versus preserving these areas both for their esthetic value as well as for their breakwater ability for more inland areas.

The number of possible uses of the coastal zone is almost endless. In some instances, multiple usage can be accomplished but this requires knowledgeable management, skill, and compromises among the different interests.

To get a better view of the demands on the coastal zone, consider below some important facts about this region in the United States; many points, however, also relate to other coastal areas in the world.

1. Seventy percent of the U.S. population lives within a day's drive of a coastal region.
2. According to the Department of the Interior, only about 6.4 percent of the U.S. shoreline that can be used for recreation is under public ownership, the rest is privately owned. The contiguous states have about 96,540 km or 60,000 mi of shore-line, only about one-third of which has some recreational potential.
3. Concerning marine recreation, in the mid-1970s over $3.9 billion in retail sales was generated for coastal zone use and there were almost 10 million boats, 80 percent of which were in the 30 coastal states. By the year 2000 the number of boat owners is predicted to more than double.

383

4. Seven of the largest cities in the United States are on the coast and over 33 percent of the U.S. population live in coastal counties, which contain only about 15 percent of the total U.S. land area. The five largest cities in the world (Tokyo, New York, London, Shanghai, and Osaka) are found in coastal areas.

5. Shoreline property has been continuously increasing in value, usually at a rate of more than 10 percent a year, and in some areas coastal shorefront property costs are excessive except for only the wealthy. Clearly, the price of this land will not decrease in the future since such land can only get scarcer.

6. About 40 percent of U.S. manufacturing takes place in coastal counties; coastal facilities handle about 350 million tons of foreign trade and 630 million tons of domestic cargo.

7. The ownership of the U.S. shoreline (excluding Alaska) is as follows: 70 percent private; 12 percent state or local government; 11 percent federal, and about 3 percent uncertain.

8. Many large energy facilities, oil refineries, and an increasing number of nuclear facilities are situated on or very near the coast. Many of these facilities, especially the nuclear plants, require large daily supplies of water.

9. Over 100 million U.S. people participate yearly in ocean-related activities, spending close to $15 billion in the process.

10. Coastal developments have already destroyed or built over a large percentage (probably greater than 20 and less than 50 percent) of our coastal habitat areas.

11. Daily waste disposal from municipal and industrial activity is about 30 billion gal. Dredging spoils are about 100 million tons per year.

12. A very large percentage of commercial fish spend a major portion of their life cycle in the coastal zone. Shellfish, like clams, oysters, and mussels, spend their entire life cycle in the more restricted waters of estuaries, bays, marshes, and the like.

13. Coastal waters receive the major portion of waste material from land, including sewage, radioactive and thermal discharge, and dredging, mining, and construction spoils.

There are authorities who feel that the destruction of the coastal zone in many regions of the world has proceeded so far that it can never be returned to its original state. Certainly, in some areas it will be impossible to restore the coastal zone to its original pristine character before people's arrival and development of this area (Figure 11-21). A more realistic goal would be to keep pollution and destruction to a minimum and to make choices whose long-term effects will be beneficial to as many people as possible. This will mean a nuclear power plant in one place and an undeveloped coastal park in another, rather than either of these everywhere.

One of the major problems of the coastal zone is the discharge of sewage and other types of effluent into it. Among the important questions about an offshore effluent system are, What happens to the material once it reaches the water? How far away can the effluent still be detected and therefore be considered a pollutant? How much effluent can be safely discharged into an area? and Does it have any effect on public health and the esthetic aspects of the nearshore waters?

FIGURE 11-21 Aerial view of Port Newark and Elizabeth Port Authority Marine Terminal—an extremely developed coastal area. (Photograph courtesy of The Port Authority of New York and New Jersey.)

An example of the size of the problem can be seen in the situation off New York City (called the *New York Bight*) in which the area received an average of 4.6 million metric tons of solid material per year during the late 1960s. About 76 percent of this material was dredge waste, 12 percent came from construction and demolition rubble, 7 percent was solids and waste chemicals, and 4 percent was from sewage sludge. The material was discharged over an area of about 160 km² (about 62 mi²) in the city's harbor and over 50 km² (over 19 mi²) offshore on the continental shelf. This is one of the largest areas of waste disposal on the east coast.

The concern for the coastal zone is clearly one of wise management and a major part of that is jurisdictional. Local governments will, of course, see only the local view whereas the national government (in theory) should be considering a broader picture. Decisions concerning the coastal zone are naturally influenced by the potential and present resources of the area. The value of ocean resources, most of which are in or closely related to the coastal zone, is immense. According to a recent study, by the year 2000 this economic potential could exceed $33 billion for the United States (Table 11-3).

There are many reasons why decisions concerning the coastal zone have

TABLE 11-3 Estimated and Projected Primary Economic Value
of Selected Ocean Resources to the United States,
by Type of Activity 1972/73–2000, in Terms of Gross
Ocean-Related Outputs (in Billions of 1973 Dollars)

Activity	1972	1973	1985	2000
Mineral resources:				
Petroleum		2.40	9.60	10.50
Natural gas		0.80	5.80	8.30
Manganese nodules			0.13	0.28
Sulfur		0.04	0.04	0.04
Fresh water		0.01	0.02	0.04
Construction materials		0.01	0.01	0.03
Magnesium		0.14	0.21	0.31
Other			0.01	0.02
Total		3.40	15.82	19.52
Living resources:				
Food fish	0.74		0.95–1.58	1.37–4.01
Industrial fish	0.05		0.05–0.08	0.05–0.14
Botanical resources	a		a	a
Total	0.79		1.00–1.66	1.42–4.15
Nonextractive uses:				
Energy			0.58–0.81	3.78–6.03
Recreation	0.70–0.97		1.12–1.50	1.64–2.53
Transportation	2.57		4.40–6.21	6.88–11.41
Communication	0.13		0.26–0.36	0.44–0.85
Receptacle for waste	b		b	b
Total	3.40–3.67		6.36–8.88	12.74–20.82
Grand total	7.59–7.86		23.18–23.36	33.68–44.49

SOURCE: U.S., Congress, Senate, Committee on Commerce, *The Economic Value of the Ocean Resources to the United States*, National Ocean Policy Study, 93rd Congress, 2nd sess., 1974.
a Insignificant.
b Potentially significant, but unmeasurable.

been delayed or been incorrect. Among the more obvious are the conflicting jurisdiction and lack of coordination among government agencies; lack of a clear plan for the development of the coastal zone (see next paragraph); lack of knowledge and data concerning the resources and marine interactions of the coastal zone; lack of funds to manage the coastal zone; and conflict between economic uses and ecological uses.

A leader in the concern for the coastal zone has been the United States, which in 1972 passed the National Coastal Zone Management Act (P.L. 92-583). This act encourages individual coastal states to exert a strong role in all matters affecting the coastal zone. An Office of Coastal Zone Management was established to make federal funds available to a state if it develops a comprehensive plan for the use of its coastal zone. As part of this plan the state must make baseline studies of its coastal resources, define problem areas, and develop the appropriate political mechanisms for a management program. When the management program of the state is completed, it must be approved by the secretary of commerce before the state starts to receive coordination of federal and state interest, especially in the areas of conservation and energy development. Under the program, states first receive planning grants to develop some sort of management program. Once these are approved, the states can then receive further grants called administrative grants. Several things are necessary for the beginning of a coastal zone management program. First, the local governments must define the area of their coastline for management purposes. Generally this extends out to 3 mi (about 5.5 km) off their coast and includes intertidal areas, beaches, dunes, and wetlands as well as inshore areas to that extent to which the shoreline would affect coastal waters. Within these regions the state has to designate areas of particular concern for one reason or another. These can be oil and gas considerations, environmental problems, recreation potentials, and so forth. The second thing a state has to do is to designate and control those uses in a coastal zone that can affect the coastal waters. As the states start to develop their management programs, they must collect information concerning the resources of the coastal zone and then develop a plan based on that information and show how the coastal zone should be used and developed. The final item is to develop a management technique to implement the plan.

In this process the states must develop a specific program, preparing maps, illustrations, and other such materials concerning the coastal zone, and must define the areas that are relevant to the management program. They must develop objectives and policies to guide public and private use of the lands and waters in the coastal zone and they must inventory the areas involved and prepare guidelines for present and future use of the areas.

Public hearings must be held as the program is developed, the program must be approved by the governor of the state, and a state agency must administer the grants for the development of the program. As these mechanisms are developed there will be much input from different user groups within a state and, in particular, within a coastal zone. This will involve adjustments and changes to the program as it is developed.

The actual future of the National Coastal Zone Management Act is somewhat in doubt due to budgetary constraints imposed by the federal government. However, in many states, plans have been sufficiently developed and may be maintained by the local government.

In the United States the interaction of the coastal zone and offshore oil and gas development can have considerable impact. It is not just the drilling that presents the problems but also that personnel are needed for drilling, and that ship facilities, support industries, and staging areas are necessary along the coast for offshore operations. Refineries, pipelines, and storage tanks may also be built in the nearshore area. All this development will put considerable strain on existing harbor facilities. Likewise, if people come to work in these areas, they will need homes, schools, and other facilities that again will make an additional impact on what probably was a relatively small community. The local areas can rightly argue that the oil companies, to get offshore concessions, pay immense amounts of money directly to the federal government with little, if any, coming directly back to the state or local government. Indeed, any oil drilling 3 mi (about 5.5 km) or more off a coast falls under federal jurisdiction and local governments do not directly receive any of the royalties associated with the drilling.

Two other important U.S. programs for the coastal zone have been developed; one is the Marine Protection Research and Sanctuaries Act of 1972. A provision of this act permits the federal government to designate specific areas of the coastal zone from the high tide line out to the edge of the continental shelf for research protection and recreational purposes. The first such national marine sanctuary was the site of the sinking of the Civil War vessel U.S.S. *Monitor* off Cape Hatteras (Figures 14-16 and 14-17). The second area is a marine park in the Florida Keys. By 1977 more than five areas covering 25,000 acres had been designated as sanctuaries.

The second important act is the Fishery Conservation and Management Act passed in April 1976 and implemented in March 1977. This act has given the United States a good start in conserving and managing the fish resources off its coast. A total of eight Fishery Management Councils have been formed, and they are developing managing plans for their geographical areas (see Chapter 12 for more details).

Summary

The coastal zone, which is situated at the boundary of earth and land, is probably the most important portion of the ocean. It can be broadly defined as that part of the ocean affected by land and that part of land affected by the ocean. Many of the features of the coastal zone are temporary, being controlled by the height of sea level, wave height, and tide conditions. This region is the site of considerable urbanization, and thus many, often conflicting, uses are placed on this environment. Development, both industrial and residential, has damaged or removed large portions of the land and parts of the coastal zone from public use. In recent years control by individual states and the federal government has resulted in improved management plans for coastal regions in the United States.

Suggested Further Readings

CSANADY, G. T., "What Drives the Waters of the Continental Shelves?" *Oceanus*, **22**, no. 2 (1979), pp. 28–35.

DAVIES, J. L., *Geographical Variation in Coastal Development* (2nd ed.). New York: Longman, 1980, 212 pp.

KAUFMAN, W., and O. PILKEY, *The Beaches Are Moving*. Garden City, N.Y.: Anchor Press/Doubleday, 1979, 326 pp.

LEATHERMAN, S. P., *Barrier Island Handbook*. National Park Service, Department of Interior, Washington, D. C., 1979, 101 pp.

PHILLIPS, R. C., "Seagrasses and The Coastal Marine Environment," *Oceanus*, **21**, no. 3 (1978), pp. 30–40.

SIMON, A. W., *The Thin Edge*. New York: Avon Books, 1979.

12

Resources of the Ocean

THE PRECEDING CHAPTERS have emphasized the various scientific aspects of ocean-ography. How big the ocean is and in some respects how little we know about it should be obvious by now. It would probably be easier if oceanographers and other scientists could study the ocean in its pristine state without having to consider any other marine uses. Unfortunately, this is not to be the case since exploitation and development of the resources of the sea are proceeding at a rapid rate. Questions of how the biological, mineral, and energy resources of the ocean are to be recovered, and at what rate, must be considered now, even in many instances without the benefit of adequate scientific study. Advanced technological developments such as offshore drilling platforms (Figure 12-1) allow almost any portion of the shelf to be drilled for oil. New approaches to resource development such as aquaculture and innovative sources of energy should lead to unique developments in the coming years (see also Chapter 14). This chapter considers the importance of the ocean from an economic standpoint. It should be emphasized that almost any use of the ocean must also consider the potential environmental damage that might result. These two aspects must be balanced by some criteria. With proper care, the activities described in this chapter can be made with minimal environmental damage. Often it is the activity that has not been adequately planned that causes the severe environmental damage.

The resources of the world can be considered in two ways. Some resources are renewable, such as food that is replaced usually on a yearly basis by the growth of plants and animals. Resources of the other type, the nonrenewable ones, are present in fixed amounts and cannot be replenished. These include most mineral deposits and especially oil, gas, and coal. These two types of resources, nonrenewable and renewable, are, respectively, equivalent to the mineral and biological resources of the ocean.

The biological resources of the ocean, such as fish, clams, and lobsters, have been harvested by people since the early days of their existence. Mineral resources, with the exception of salt obtained from evaporated seawater, are a fairly new development. The following sections consider the important aspects of these different types of resources.

FIGURE 12-1 A very advanced design of a deep-water, semisubmersible drilling rig. This vessel can work in water depths of about 610 m or 2,000 ft and drill to depths of 7,620 m or 25,000 feet. (Photograph courtesy of Zapata Corporation.)

Mineral Resources of the Ocean

Over the past few years much interest has developed among marine scientists, industry, and governments in the mineral wealth of the ocean (Table 12-1). Some early estimates of the value of these marine mineral resources have been staggering—values in the billions or even trillions of dollars. For some resources, these estimates are ridiculously high; for some they may turn out to be conservative. It should be noted that many industrialized countries, like the United States,

TABLE 12-1 Some Possible Marine Mineral Resources and Their Sources

Mineral resource	Source
Boron, Bromine, Calcium, Magnesium, Potassium, Sodium, Sulfur, and Uranium	Seawater
Sand and Gravel, Phosphorite, Glauconite, Lime and Silica, Sand, and Heavy Minerals (Magnetite, Rutile, Zircon, Cassiterite, Chromate, Monazite, Gold)	Sediment (continental shelf and slope)
Copper, Lead, Silver, and Zinc	Heavy-metal muds
Oil, Gas, and Sulfur	Subsurface (continental shelf, slope, and rise)
Manganese nodules (Copper, Nickel, Cobalt, and Manganese)	Deep sea

are deficient in most of the resources listed in Table 12-1. The most valuable marine mineral resources, based on our present knowledge, are the hydrocarbons, such as oil and gas, found buried within the sediments of the continental shelf and probably in the deeper parts of the continental margin.

One difficulty in assessing the value of marine mineral resources is that they are underwater and thus conventional techniques of mineral exploration and evaluation cannot be used. Exploitation of some potential marine resources must await technological advances. Other resources may never be developed because transportation or pollution prevention costs are excessive or because competitive land resources are cheaper.

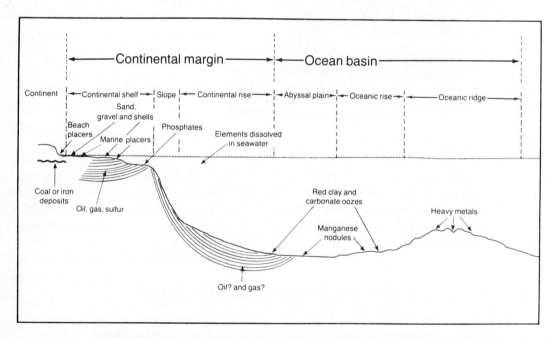

FIGURE 12-2   A generalized view of the principal mineral resources potentially available from the ocean.

Perhaps the best way to discuss the mineral resources of the ocean is to consider which main geographic area they come from: either the continental margin (including coastal zone) or the ocean basin (Figure 12-2). Since many resources are common to both the coastal zone and the continental margin, they are included together. One reason for this division is that the distance from land and ownership of a mineral deposit (see Chapter 15) are vital considerations in its exploitation.

Mineral Resources of the Continental Margin
Including Those from the Coastal Zone

The continental margin, which includes the continental shelf, the continental slope, and (if present) the continental rise or continental borderland covers an immense area, equal to about 50 percent of the total land area of the world. Geologically, the continental margin often marks the transition from continental structure, with a thick crust mainly of granitic rock, to the thin gabbroic crust of the ocean.

The shelf and rise, where present, are usually areas of thick sediment accumulations. Shelf widths may vary from hundreds of kilometers off some countries to only a few off others. The United States is fortunate to have an extensive continental shelf; the shelf area out to a depth of 100 fathoms (600 ft, or 183 m) is equivalent to 23.8 percent of the total U.S. land; out to 1,000 fathoms (6,000 ft, or 1,830 m), the area is equal to 36 percent of the U.S. land area.

The minerals of the continental margin can be divided into several categories: elements in solution; minerals recoverable from the underlying bedrock; minerals on the ocean bottom; and oil, gas, and sulfur.

ELEMENTS IN SOLUTION

Although seawater is the most abundant fluid on earth, freshwater is in short supply in many localities. Thus, freshwater can be considered as a marine resource obtainable from the sea by desalination (see pages 478–479 for further discussion). The need for water will probably increase even more in the future because of energy problems. Many countries are developing plants to produce natural gas from coal or to develop oil shale deposits. A plant producing 1 million bbl of oil (42 million gal) from oil shale could use somewhere between 100 and 170 million gal of water per day.

One difficulty with desalination is that most processes yield a hot, salty residue. If the plant is in a restricted coastal environment, the disposal of this residue can have a detrimental effect on the nearshore ecology. Because of this and the generally high cost of desalination, some unique ideas for obtaining freshwater have been suggested. Perhaps the most imaginative is to tow icebergs from polar regions to an area where they could be used. Another innovative method is to extract water vapor from the atmosphere in tropical areas by cooling it with cold ocean water brought from depths to the surface. This system can also be used as part of an aquaculture system; both ideas are treated further in Chapter 14.

Seawater contains almost all known elements. The main elements or compounds presently being economically removed from seawater are common salt, bromine, and magnesium. There is potential for the recovery of other elements if the technology were available and the economic return sufficient.

The ocean, because of its immense volume ($1,350 \times 10^6$ km^3 or 318×10^6 mi^3), contains a virtually inexhaustible supply of many elements. Seawater has an average salinity of 35 ‰, or 3.5 percent of the water is elements in solution. A cubic mile of seawater weighs about 4.7 billion tons and thus contains about 165 million tons of dissolved elements. The ocean contains over 5 billion tons of uranium and copper, 500 million tons of silver, and as much as 10 million tons of gold. For gold alone, this works out to about 5 lb for every person on earth, which is about $56,000 per person (at $700 per ounce). One must bear in mind that these estimates are not very realistic when one considers the cost of extraction of some of these elements. For example, still considering gold, its average concentration ranges from 0.000004 to 0.000006 mg per liter, or about 50 lb in a cubic mile of water. This would, even at today's high price, be about 0.001¢ per ton of seawater. However, other elements, like bromine and magnesium, which have a value of $0.02 and $1.00, respectively, per ton of seawater, can be profitably recovered.

MINERALS RECOVERABLE FROM THE UNDERLYING BEDROCK ON THE CONTINENTAL MARGIN

Some land-based mining operations have been extended under the ocean. These include offshore coal mining in Japan, the United Kingdom, and Nova Scotia; iron ore mined off Newfoundland; and tin mined off the United Kingdom. In most instances, these operations differ little from similar operations on land and probably developed only because the ores were originally found on land.

In some areas phosphorite rocks and phosphorite-rich sands are exposed on the sea floor (Figure 12-3) and could be mined and used for fertilizer. However, the water depth, difficulty of recovery, transportation costs, and better land sources prevent the present mining of most marine phosphorite. In some areas, such as near agricultural regions like Georgia, North Carolina, Florida, and California or off countries that are phosphorite importers, marine phosphorite mining operations are being seriously considered. As the need for and cost of fertilizers increases in the future, it is possible that marine phosphorite will supply a significant portion of that need. Phosphate deposits off the southeastern United States are estimated to cover an area of over 125,000 km^2 (about 48,200 mi^2) and to contain several billion tons of phosphate.

Phosphorite deposition tends to occur in areas of upwelling, where cold nutrient-rich waters (with dissolved phosphate) are brought to the surface. As these waters are warmed and their pH increases, phosphate is precipitated. Rich accumulations occur where the phosphate is not diluted by land-derived sediments; topographic highs near cold currents are especially favorable areas.

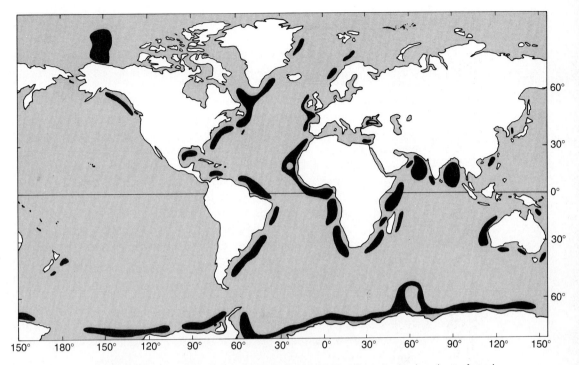

FIGURE 12-3 Offshore areas (black lines) where phosphorite has been found. (Adapted from McKelvey and Wang, 1970.)

MINERALS ON THE OCEAN BOTTOM IN THE CONTINENTAL MARGIN REGION

Many of the mineral resources on the continental margin are related to three main factors: first, the large amount of sediments present; second, the reworking effect of waves and currents combined with past changes in sea level due to recent ice ages; third, the development of the margin throughout geologic time, which involves the effects of sea-floor spreading.

Most continental margins of the world have a large thickness of sediment on their continental shelf, considerably less on the continental slope, and sometimes a very thick mass of sediment on the continental rise. These accumulations are due to the large amount of material eroded from land and carried to the ocean by rivers, wind, and coastal erosion. In some areas, the sediment mass has been accumulating for as much as 200 million years and is almost 20 km (about 12.5 mi) thick. This thickness, combined with a relatively high content of organic material from life processes in the overlying water during sediment deposition, can result in oil and gas deposits. This is discussed in more detail in the following section.

Sea level has risen and fallen several times over the past million years in response to extensive periods of glacial growth and melting. The most recent such glaciation occurred about 15,000 years ago (Figure 4-4). During this glaciation, sea level dropped by as much as 130 m (about 426 ft) because water was removed from the ocean and incorporated into glaciers. As the glaciers melted, the water returned to the ocean and sea level slowly rose. Thus, 15,000 years ago sea level was 130 m lower than present and the shoreline was much farther seaward than now (Figure 4-5). Subsequently, the shoreline has been migrating landward over the previously exposed portion of the continental shelf (the shoreline moved seaward when the glaciers were forming). As the shoreline moved, so did the beaches, marshes, and other nearshore features. If the shoreline movement was very fast, the beach could have been left behind resulting in an ancient submerged beach or ridge. Associated with any shoreline are breaking waves and nearshore currents (Figure 9-23) that can remove the finer-sized particles from the sediment and carry them seaward, leaving a coarser-grained sand or beach deposit behind. If the shelf is narrow and has a modest slope, reworking by waves and currents will be more intense, resulting in well-sorted (sediment particles generally of the same size) beach sands over the continental shelf [Figure 12-4(b)]. If sea level paused at a particular level for a period of time, the lighter sand particles could be removed leaving behind a valuable deposit enriched in the relatively heavier minerals. This type of deposit is called a *placer*.

Since reworking is generally more effective on beaches, they are generally the areas where economic concentrations of heavy mineral deposits or placers are found. Ancient beaches, formed during a lower or higher stand of the sea,

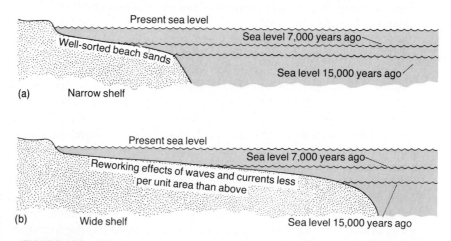

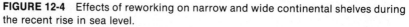

FIGURE 12-4 Effects of reworking on narrow and wide continental shelves during the recent rise in sea level.

can also be enriched in heavy minerals. Present beaches generally can better be used as recreational areas, so they are rarely mined. It is generally more acceptable to mine ancient offshore beaches, however, and they frequently are mined for their sand and gravel, which is used in the building industry. The more common minerals found in placers are listed in Table 12-2.

Sand and gravel is actually the second most valuable mineral deposit in the ocean after oil and gas. Its extensive occurrence, ease of mining, and need in the building industry has made it a most valuable resource. Estimates are that as much as 3 billion tons will be mined by the year 2000. Most people do not realize how extensively sand and gravel is used; for example, a 30- by 40-ft (9.1 × 12.2 m) concrete basement for a house needs 80 tons of sand and gravel aggregate, and 1 mi (1.6 km) of a four-lane highway could use 60,000 to 100,000 tons of mineral aggregate.

The value of sand and gravel will undoubtedly increase in the future as present sources on land are used up or covered by the intense building being done in coastal areas. Since transportation is a major cost in a sand or gravel mining operation, the vast deposits of sand and gravel in the marine environment will continue to become more appealing. Manheim in 1972 noted that the sand reserves off the northeastern United States are more than 400 billion tons, or several hundred years' supply, and he included only the upper 3 m (about 10 ft) in his calculation.

Offshore mining of sand and gravel can cause environmental problems. These may, in some instances, be less severe than those caused by land quarries near residential sections where large trucks are needed to transport the material to the areas of usage. Offshore problems concern erosion of beaches (since some of the offshore sand may be involved in the seasonal beach cycle) and interference with bottom ecology. The actual dredging of the material (Figure 12-5) can also introduce large quantities of suspended material into the water, which can have environmental effects (Figure 13-15).

Other minerals mined on the continental margin include calcium carbonate, barite, and (in the past) diamonds off South Africa. The diamond operation was hampered by extreme sea conditions and relatively low concentrations of diamonds on the shelf. Although this operation received considerable publicity, it appears to have never really been profitable. A relatively pure form of calcium carbonate, called aragonite, has been mined by the Dillingham Corporation in the Bahamas. The carbonate can be a source of lime for agriculture or used for cement. Barite is mined off the Alaskan coast and used by the petroleum industry as a major ingredient in drilling mud. One potential mineral that could be mined is glauconite—an iron- and potassium-rich clay mineral. It contains between 4 and 9 percent potassium oxide and could be used as a source of potassium for fertilizers.

Before discussing oil, gas, and sulfur, which are the main resources of the continental margin, it is appropriate to review briefly some aspects of the concept

TABLE 12-2 Mineral Deposits on the Sea Floor of the Continental Margin

Minerals	Use	Possible minable marine areas[a]	Value[b]
Marine placers			
Gold	Jewelry, electronics	Alaska, Oregon, California, Philippines, Australia	$350 or more/oz
Platinum	Jewelry, industry	Alaska	$500 or more/oz
Magnetite	Iron ore	Black Sea, U.S.S.R., Japan, Philippines	$6–11/ton
Ilmenite	Source of titanite	Baltic, U.S.S.R., Australia	$25–35/ton
Zircon	Source of zirconium	Black Sea, Baltic, Australia	$45/ton
Rutile	Source of titanite	Australia, U.S.S.R.	$100/ton
Cassiterite	Source of tin	Malaysia, Thailand, Indonesia, Australia, England, U.S.S.R.	
Monazite	Source of rare earth elements	Australia, United States	$170/ton
Chromite	Source of chromium	Australia	$25/ton
Sand and gravel	Construction	Most continental shelves	
Calcium carbonate (aragonite)	Construction cement, agriculture	Bahamas, Iceland, southeastern United States	
Barium sulfate	Drilling mud, glass, and paint	Alaska	
Diamonds	Jewelry, industry	Southwest Africa	
Phosphorite	Fertilizer	United States, Japan, Australia, Spain, South America, South Africa, India, Mexico	$6–12/ton
Glauconite	Source of potassium fertilizer		
Potash	Source of potassium	England, Alaska	

Source: From Ross, 1980.

[a] Not necessarily including all areas, because in many localities exploration has been nil or minimal.

[b] Can vary depending upon degree of refinement; data from various sources and may be in error because of changing economic conditions.

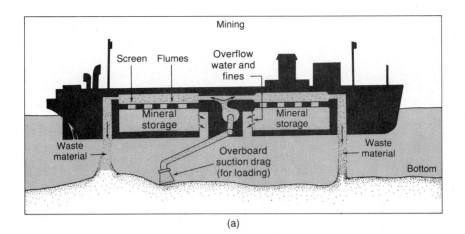

(a)

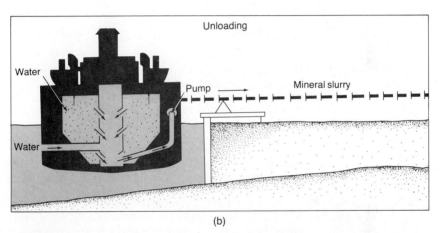

(b)

FIGURE 12-5 Cross section of (a) a dredge used in mining sand and gravel and (b) the unloading procedure. (Figure courtesy of Construction Aggregates Corporation, Chicago, Illinois.)

of sea-floor spreading that explain why some of these resources are in one locality and not in another.

Sea-Floor Spreading and Mineral Resources

The basic aspect of sea-floor spreading is that new sea floor forms along the mid-ocean ridges and slowly moves away from the ridge to become part of an expanding ocean floor and perhaps eventually to be consumed or subducted at the edge of the continents (Figure 12-6). This concept is discussed in detail in Chapter 6.

One especially interesting aspect of the sea-floor spreading concept is that

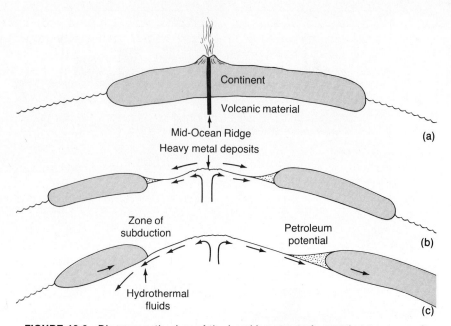

FIGURE 12-6 Diagrammatic view of the breaking apart of a continent as a result of sea-floor spreading. In the first phase (a) an initial large continental mass is intruded by volcanic material along zones of weakness. The continental mass eventually breaks apart (b) forming two continents. During this phase and the next, heavy metal deposits can be formed along the ridge axis by volcanic activity. A new ocean floor forms between the two continents and sediment is deposited along the margins of the continents. Sometimes the initially small ocean basin can become evaporitic resulting in thick sequences of salt and anhydrite deposits. In the last phase (c) the sea floor has underthrust the continent on the left side of the figure forming a zone of subduction and removing the sedimentary material either by carrying it under the continent or by accreting it onto the continent. This is similar to what is happening on the west coast of South America and Central America and much of the western Pacific. On the right side of the figure the continent is still coupled to the movement of the sea floor, and sediment accumulation on the continental margin continues. This side is similar to the margins around the Atlantic Ocean and the east coast of Africa.

there is mineralization due to rising magma on some of the mid-ocean ridges where the new sea floor is created. This mineralization has included deposits of copper, zinc, iron, silver, gold, nickel, vanadium, lead, chromium, cobalt, manganese, and other minerals. Among the most dramatic examples are heavy-metal deposits of the Red Sea and the mineral vents of the Pacific described in later sections (Figure 7-18). Our present knowledge of the processes forming these minerals is small and it is too early to know whether these mineralized areas could ever be mined. Although most of the mineralization occurs at the ridge axis, some minerals may be found in other areas of the sea floor since, with time, material moves away from the ridge crest across the ocean floor to eventually become accreted onto a continental landmass or be depressed or subducted below it. In addition, there appear to be mechanisms for mineral enrichment near the zone of subduction by the movement of hydrothermal fluids (hot magmas enriched in water) into the overlying rocks and sediments. Although this process is poorly understood, there has been increased prospecting activity in regions of ancient subduction or downthrusting.

The concept of sea-floor spreading can also be used to find mineral deposits by extrapolating backward to determine the position of the continents at different times in the past (Figure 6-2). If a mineral deposit that predates the actual separation of two then-joined continents is found near the edge of one of the continents, it may be that part of the same deposit can be found on the other continent. In other words, the original deposit may have been split in two.

There are portions of the ocean floor where, although sea-floor spreading has occurred, the ocean crust and continents have moved together without developing a zone of subduction. A good example of this is the Atlantic Ocean, especially the east coasts of South America, North America, and Africa. One result is that the continental margins, and especially the continental rise in this area, have been available to receive and accumulate sediments for at least the last 200 million years. As a result, these areas have very thick sediment sequences and considerable potential for oil and gas accumulations.

OIL, GAS, AND SULFUR DEPOSITS

Oil, gas, and sulfur easily qualify as the most valuable marine resources. The present value of the oil and gas obtained yearly from the marine environment exceeds $60 billion and is higher in value than marine biological resources. With the continuing demand for energy, there will be even further future exploration for oil and gas in the ocean.

Petroleum (oil is petroleum in the liquid state) originates from the organic remains of animals and plants that once lived in the sea or in rivers and after death settled to the ocean bottom and were buried by sediments. The actual conversion process whereby the organic material changes into petroleum is extremely complex and not completely understood. Some aspects are known; for example, if the organic material decays or is oxidized, it will not form petroleum. If the basin where the material settles is oxygen poor, however, the organic material can be easily preserved and can accumulate. It can also be preserved if the sedimentation rate is sufficient to bury the organic material before it is oxidized. About 95 percent of the world's oil is found in sediments originally deposited in a marine environment. Once the organic material reaches the bottom, the conversion process starts when bacteria or other organisms digest the organic material and redeposit it as fecal material. As the sediments are buried, further chemical changes occur, in large part influenced by the heat and pressures associated with burial. The time needed for the formation of petroleum or gas deposits is unknown but they are rarely found in rocks that are less than 2 or 3 million years old. The conditions necessary for an oil deposit are summarized in Table 12-3.

TABLE 12-3 Essential Requirements for Favorable Petroleum Prospects

1. A sufficient source of the proper organic matter.
2. Conditions favorable for the preservation of the organic matter—rapid burial or a reducing environment.
3. An adequate blanket of sediments to produce the necessary temperatures for the conversion of the organic matter to fluid petroleum.
4. Favorable conditions for the movement of the petroleum from the source rocks and the migration to porous and permeable reservoir rocks.
5. Presence of accumulation traps, either structural or stratigraphic (Figure 12-7).
6. Adequate cover rocks to prevent loss of petroleum fluids.
7. Proper timing in the development of these essentials for accumulation and a postaccumulation history favorable for preservation.

SOURCE: Adapted from Hedberg et al., 1979.

In searching for oil or gas deposits, a thick sequence of marine sediments with occasional coarse sediment layers (reservoir beds) is usually considered as having high potential. Ancient coral reefs can also have good reservoir potential, and many have hydrocarbon deposits. But it should be emphasized that drilling is always necessary to determine if oil or gas is actually present (Figure 12-7).

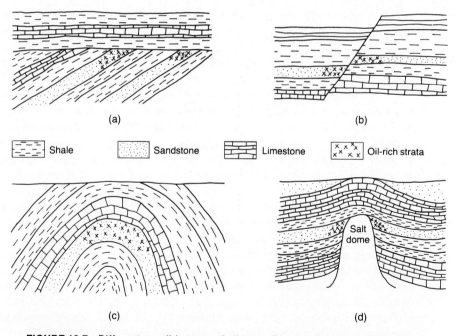

FIGURE 12-7 Different possible types of oil traps. For simplicity, the reservoir beds are always indicated as sands. (a) Stratigraphic trap, (b) structural trap, (c) anticline, (d) salt dome, or diapir.

There are several technical limitations to drilling in the marine environment. The main one is the actual drilling platform, which can be fixed, semifixed (can move after drilling), or floating relative to the sea floor. The first two types presently limit drilling to depths of less than 183 m or 600 ft. Floating platforms (Figure 12-1) can be used to drill in deeper water while maintaining their position over the drill site by dynamic positioning on subsurface buoys (Figure 3-13). The record for drilling for oil belongs to a vessel called *Discoverer Seven Seas* that drilled in a water depth of 1,486 m (4,876 ft) off Newfoundland. The previous record (held by the same ship) was 1,353 m (4,441 ft) in the Mediterranean in February 1979. Ships that drill for scientific purposes, such as the *Glomar Challenger* (Figures 3-11 and 3-13), can drill in waters up to about 6,100 m or 20,000 ft deep but are not equipped to exploit an oil deposit if one is found. Another limitation to drilling in the ocean is that costs are considerably more than for a land operation; the cost of drilling a well in 200 m of water can easily exceed $1 million.

At present, the Middle East area (the Persian Gulf) is the biggest producer of offshore oil. By the mid-1980s as much as 35 percent of the total oil produced in the world could come from the ocean. The ultimate amount of oil obtained from the marine environment could eventually be very considerable when compared with that found on land. This in part results from the fact that most land areas have been explored and promising areas already drilled, whereas exploration in the ocean has been restricted to relatively shallow waters. Exploration of some sort has already taken place off the coasts of most countries that have marine boundaries. There are several potential favorable areas off the United States that have yet to be fully explored (Figure 12-8).

Marine areas especially favorable for oil or gas are deltas, where basins of thick sediments extend from land into the offshore area, and other localities where thick sedimentary sequences occur. One major source of petroleum in the future may be the continental rise, which, where present, is a thick wedge of sediment at the base of the continental slope. The distribution of continental rises (Figure 5-26) is strongly controlled by sea-floor spreading. Where oceanic plates are being thrust under the continents, rises do not form since the sediment is removed (Figure 12-6). Continental rises and other marine sedimentary basins cover an area of about 19 million km^2 (7.3 million mi^2), which is equal to about 33 percent of the total area of the continental shelf and slope.

In the late 1970s and early 1980s there was a clear trend toward a reduction in oil consumption among developing countries. Uncertainties of oil supply due to actions like the Iran–Iraq war have led many countries to look for alternative sources of energy, such as nuclear power, coal, and more exotic things like marine biomass and wave and tidal energy (see Chapter 14).

Sulfur can sometimes be found associated with a petroleum deposit as part of the cap rock of salt domes that are occasionally found beneath the sediments of the continental shelf or continental rise (Figure 12-9). The salt was deposited when the basin was under evaporitic conditions and later flowed through the

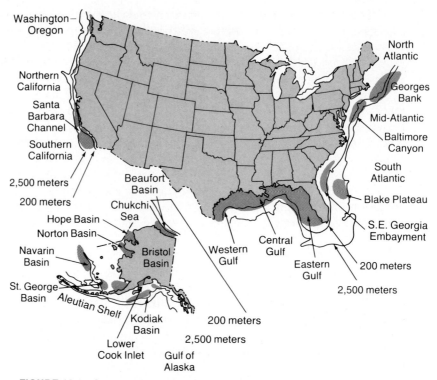

FIGURE 12-8 Some regions of good hydrocarbon potential (striped areas) off the United States.

FIGURE 12-9 A seismic reflection profile showing salt domes, or diapirs, off the coast of Angola [compare with Figure 12-7(d)]. The scale on the left is reflection time and 1 second is about 1 km or 0.6 mi. The scale on the top is time aboard the ship as the record was being made and 1 hr is essentially equivalent to 15 km or 8 n mi, of steaming. (Photograph courtesy of K. O. Emery.)

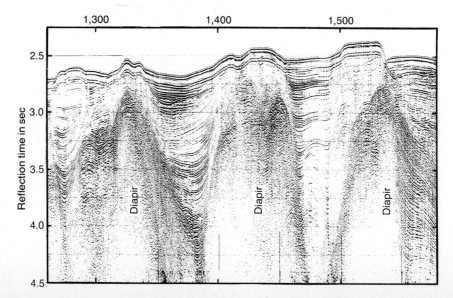

overlying heavier sediments, moving upward as a long, thin plug, and produced a domelike structure; hence the name salt dome, or diapir. These features are extremely common in the Gulf of Mexico and have formed traps for much of the oil recovered from that area. As the salt nears the surface, some of it is dissolved by water and a residue or cap rock of relatively insoluble anhydrite remains. Anhydrite (calcium sulfate) can react with petroleum and bacteria to form hydrogen sulfide gas, water, and the mineral calcite (calcium carbonate). If oxygen then interacts with the hydrogen sulfide, water and elemental sulfur can be produced and concentrated within the cap rock. The sulfur can be recovered fairly simply by the so-called Frasch process in which superheated water and air are piped down the drill hole. This melts the sulfur and allows the liquid sulfur and water to be pumped to the surface. Salts domes off Louisiana have yielded 200 million tons of sulfur.

Resources of the Deep Sea

The deep sea includes the main ocean basins, trenches, rises, and oceanic ridges. These provinces total about 79.4 percent of the ocean and cover an area of about 287 million km^2 (about 110 million mi^2), which is about four times the area of the continental margin. The resource potential of the deep sea is considerably less understood, however, than that of the continental margin. The major resource of the deep sea is probably manganese nodules; of substantially lesser potential are deep-sea muds and oozes.

Some recent scientific discoveries concerning sea-floor spreading may lead to additional resources in the deep sea. The recent findings of heavy-metal accumulations (see discussions in Chapters 6 and 7) along zones of active spreading and zones of subduction may eventually be economically significant.

DEEP-SEA MUDS AND OOZES

Much of the deep-ocean floor is covered by slowly deposited, fine-grained mud deposits. The most common type is called brown clay, sometimes called red clay, which accumulates at a rate of a few millimeters per 1,000 years (Figure 5-42). Areas of brown clay deposition cover an area of about 100 million km^2 or 38 million mi^2. If one assumes an average thickness of 300 m (984 feet), the deposits have a volume of about 30 million km^3 (about 7.1 million mi^3). Chemical analyses of these sediments show as much as 9 percent aluminum and 6 percent iron, plus modest amounts of copper, nickel, cobalt, and titanium. Some of these metals are more enriched in the brown clay than they are in rocks mined on land. John Mero in 1965 estimated that brown clays contain enough aluminum and copper, if they could be economically mined, to last over 1 million years at the present rate of use. There are several problems that have to be overcome before these clays could be economically mined, however, including recovery from depths of about 6,100 m or 20,000 ft or more below sea level and transportation and refining of the fine-grained material. One advantage of the deposit is that it lies uncon-solidated directly on the sea floor with no overlying rocks.

Oozes are muds that contain relatively large amounts (usually more than 30 percent) of shells, or tests, of dead organisms (Figure 5-42). These oozes can be of two main types: calcareous or siliceous, depending on whether they are com-posed mainly of calcium carbonate or siliceous shells. Calcareous oozes can be 95 percent pure carbonate, which could be used as a source of limestone for cement. As in the case of brown clays, the volume of this material is awesome, and if ever mined it could supply limestone at a rate equal to several million years of consumption. Siliceous oozes could be mined for their silica content, which can be used for insulation and soil conditioners. This deposit also covers an extensive area of the sea floor, although less than that covered by calcareous ooze. Both are deposited at a rate of a few centimeters per thousand years.

Oozes and muds must be considered only as potential resources because of the technological difficulties in mining them. Even so, numerous people have been impressed by their vast extent and have suggested that they could become a very important potential resource in the near future. One especially impressive feature of these potential resources is that their rate of accumulation, which although amazingly small extends over such a large area that the gross rate of accumulation of several elements in the sediment is considerably higher than the present rate of their consumption on land. One type of deep-sea mud, the so-called Red Sea metalliferous (or metal-rich) muds, may be mined by the mid-1980s.

Among the more interesting discoveries of recent years has been the finding of sediments enriched in certain metals in various areas of the sea floor, especially those associated with present or past spreading centers. Discovery of these de-posits was first made in the early 1960s along the East Pacific Rise and later in the Red Sea (see below). The sediments can be divided into three different types: sulfide deposits such as those found in the brine pools of the Red Sea, localized areas of sediments on some of the mid-ocean ridges, and dispersant deposits mainly enriched in iron and manganese found on, adjacent to, or near some areas of sea-floor spreading. The consensus is that these deposits, although different, come from a similar process of precipitation from hydrothermal solutions that result from the interaction of seawater with the oceanic crust. When these hy-drothermal waters return to the ocean floor, precipitates are formed by various types of mechanisms. Precipitates are restricted to vent areas such as those in the Galapagos (Figure 7-18), or to isolated pools where dense hydrothermal water can accumulate, or can occur uniformly over a large area if hydrothermal water is mixed with the overlying water. Buried deposits of this type have been pene-trated by cores far from present spreading centers, which suggests that this mech-anism has been active in the past and is not just restricted to present times and may explain the formation of many of our known mineral deposits on land—in particular, Cyprus.

RED SEA METALLIFEROUS MUDS

One of the most interesting potential mineral resources of the deep-sea floor was accidently discovered in 1948 in the Red Sea, in an area of active sea-floor spreading where Africa and the Saudi Arabian Peninsula are slowly moving away from each other (Figures 6-12 to 6-15). During a routine hydrocast (collecting water samples and measuring its salinity and temperature), the Swedish research vessel *Albatross* noted slightly anomalous values for temperature and salinity of water collected from the bottom of the central ridge of the Red Sea. At the time, they did not put much faith in these observations and assumed that the anomalies were due to instrument error. Further study in this area was done by the British R.R.S. *Discovery* and ships from the Woods Hole Oceanographic Institution that finally led to the discovery in the early 1960s of three pools of hot, saline water at a depth of about 2,000 m (about 6,100 ft) on the bottom of the Red Sea. Salinity of these strange waters reached a maximum of 257 ‰[1] and temperatures were then as high as 56°C (138°F). It was not until a major expedition to the area by the Woods Hole Oceanographic Institution in 1966 that the area was mapped and numerous sediment samples, to a depth of 10 m (about 33 ft), were taken that the economic implications of the area were first realized. On the basis of a few chemical analyses of the sediment underlying one of these pools, it was estimated that their *in situ* value was about $2 billion. A later report by Bischoff and Manheim (1969) gave estimates of as high as $2.4 billion for the top 10 m of sediment (Table 12-4). This value does not consider the cost of raising the sediments from the bottom and refining and marketing the minerals. In any case, the governments of Saudi Arabia and Sudan have joined forces in an attempt to mine these deposits. An optimistic view is that these sediments may be mined by the mid-1980s.

What makes the Red Sea deposits so interesting is not necessarily their high content of copper, zinc, and silver, but the fact that oceanographers can actually "see" a mineral deposit forming on the sea floor. The Red Sea deposits bear some similarity to land deposits but these are usually hundreds of millions of years old. It follows that if such a deposit is actively forming now on the Red Sea floor, in an area of sea-floor spreading, that similar deposits should occur in other similar parts of the ocean. Early, but limited, sampling along the oceanic ridges showed definite suggestions that hydrothermal and volcanic activity are concentrating metals in some areas within the bottom sediments and rocks. Enrichments of elements such as cobalt, copper, chromium, iron, manganese, nickel, uranium, and mercury with lesser amounts of bismuth, cadmium, and vanadium have already been found. These concentrations are generally not of economic grade but occasionally some especially high concentrations are found. Small veins of pure copper were recovered by the Deep Sea Drilling Project at two locations: in the

---

[1] This is not salinity in the true sense of the word since the ratio of the major elements in the Red Sea is not similar to the ratios found in normal seawater, and some elements such as iron are enriched thousands of times more, and others are less abundant than in normal seawater.

TABLE 12-4   Gross Value of Metals in Upper 10 m of Sediments
Collected from *Atlantis II* Deep in the Red Sea,
Based on 1967 Metal Prices

Metal	Average assay (%)	Tons	Value ($)
Zinc	3.4	2,900,000	860,000,000
Copper	1.3	1,060,000	1,270,000,000
Lead	0.1	80,000	20,000,000
Silver	0.0054	4,500	280,000,000
Gold	0.0000005	45	50,000,000
Total			2,480,000,000

Source: Adapted from Bischoff and Manheim, 1969.

Indian Ocean near the equator and some 650 km or about 350 mi southeast of New York in sediment under the lower continental rise in a water depth of about 5,200 m or 17,060 ft. None of the above has yet attained the economic potential of the Red Sea, but this just may be due to inadequate sampling or because we have not yet looked in the right localities.

It was not until the late 1960s that the hydrothermal activity of spreading centers was recognized. The most dramatic results came from the Galapagos region, where waters as hot as 350°C were observed forming vents and mineral deposits (Figure 7-18). At this time the economic implications of these sediments are unknown.

MANGANESE NODULES

One of the more interesting resources of the deep-sea floor, especially from a scientific point of view, is manganese nodules (sometimes called iron-manganese or ferromanganese deposits). These deposits, first discovered during the Challenger Expedition (1872–1876), commonly occur as round spheres of about 1 to 20 cm (about 0.4 to 8 in) in diameter (Figures 7-5 and 7-17) but can also form as coatings on rocks and other objects or as long slabs called manganese pavements. See Chapter 7, pages 205–207, for a discussion of their origin. The principal economic interest in the nodules is their accessory elements—copper, nickel, and cobalt—and their main component, manganese, combined with their supposed ease of recovery from the sea floor.

Manganese nodules are common to all the oceans (Figure 12-10), especially in areas of low sedimentation rates, such as abyssal plains. This is because the nodules form at a rate of about 1 mm per million years and could be easily buried where the sedimentation rate is high. It has been estimated that more than 25 percent of the sea floor is covered by nodules and that over 1.5 trillion (1.5×10^{12}) tons are in the Pacific Ocean alone.

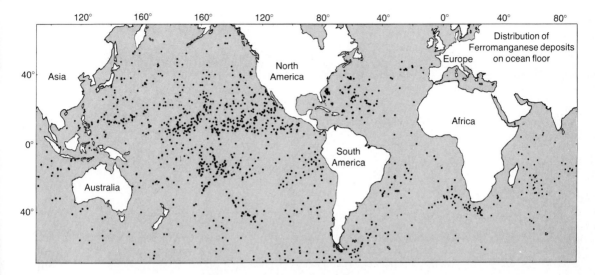

FIGURE 12-10 Location of surface stations where ferromanganese nodules were obtained from the ocean floor. (Data from Horn et al., 1973.)

Numerous analyses have been made of manganese nodules (Table 12-5). These generally show differences among oceans and some general similarities within certain physiographic provinces in an ocean. The distribution of the most economically important elements—manganese, copper, nickel, and cobalt—has been plotted for each ocean as part of an IDOE study (Figure 12-11). The same studies have shown that a red clay and siliceous ooze area of the North Pacific looks most favorable (especially the siliceous ooze area) for mining because of its relatively high metal content (Table 12-5). These deposits lie north of the

TABLE 12-5 Average Composition of Manganese Nodules from the Different Oceans

Minerals	South Pacific[a]	North Pacific[a]	West Indian[a]	Atlantic[b]	Favorable North Pacific area[c] Red clay	Favorable North Pacific area[c] Siliceous oozes
Manganese	16.61	12.29	13.56	16.1	17.43	22.36
Iron	13.92	12.00	15.75	21.82	11.45	8.15
Nickel	0.433	0.422	0.322	0.297	0.76	1.16
Copper	0.185	0.294	0.102	0.109	0.50	1.02
Cobalt	0.595	0.144	0.358	0.309	0.28	0.25

[a] Data from Cronan, 1967, and Cronan and Tooms, 1969, in weight percent, air-dried weight.
[b] Data from Cronan, 1972, in weight percent, air-dried weight.
[c] Data from Horn et al., 1972.

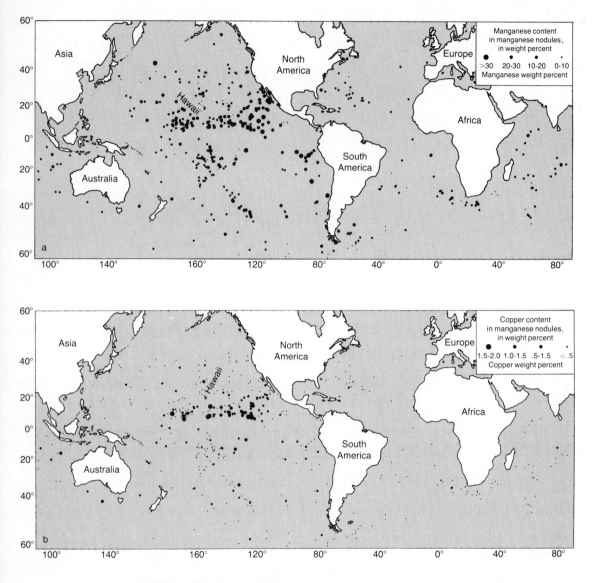

FIGURE 12-11 Concentration of manganese (a), copper (b), nickel (c), and cobalt (d) in manganese nodules found on the ocean floor. (Data from Horn et al., 1972.)

equator in a broad band between 6°30′ N and 20° N, extend from 110° W to 180° W, and cover an area of 800 by 4,200 n mi or about 3.4 million n mi² (about 11.5 million km²). They are in a relatively smooth portion of the sea floor, which would simplify recovery. The nodules of the Atlantic and Indian Oceans are generally characterized by having copper, nickel, and cobalt values below those generally considered necessary for recovery.

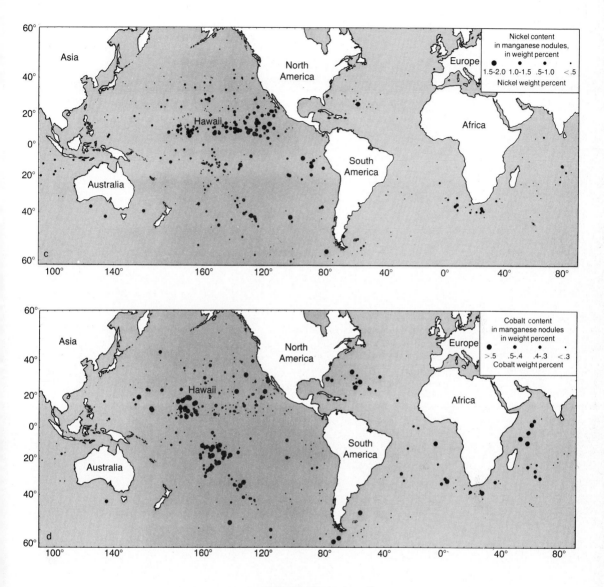

FIGURE 12-11 (*cont.*)

The potential value of the manganese nodules has excited many about the riches of the ocean, as well as increased the intensity of the international discussions concerning ownership of the sea floor (see Chapter 15). Within recent years several companies or international consortiums have been moving at a rapid rate toward actually mining manganese nodules. These groups at one time included the Summa Corporation, owned by Howard Hughes, but more about this shortly. Deepsea Ventures, Inc. was probably the first to develop an operational

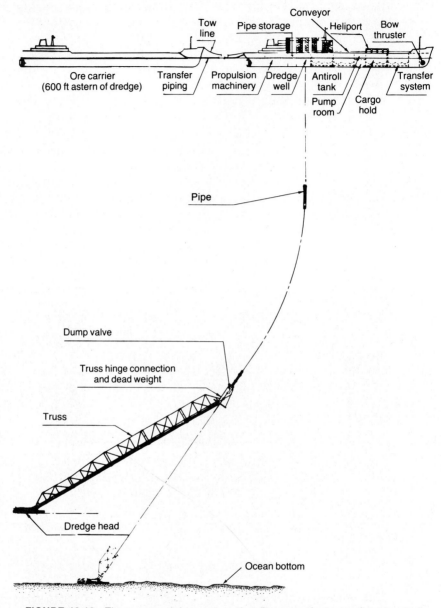

FIGURE 12-12 The ocean mining system that Deepsea Ventures plans to use to mine manganese nodules from the ocean floor. (Courtesy of Deepsea Ventures, Inc.)

prototype hydraulic dredge system for the recovery of nodules (Figure 12-12). In 1970 the system was successfully tested in a water depth of about 762 m or 2,500 ft. A specially designed ship, using an airlift hydraulic suction dredge, was able to lift large amounts of manganese nodules off the sea floor (Figure 12-13). The success of the prototype system indicated to the company that recovery from about 6,100 m or 20,000 ft is possible.

For several years it was thought that the Summa Corporation was going to be in the manganese nodules business. They had built a 618-ft (188 m) long, 115.5-ft (35.2 m) wide, 36,000-ton experimental ship called the Hughes's *Glomar Explorer* (Figure 12-14). Numerous press releases (although always with some element of mystery) told how the ship would bring nodules up from the bottom by air suction and store them in a 324 ft (98.7 m) barge submerged below the Hughes's *Glomar Explorer*. In early 1975, however, it was learned that the Hughes venture was really a cover for a CIA attempt to raise a sunken Soviet submarine from the Pacific. The cover story was very successful although it is not yet known how successful the recovery was. The operation showed several things: one was that there was technology available to recover large objects from any portion of the sea floor. It also indicated to the rest of the world, and especially less-developed countries, that what appears to be a peaceful operation in the deep sea can have military implications. This point was not missed by the delegates meeting at the 1975 Law of the Sea Conference in Geneva. Ironically, this meeting was being

FIGURE 12-13 Manganese nodules on a conveyor belt after recovery from the sea floor and discharge from a nodule/water separator. (Photograph by B. J. Nixon, Deepsea Ventures, Inc.)

FIGURE 12-14 The Hughes's *Glomar Explorer,* which for many years was thought to be an experimental deep-sea manganese nodules mining ship; in 1975 it was revealed that one of its main objectives was salvage of large objects from the sea floor. More recently the vessel was used in a prototype mining operation in the Pacific. This vessel may also be used in the Ocean Margin Drilling Project. (Photograph courtesy of Summa Corporation.)

held when the Hughes story broke. Even stranger, the *Glomar Explorer* was actually used in later years in a manganese nodule mining operation.

The Ocean Minerals Company (a consortium of several companies) chartered *Glomar Explorer* and used it to test a prototype mining device in water depths of about 5,000 m (16,405 ft) in the Pacific (Figure 12-15). The device, only about one-tenth scale of an operational system, is still almost 14 m (46 ft) long and 9 m (29 ft) wide. As it moves along the bottom it rakes in the nodules onto a conveyor belt where they are washed, crushed, and transported to the surface.

There are several problems with a deep-sea manganese nodule program. These especially include economic feasibility, environmental effects (see Chapter 13, pages 457–458), and legal implications. Probably the main problem concerning the mining of manganese nodules is the ownership of these deep-sea resources. This point was indirectly covered in the 1958 U.N. Conferences on the Law of the Sea held in Geneva (see Chapter 15). Article 1 from the Convention on the Continental Shelf states that the continental shelf is "the seabed and subsoil of the submarine areas adjacent to the coast but outside of the area of the territorial

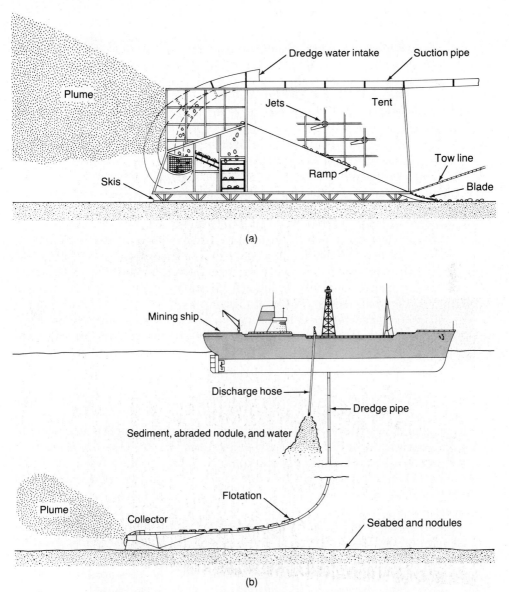

FIGURE 12-15 A prototype manganese nodule mining device developed by Ocean Minerals Company. (a) Is a side view of the device; (b) as it looks being towed by a surface ship. (From J. E. Flipse, 1980.)

sea, to a depth of 200 m, or beyond that limit, to where the depth of the superadjacent waters admits the exploitation of the said areas and to the seabed and subsoil of similar submarine areas adjacent to the coasts of islands.'' This point of exploitability would theoretically apply to manganese nodule deposits in the deep sea. The recent desire, however, on many countries to expand their territorial claims and the suggestion that the resources of the ocean are the ''common heritage of all mankind'' suggest an element of risk or uncertainty to a deep-sea mining venture. The present Law of the Sea Conference should establish international and legal guidelines to the resources of the deep sea.

As of early 1981 the legal situation concerning the mining of manganese nodules was still unclear. The complex regime proposed by the Law of the Sea Conference is unacceptable to many mining consortiums and several have indicated that they will not continue their development activities. In 1980 the U.S. Congress passed the Deep Seabed Hard Minerals Resources Act that will allow U.S. mining companies to begin commercial mining after January 1, 1988. This legislation was not intended to interfere with the Law of the Sea Conference, but rather to establish the right of companies to proceed if the conference fails. If a treaty is passed and approved it would supersede the act.

In summary, the mining of manganese nodules from the deep-sea floor is technologically feasible and mining could begin before the end of the 1980s. At present, there are nearly sufficient commercial interest and techniques to ensure that a mining operation will occur, although questions concerning legal ownership and profitability still remain. No legal regime presently exists to define and adequately protect a deep-sea mining operation, and this fact has affected the plans of some companies or consortiums concerning their exploitation and development activities.

Biological Resources

The biological resources of the sea are one of the most important aspects of oceanography. Many believe that the ocean must be used to help solve the world's food problems—problems that will become important in future years. It would be naive to think that the ocean provides an easy solution; even if the fish harvest from the ocean were doubled, it would not meet the global deficiency of animal protein, but it would help.

The food problem is such that almost 2 billion people—about half the population of the world—suffer from a protein-deficient diet. The deficiency is compounded in less developed countries because their population is growing faster than their ability to produce adequate food supplies. The main need is for animal protein, a commodity present in large quantities in the ocean. A daily supplement of 10 to 20 g (0.3 to 0.6 oz) of animal protein is considered sufficient to prevent a protein diet deficiency; this is equivalent to 3.6 to 7.2 kg or about 8 to 16 lb of protein per year. Some have suggested that large amounts of protein could be obtained by harvesting plankton, which comprise the largest group of organisms in the ocean. As individual organisms, most are microscopic in size. Therefore, to obtain large quantities of plankton an immense quantity of water must be filtered (as much as 1 million lbs of seawater to obtain 1 lb of plankton). There are other reasons, such as taste, why plankton are not a suitable source of protein. To find a usable source of protein, one must look to animals, such as fish.

The biological resources of the ocean, like those of land, are ultimately directly related to the production of organic matter by plants. As discussed in Chapter 8, the plants in the ocean are primarily floating plants or phytoplankton that live in the upper layers of the ocean because of their need for light to photosynthesize. Nutrients are another vital ingredient for photosynthesis. The supply of nutrients is controlled, in large part, by water circulation. Nutrients are especially abundant in areas of upwelling, where deeper, nutrient-rich waters are brought to the surface. The geographic distribution of phytoplankton production of organic material (Figure 12-16) shows four main zones within the ocean.

1. The deep-ocean areas far from land that are essentially biological deserts.
2. The Antarctic, Arctic, and equatorial waters that are moderately productive due to water mixing by currents and winds.
3. The shallow waters of most continental shelves that generally have high production.
4. Upwelling areas such as those off Peru, California, and western Africa, where organic production is extremely high.

In general, the location of the major fish areas of the world follow the pattern of the phytoplankton production of organic material, especially in the nearshore parts of the ocean (Figure 12-17). This pattern is reasonable since fish either

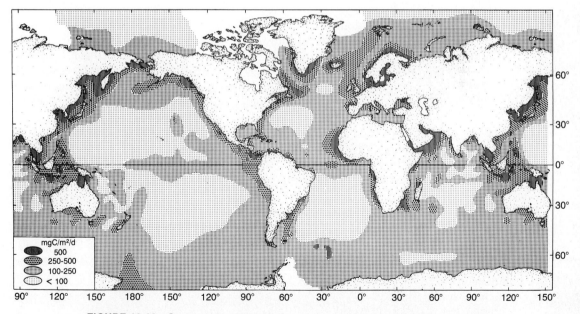

FIGURE 12-16   Geographical distribution of phytoplankton production of organic material (in mg of carbon per square meter per day). [Adapted from Food and Agriculture Organization (FAO) of the United Nations, *Atlas of the Living Resources of the Sea,* 1972.]

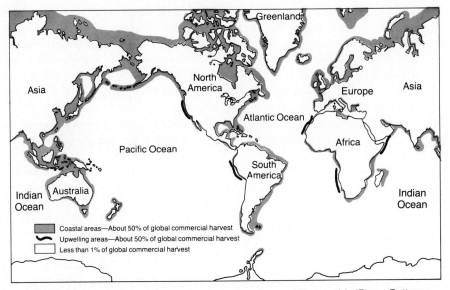

FIGURE 12-17 Location of the major fishing areas of the world. (From *Patterns and Perspectives in Environmental Science:* Report Prepared for the National Science Board, National Science Foundation, 1972.)

directly, or indirectly by eating zooplankton, feed on the organic material produced by phytoplankton.

The ocean can be divided into three provinces (Table 12-6) based on production of organic material (zones 1 and 2 above are lumped together). It can be noted that although the open ocean makes up about 90 percent of the ocean, it produces less than 1 percent of the fish caught and has little potential for increased production. The coastal zone and upwelling provinces occupy 9.9 and 0.1 percent,

TABLE 12-6 Production of Organic Material in the Ocean and Estimated Amounts of Fish Production (See text for more explanation)

Province	Percentage of ocean	Mean productivity of organic material (g C/m²/yr)	Total primary production (tons organic carbon)	Feeding levels	Efficiency of conversion	Total fish production (tons fresh weight)
Open ocean	90	50	16.3 billion	5	10	1,600,000
Coastal zone (including offshore areas of high production)	9.9	100	3.6 billion	3	15	120,000,000
Upwelling zone	0.1	300	0.1 billion	1.5	20	120,000,000

SOURCE: Adapted from Ryther, 1969.

respectively, of the ocean but each produces about 50 percent of the world's fish catch. One of the main reasons for this is that in the open ocean there are more different feeding levels necessary to go from a microscopic phytoplankton to an animal large enough to be used by man, than there are in the other two provinces. In going from one feeding level to another (Figure 8-51), as much as 90 percent of the organic material may be lost. In addition, the efficiency of conversion (rate of growth of the animal relative to the amount of food consumed) is greater in the coastal zone and upwelling provinces than in the open ocean.

The total amount of fish, crustaceans, mollusks, and other aquatic plants and animals that can be harvested from the sea is subject to considerable debate. In 1978, the total world catch was 72,400,000 metric tons; about 90 percent was fish. This number is 2 percent higher than the 1977 catch. Some marine scientists have suggested that this number is near the maximum sustainable yield from the sea. Others, perhaps too optimistically, think yields of 3 to 30 times this number are possible. Further increases in the yield from the sea, if possible, will require more sophisticated methods of fishing. It will be necessary for people who fish to change from "hunters," which they essentially are now, to "herders." In other words, it will be necessary for them to be able to control or influence the movement of fish. Aquaculture and innovative ways to increase biological productivity are an attempt at such control (see Chapter 14). Sophisticated electronic devices (Figure 12-18) permit modern fishermen to find schools of fish. Some countries, especially those fishing far from their home ports like the Soviet Union, use modern factory ships to process their catch at sea (Figure 12-19). These fish are generally still caught by conventional techniques, however, using nets from small ships (Figures 12-20 and 12-21).

Fishing within the United States is not so impressive a business as might be expected but has increased in recent years. In 1978 the United States ranked fourth among the countries of the world in total weight of fish caught, behind Japan, U.S.S.R., and China (Table 12-7).

The total U.S. commercial harvest of fish, shellfish, and other aquatic life was 6.3 billion lb (2.85 billion kg) in 1979. This was worth $2.233 billion to the fishermen, or an average of about 35¢ per pound (or per 0.45 kg). Menhaden made up 42 percent of the U.S. commerical catch. Salmon was the second most important, both in quantity and value, followed by crab and tuna. Shrimp was the fifth most important in quantity but had the highest value. Louisiana was the leading state in volume of fish followed by Alaska, California, Virginia, and North Carolina. Alaska was clearly the leader in the value of catch followed by California, Louisiana, Massachusetts, and Texas.

Of the total U.S. catch about half is used directly for human consumption, the remainder is for industrial purposes such as chicken and pet feed. The per capita consumption of fishery products was about 13 lb (5.9 kg) per person in 1979, an especially low value considering the nutritional value of fish. In contrast, the consumption of beef, pork, lamb, veal, and poultry in the United States was about 250 lb (113.4 kg) per capita.

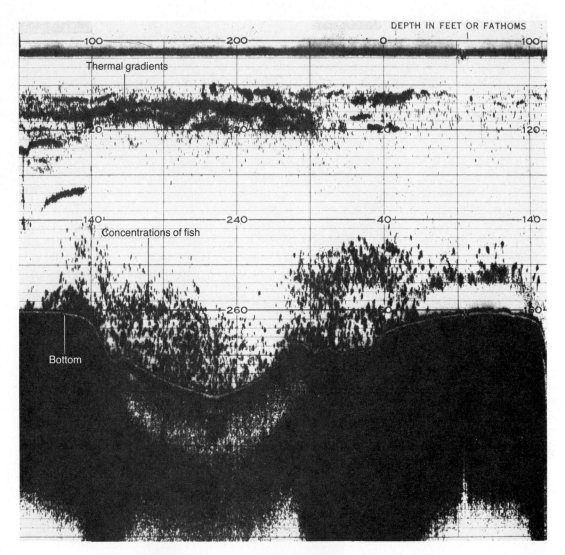

DEPTH IN FEET OR FATHOMS

Thermal gradients

Concentrations of fish

Bottom

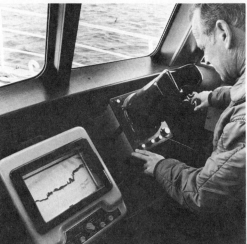

FIGURE 12-18 Echo-sounding record (above) that is used by fishermen (below) to locate concentrations of fish. A moderately large school of fish is seen near the bottom of the record. (Photograph courtesy of Raytheon Marine Company.)

FIGURE 12-19 Soviet stern trawler (on the left) about to offload its catch onto a large factory ship. Several other ships can be seen waiting in the background. (Photograph courtesy of National Marine Fisheries Service.)

FIGURE 12-20 Empyting of a fishing net.

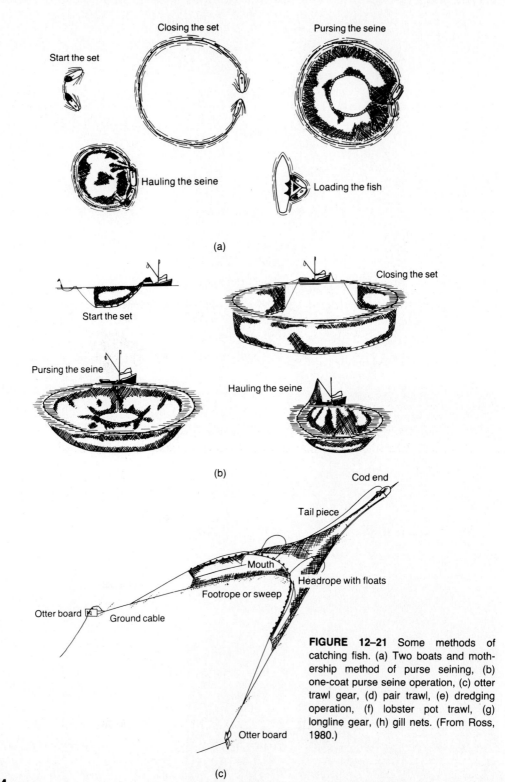

Start the set

Closing the set

Pursing the seine

Hauling the seine

Loading the fish

(a)

Start the set

Closing the set

Pursing the seine

Hauling the seine

(b)

Cod end

Tail piece

Mouth

Headrope with floats

Footrope or sweep

Otter board

Ground cable

Otter board

FIGURE 12–21 Some methods of catching fish. (a) Two boats and mothership method of purse seining, (b) one-coat purse seine operation, (c) otter trawl gear, (d) pair trawl, (e) dredging operation, (f) lobster pot trawl, (g) longline gear, (h) gill nets. (From Ross, 1980.)

(c)

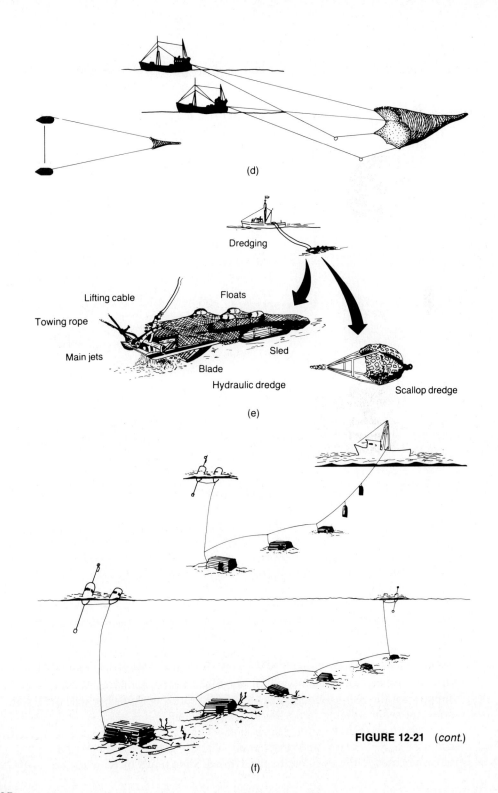

(d)

Dredging

Lifting cable

Towing rope

Floats

Main jets

Sled

Blade

Hydraulic dredge

Scallop dredge

(e)

FIGURE 12-21 (cont.)

(f)

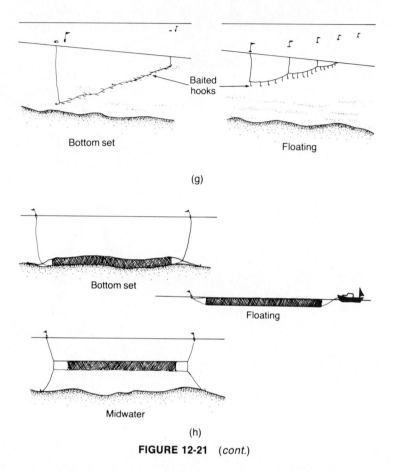

Baited
hooks

Bottom set

Floating

(g)

Bottom set

Floating

Midwater

(h)

FIGURE 12-21   (*cont.*)

A surprising statistic is that the American sport fisherman, whose catch is not included in the above figures, spent $1.4 billion in 1970 (last year that data is available) on this hobby and caught about 1.6 billion lb (0.73 billion kg) of edible (saltwater) fish. This number is approximately equivalent to the value of the total U.S. 1973 fishing catch.

There are many reasons why the U.S. fishing industry has not done so well as it might. For example, in the past foreign fishermen used to take about twice the amount of fish that American fishermen caught from the waters adjacent to the United States. This was perfectly legal, but recent legislation (see below) has changed this. Another reason is the lack of financial incentive to the fisherman. In many other countries a food shortage has encouraged increased government support for the fishing industry. The United States, perhaps because of its wealth, does not seem to mind importing the fish it needs.

There are other factors, such as laws and customs, that have restricted the catch of U.S. fishermen. Many fishing boats (which by law must be built in the United States if they unload in the United States) are inferior to those of other

TABLE 12-7 World Commercial Catch of Fish, Crustaceans, Mollusks, and Other Aquatic Plants and Animals (Except Whales and Seals), by Countries 1974–1978

Country[a]	1974	1975	1976	1977	1978
	.... (Thousand metric tons)				
		Live weight			
Japan	10,805	10,524	10,662	10,763	10,752
U.S.S.R.	9,257	9,975	10,134	9,352	8,930
China, mainland	4,400	4,500	4,600	4,700	4,660
United States	2,929	2,920	3,160	3,085	3,512
Peru	4,145	3,447	4,343	2,541	3,365
Norway	2,668	2,542	3,416	3,460	2,647
India	2,255	2,266	2,174	2,312	2,368
Republic of Korea	2,024	2,134	2,405	2,419	2,351
Thailand	1,516	1,553	1,660	2,190	2,264
Denmark	1,835	1,767	1,912	1,807	1,745
Chile	1,158	929	1,409	1,349	1,698
Indonesia	1,336	1,390	1,483	1,572	1,655
North Korea	1,400	1,500	1,600	1,600	1,600
Iceland	945	996	992	1,378	1,579
Philippines	1,371	1,443	1,393	1,511	1,558
Canada	1,042	1,033	1,133	1,270	1,407
Spain	1,510	1,518	1,475	1,394	1,380
Vietnam	1,014	1,014	1,014	1,014	1,014
Brazil	740	772	659	748	858
France	808	806	806	760	796
Mexico	442	499	572	670	752
Malaysia	526	474	517	619	685
Bangladesh	822	823	826	835	640
Republic of S. Africa	648	642	640	603	628
Poland	679	801	750	655	571
England and Wales	534	497	520	525	548
Burma	434	485	502	519	540
Argentina	296	229	281	392	537
Nigeria	473	466	497	504	519
Scotland	538	468	503	468	479
Ecuador	174	263	315	476	476
Namibia (S.W. Africa)	840	761	574	404	418
Fed. Rep. of Germany	526	442	454	432	412
Italy	426	406	420	380	402
Senegal	357	363	362	289	346
Netherlands	326	351	285	313	324
Faeroe Islands	246	286	342	310	318
All others	7,450	7,323	7,323	7,594	7,646
Total	68,895	68,608	72,113	71,213	72,380

Source: Food and Agriculture Organization of the United Nations (FAO), *Yearbook of Fishery Statistics*, Vol. 46, 1978.

[a] Statistics for mariculture, aquaculture, and other kinds of fish farming, seaweed harvesting, and so on, are included in country totals. Statistics on quantities caught by recreational fishermen are excluded.

countries. Fishing regulations, although usually imposed with good intentions, can in some instances be very cumbersome and ineffective. Some regulations are necessary to prevent overfishing and depletion of a particular species of fish.

Successful fishing is also influenced by factors beyond the control of fishermen or politicians. In particular, there are changing oceanographic or atmospheric variables that can dramatically influence fishing success. Perhaps the best example of this occurs off Peru, where for many years one of the largest fisheries of the world has existed. This is an area of intense upwelling mainly due to prevailing southerly winds. Occasionally, however, atmospheric conditions change, causing more northerly winds (from the north). This results in the El Niño condition (see Chapter 10, page 353) and the warming of surface waters and a slackening of the upwelling. The available nutrients are quickly used up and organic production by phytoplankton and the associated marine and bird life dependent on it disappear or die. El Niño can have a disastrous effect on Peruvian fisheries. In 1971 Peru had an annual anchovy catch of over 10 million metric tons, which made it the world leader in fishing. An El Niño condition occurred in 1972 and Peru's catch dropped to 4.5 million tons and further in 1973 to 1.8 million tons.

The problem of control of fishing areas is an especially complicated one. In recent years there has almost been a miniwar between Great Britain and Iceland over Iceland's claim to a 50-mi fishing limit off its coast that was imposed, in its words, to protect its stocks of herring, cod, and haddock. The United States and Peru, Ecuador, and Chile have had numerous confrontations resulting from the claim by these countries in 1952 that fishing within 200 mi of their coast be limited to nationals and properly licensed foreign vessels. The United States for many years did not recognize such claims, but things changed in the mid-1970s.

In 1976 the U.S. Congress passed the Fishery Conservation and Management Act (effective March 1, 1977). This act gives the United States a 200-n mi offshore zone over which it has control over its fisheries. The zone does not presently restrict foreign fishing, but permission must be obtained before foreign vessels can enter and fish. The act established eight Regional Fishery Management Councils that prepare the rules and regulations for U.S. fishermen within each region. The act has forced the United States to take a position concerning the allocation and management of its offshore resources. The act has not been in force long enough for a full evaluation of its goals, but U.S. fishing catch from its waters has slowly been increasing. In 1979 the U.S. caught 33 percent of the total fish catch within its 200-mi waters, up from 27 percent the previous year.

Regardless of the amount of fish caught from the ocean or the legal restraints imposed on fishing, it is still questionable if the food from the ocean could ever be sufficient or used to feed the world's population. In many countries there are social customs and religious barriers against fish consumption. Among Catholics in the United States and elsewhere, it was a custom to eat fish on Friday. Many people thus ate fish only on Friday and as this custom was modified, the consumption of fish dropped considerably. In some countries, problems of storage and transportation limit fish consumption.

Another problem facing the fishing industry is that of pollution. Many fish and other biological resources, such as clams, mussels, and some crabs, breed in estuaries, which are areas especially vulnerable to pollution. In addition, increased levels of mercury, DDT, and other pollutants have been found in some marine organisms. An example was the 1971 public warning concerning consumption of swordfish because 50 percent of the swordfish examined within a given period of time were found to contain over 1 ppm mercury, which is twice the federal Department of Agriculture's guideline for mercury content. It is still not clear how swordfish obtained such high concentrations or how dangerous it is to eat them.

Some marine species have been so overfished that they are in danger of extinction. Especially vulnerable are several species of whales, some of which are on the U.S. endangered species list. This prevents the importation into the United States of any whale products. Several international conventions have been held and treaties signed to protect these magnificent mammals, but their slaughter continues.

OTHER BIOLOGICAL RESOURCES

Aquaculture may be one of the future methods of obtaining biological resources from the ocean. It aims at providing an environment in which certain desirable species of life will grow rapidly and can be harvested. At present aquaculture is generally restricted to nearshore areas, like bays and estuaries, and is applied only to such animals as clams, oysters, and shrimp. It may soon be possible to extend this technique to the open ocean (see Chapter 14) as well as to other animals. Kelp is a marine biological resource that is harvested in many areas of the world and holds promise for some innovative uses (see Chapter 14 and Figure 14-10).

Oyster farming is potentially a very important industry. The baby oysters, called *spat*, float and move with the ocean currents until they encounter a shell on the bottom to which they can become attached. Oyster fishermen—actually, oyster farmers is a better term—place shells on the bottom in environments suitable for growth. These environments are certain bays, inlets, and other nearshore areas. The spat, after attaching to the shells, grow until large enough to be harvested by the "farmers."

Another method (new to the United States; it has been used in Japan and Europe for decades) is to suspend shells on wires throughout the water (Figure 12-22). Thus more spat can grow in a particular area. This method, a sort of three-dimensional farming, has several other advantages: it removes the baby oysters from their natural enemies, the starfish, and places the oysters where more food

FIGURE 12-22 Scallop shells (attached to strings) being used for the settlement of oyster spat. (Photograph courtesy of Woods Hole Oceanographic Institution.)

is available. Bottom-living oysters can feed only on the plankton that settle to the bottom, while oysters living on suspended shells can feed on the much larger quantity of plankton that float by in the currents.

Lobsters can also do very well in an aquaculture system, especially ones that have heated water. Selected lobsters grown in warm seawater in the Massachusetts Lobster Hatchery reached sexual maturity in 2 years, whereas it usually took 8 years in their natural environment (Figure 8-30). This result raises the possibility of using heated water from perhaps a nuclear power plant to increase growth of selected organisms.

Aquaculture is not a major industry at present in the United States. If, however, we attempt to farm the sea with a vigor similar to that which we apply to land farming, tremendous potential awaits us. An advantage of aquaculture is that it can be carried out in the nearshore areas of countries that most need the protein.

A new product (to the United States) called fish protein concentrate (FPC) could also help in fuller utilization of the harvest from the sea. FPC is an odorless and tasteless protein concentrate powder that can be made from almost any kind of fish. It is an absolutely safe and stable product that can be stored without refrigeration. It can easily be added to the diet, perhaps even baked into bread; ten grams of FPC provide the daily protein needs of a child at a cost of about 1¢ per day. The FPC method allows the utilization of species of fish, like hake, that would not usually be consumed by humans. There is nothing wrong with these fish except that they have not been culturally and socially accepted as food. Because FPC does not spoil, most shipping and storage problems are eliminated. If this product becomes accepted in the undernourished countries, it can solve some of the food problems of the world, at least for a few years.

A new, potentially valuable biological resource is krill, a crustacean common to the Antarctic (Figure 8-31). Some feel that a harvest of this organism could exceed that of all other fish combined; others take a more conservative view. This subject is discussed in more detail in Chapter 14, page 482.

It may be possible in the future to use marine organisms to extract elements from seawater. Some marine animals concentrate trace elements from the seawater into their skeletons or tissues by factors of over 10,000 times. If these animals could be cultivated, they could provide a source of certain elements. One cannot help but wonder if there is an animal in the sea that likes gold.

Physical Resources of the Ocean

The physical resources of the ocean are perhaps not so obvious as other resources. One such resource is the large amount of energy contained within the ocean; tidal energy has special fascination for many scientists. Various ideas and schemes have been devised whereby the rise and fall of the tides or temperature differences could be used to drive electrical turbines or other devices; these are discussed in Chapter 14.

Another physical use of the ocean is for transport of commerce, especially oil. Recent technological developments such as more mechanized cargo handling, more efficient ship design, and increased size of freighters have increased the efficiency of ocean transport. Hydrofoils, which reach speeds of up to 60 knots, are being used in some areas as ferryboats. The Hovercraft, a vehicle that rides on a cushion of air, can be used for short trips.

Other resources of the ocean include its use as a recreational area. Many ocean activities such as sailing, fishing, skin diving, surfing, and swimming are becoming increasingly popular. In many areas, however, there is a conflict of interest between using nearshore areas for industrial development or for recreational sites. The ocean is also used for communication; underwater telephone cables link many of the major cities of the world. In the future, direct underwater communications may be possible using lasers.

The ocean presents some very interesting engineering problems. One, corrosion, affects almost anything placed in the ocean. Seawater, because of its dissolved salts, can conduct electricity, and when a metal object is placed in the sea, electrical current is generated similar to the way it is in an automobile battery. The current conducts particles from the metal, slowly dissolving it.

The future uses of the ocean are probably limited only by people's imaginations, with financial and legal aspects also playing important roles. Chapter 15, concerning the law of the sea, discusses what may be one of the major problems facing the use of the ocean in the future.

Summary

The resources of the ocean can be categorized either as renewable (essentially the biological resources) and nonrenewable (mineral resources). Many of the mineral resources that could be exploited from the ocean are in short supply in industrialized countries, especially the United States. Among the important mineral resources presently being obtained from the ocean are freshwater from desalination techniques, some elements like bromine and magnesium dissolved in seawater, sand and gravel, marine placer deposits, oil, sulfur, and gas. Hydrocarbons, in terms of dollars, are by far the most valuable resource from the ocean. Oil and gas exploration and exploitation from the marine environment will very likely increase in coming years. Potential marine mineral resources include phosphorite, manganese nodules, heavy metal muds, and possibly minerals associated with sea-floor spreading activities at the ocean ridges. The mining of manganese nodules, a very impressive potential resource in terms of areal distribution, is ensnarled in legal problems resulting from the Law of the Sea activities.

Biological resources are extremely important in that they supply a major source of animal protein in several areas of the world. However, many feel that the exploitation of biological resources may be near or even over its sustained capacity. The exploitation of new types of organisms, like krill, or using new techniques such as aquaculture or electronic methods for harvesting fish could increase the yield from the ocean. Other marine resources include energy from the sea and the use of the ocean for transportation, communication, and recreation.

Suggested Further Readings

Bonatti, E., "The Origin of Metal Deposits in the Oceanic Lithosphere," *Scientific American,* **238,** no. 2 (1978), pp. 54–61.

El-Sayed, S. Z., and M. A. McWhinnie, "Antarctic Krill: Protein of the Last Frontier," *Oceanus,* **22,** no. 1 (1979), pp. 13–20.

Flipse, J. E., *The Potential Cost of Deep Ocean Mining Environmental Regulation.* Texas A & M Sea Grant Publication 80-205, 1980, 47 pp.

Food and Agricultural Organization, Department of Fisheries, *Atlas of the Living Resources of the Seas.* Rome, 1972.

Food and Agricultural Organization, *Yearbook of Fishery Statistics,* **46,** 1978.

FRALICK, R. A., and J. H. RYTHER, "Uses and Cultivation of Seaweeds," *Oceanus,* **19,** no. 4 (1976), pp. 32–39.

GLANTZ, M. H., "El Niño: Lessons for Coastal Fisheries in Africa?" *Oceanus,* **23,** no. 2 (1980), pp. 9–17.

HEATH, G. R., "Deep-Sea Manganese Nodules," *Oceanus,* **21,** no. 1 (1978), pp. 60–68.

HEDBERG, H. D., J. D. MOODY, and R. M. HEDBERG, "Petroleum Prospects of the Deep Offshore," *American Association of Petroleum Geologists,* **63,** no. 3 (1979), pp. 286–300.

HOLT, S. J., "The Food Resources of the Ocean," *Scientific American,* **221,** no. 3 (1969), pp. 178–94.

MACLEISCH, W. H., ed., "Harvesting the Sea," *Oceanus,* **22,** no. 1 (1979), 72 pp.

MACLEISH, W. H., ed., "Ocean Energy," *Oceanus,* **22,** no. 4 (1979/80), 68 pp.

MANHEIM, F. T., *Mineral Resources of the Northeastern Coast of the United States,* U. S. Geological Survey Circular 669, Washington, D.C.: U.S. Government Printing Office, 1970.

MCKELVEY, V. E., "Seabed Minerals and the Law of the Sea," *Science,* **209,** (1980), pp. 464–72.

MCKELVEY, V. E., and F. H. WANG, *World Subsea Mineral Resources.* U.S. Geological Survey Miscellaneous Geological Investigation Map I-632, Washington, D.C.: U.S. Government Printing Office, 1970.

MERO, J. L., *The Mineral Resources of the Sea.* New York: Elsevier, 1965.

MOTTLE, M. J., "Submarine Hydrothermal Ore Deposits," *Oceanus,* **23,** no. 2 (1980), pp. 18–27.

RONA, P. A., "Plate Tectonics and Mineral Resources," *Scientific American,* **229,** no. 1 (1973), pp. 86–95.

ROSS, D. A., *Opportunities and Uses of the Ocean.* New York: Springer-Verlag, 1980, 320 pp.

U.S., Department of Commerce, National Oceanic and Atmospheric Administration, *Fisheries of the United States, 1979,* Current Fishery Statistics No. 8000. Washington, D.C.: U.S. Government Printing Office, 1980.

13

Marine Pollution

POLLUTION OF THE OCEAN is often a controversial subject. Some sources of pollution, often dramatized in the media, are relatively not too critical whereas others, perhaps not so well known, can have awesome effects. In a few instances so-called pollutants may actually be beneficial to marine life—for example, nutrients discharged into well-circulated waters. Alternatively, the same nutrients discharged into a restricted bay or small estuary may have deleterious effects. Offshore drilling for oil is often considered an extreme case of pollution potential. The shipping and transportation of oil, however, introduces many times (perhaps more than 20 times) the oil into the environment that drilling does, but often gets much less publicity.

The preceding chapters have shown that the earth is really a water planet rather than a land planet. It should also be evident to even the most casual reader that the ocean has a considerable influence on our environment, climate, and future on earth. The waters of the ocean, which cover about 72 percent of the earth, touch the shores of over 100 nations, and a major portion of the world's population lives either on or very near the water.

The ocean is an extremely valuable source of energy, food, and mineral resources that could accommodate much of the future needs of the earth (see Chapter 12). These uses could also put considerable environmental pressures on certain marine areas, especially the nearshore zone. The ocean is also the main thoroughfare for commerce and, unfortunately, also the principal area for disposal of many of our industrial and domestic waste products. All these demands upon the ocean have greatly increased the possibility of potential damage to it by pollution.

There is a tendency to think that humans are the only creatures dependent on the ocean. This is obviously not true; as Chapter 8 has shown, the oceans contain a vast cornucopia of life, most of which are intimately dependent on each other. This marine life has evolved over many hundreds of millions of years and in most instances cannot adjust very rapidly to profound environmental changes. The decrease or elimination of one variety of life within a particular ecological level could have serious effects on higher or lower orders of life. Likewise, human's increased use and pollution of estuaries and coastal areas may damage extensive breeding areas for plants and animals on which other forms of life are dependent (Figure 13-1).

It is apparent that the marine environment is an important resource of the United States, especially since over 70 percent of its population live within a day's drive of a coastal or Great Lakes area. It is estimated that over 112 million Americans each year participate in ocean-related activities and spend over $14 billion in the process. These activities, such as fishing, swimming, and boating, which in the past have been taken so casually, are threatened by the ever-increasing pollution of our nearshore waters. In fact, most marine scientists would agree that the most serious potential problem facing the ocean is its pollution. This problem is often made more complex because of the lack of background data—we just do not know to what level the ocean is already polluted, what pollutants were present in it before the Industrial Revolution, and what levels of the pollution the ocean can tolerate. In addition, the vast size of the marine environment does not lend itself to experimentation nor can we often afford to wait for years to see what results from our present actions. Nevertheless, the monies spent in the United States for ocean pollution research, monitoring, and development are considerable, over $180 million in 1980. About 80 percent of the total goes toward research and monitoring, and a major part of this is directed toward petroleum and petroleum products.

The amount of waste materials introduced into the ocean yearly from factories, power plants, rivers, the atmosphere, and shipping are immense. At what rate these potential pollutants degrade, or change the ocean, however, is generally

FIGURE 13-1   A 15 cm (about 6 in.) living brown trout that has grown around the pop top of a beverage can. (Photograph courtesy of the California Department of Fish and Game.)

not known. In many instances if the waste material is not put into the ocean it will end up on a land site causing perhaps even more serious damage. Some of these "pollutants" are entering the ocean independent of people's activities by natural weathering and erosion of rock and soil, volcanic activity, and even by processes associated with sea-floor spreading. We are also seeing increased use of the ocean for oil and gas exploration and sand and gravel mining and, in a few years, for recovery of mineral resources such as manganese nodules. Again, the effect of these uses may not be known until long after the exploitation process has begun.

Perhaps our limited awareness of ocean problems is understandable. For many centuries it has been thought that the ocean was essentially unlimited in size and could not be harmed or overfished. This view was probably reasonable considering our knowledge at that time. We have since found, however, that lakes, including the Great Lakes, and rivers can easily be polluted by human activity—the effect of acid rain on lakes is an all-too-perfect example—and we are detecting more and more indications of pollution in the nearshore waters and in the deep sea. Ironically, one of the problems is to determine if certain pollutants in the ocean are the result of human activity or natural causes. It is conceivable that in some instances the input of a potential pollutant from natural causes can be equal to or exceed human input; but since little background data are available, especially prior to the industrial and urbanization revolution of the human race, it sometimes is difficult to evaluate exactly people's impact (see Table 13-1).

TABLE 13-1   Pollutants that are Introduced into the Ocean both by Natural Causes as well as by Human Activity

Ocean pollutant	Natural cause or input	Human input
Heavy metals	River runoff (erosion of rocks), volcanic activity from subduction zones (Figure 6-8), decay of organic matter	Industrial and municipal discharges
Hydrocarbons	Natural oil and gas seeps (Figure 13-8), river runoff, bacteria and other organisms in water, volcanoes	Shipping and drilling activities, runoff, atmospheric input
Nutrients	River runoff, reworking of bottom sediments, upwelling, and biological activity	Industrial and municipal discharge, agricultural effluents
Radioactive substances	River runoff, (rocks contain radioactive elements), volcanic activity, atmospheric interactions (^{14}C for example is formed in atmosphere)	Nuclear power plants, nuclear weapon testing, industrial and municipal discharge
Particulate matter	River runoff, biological activity, mixing of bottom sediments, atmosphere, turbidity, currents	Fishing activity, mining, drilling, municipal and industrial discharge

Pollution is clearly a very emotional subject and few, obviously, are in favor of it. Eliminating some forms of pollution is not always a possible alternative, and some risks just cannot be avoided. In some cases society must be willing either to pay the necessary costs to maintain the present quality of the environment or to accept the effects of the pollution. An example is the vast daily accumulation of garbage by large cities like New York where land disposal sites are rapidly being filled. Either disposal at sea and its consequent problems or new land sites are possible solutions. Additional land sites will involve increased costs for the acquisition of the land and transportation as well as environmental problems of their own. Innovative ideas such as using garbage or dredged material for offshore islands (airports, nuclear power plants, or just for recreation) could be alternatives, but they will be expensive.

Certainly one of the key difficulties in combating marine pollution is to define it. Exactly what is a pollutant, and can a pollutant in one situation actually be a nonproblem or even beneficial in another situation? A U.N. report (*The Sea: Prevention and Control of Marine Pollution, Report of the Secretary General,* 1971) defined marine pollution as ''the introduction by man, directly or indirectly, of substances or energy into the marine environment (including estuaries) resulting in such deleterious effects as harm to living resources; hazards to human health; hindrance to marine activities, including fishing; impairment of quality for use of seawater; and reduction of amenities.'' Note that this definition, compiled by the U.N. advisory experts, does not consider pollution from other than humans, such as natural oil leaks, volcanic eruptions, and the like.

The various types of pollutants can be as varied as our technological achievements but can be categorized (according to the same U.N. report) as follows:

1. Disposal of domestic sewage, industrial, and agricultural wastes.
2. Deliberate and operational discharge of shipborne pollutants.
3. Interference with the marine environment from the exploration and exploitation of marine minerals.
4. Disposal of radioactive waste resulting from the peaceful uses of nuclear energy.
5. Military uses of the ocean.

Individual pollutants within these groups have several ways or pathways of reaching the sea, including direct input by rivers, runoff, sewer outfalls, dumping into the sea (either deliberate or accidental), and from the atmosphere. Considering the atmosphere, as much as 40,000 tons of industrial lead may get to the oceans each year by rainfall and atmospheric fallout, whereas over a quarter million tons may enter via rivers and outfalls. However, the rain and atmospheric input are spread over the entire ocean whereas the river and outfalls are mainly in coastal regions (where much ends up in the sediments). In this manner then, the atmosphere can be the principal source of offshore pollution. Another example is atmospheric input of radioactive materials from atomic bomb testing. Alternatively, material from the ocean can get into the atmosphere and end up on land.

For example, many coastal areas, especially with onshore winds, have relatively high contents of chlorine and sodium in their soil that have come from the ocean.

In 1978 the National Ocean Pollution Research and Development and Monitoring Planning Act was signed. The act requires NOAA to develop a comprehensive 5-year plan, which they did with the help of an interagency committee. Among the things the committee did was establish a list of priorities (not ranked within priorities) or areas of immediate concern:

> *High priority*. Land use practices, municipal sewage outfalls, industrial waste disposal, radioactive waste disposal, dredge material disposal, steam electrical power plants, oil and gas development, oil transportation, hazardous materials transportation, accidental pollution discharges, pollution incident response and cleanup.
>
> *Medium priority*. Ocean thermal energy conversion, deep seabed mining, nearshore mining, brine producing activities, at-sea chemical incineration, recreation (including small craft activity),
>
> *Low priority*. Sewage sludge dumping, developmental ocean energy technologies, fish and shellfish processing, hatcheries and aquaculture.

It is generally thought that the disposal of domestic sewage and agricultural and industrial wastes is the most serious present form of pollution, especially for the nearshore area. Pollution from oil occasionally receives more attention, however, perhaps because it is so widespread or because energy is such an emotional issue.

The following sections discuss some of the categories of pollutants and how they affect the marine environment.

Domestic, Industrial, and Agricultural Pollution

Domestic, industrial, and agricultural pollution is an extremely large category of pollutants that includes human sewage, chemicals, organic material, and pesticides, and each can reach the ocean by different pathways, including rivers, runoff, and outfalls (Figure 13-2).

The problems of and genetic damage to people living in the Love Canal chemical dump area of upstate New York has shown to many the dangers from the uncontrolled or unmonitored disposal of chemical waste. Many chemicals thought to be safe are found to be dangerous, even carcinogenic. It is estimated that over 60 million tons of hazardous wastes are produced each year in the United States; that is about 227 kg or 500 pounds for each citizen, and only a small portion is disposed of adequately. In spite of various laws very little effective action has yet resulted. The cleanup bill for the Love Canal has been estimated at about $125 million and there may be another 2,000 sites in the United States that are also hazardous. The following are some of the problems that can result from such pollution:

FIGURE 13-2 Wastewater from a pulp mill being discharged into Puget Sound, Washington. (Photograph courtesy of U.S. Environmental Protection Agency.)

Disease from infectious organisms carried within the water. Disease can generally be prevented by disinfecting techniques but many municipalities are finding potentially dangerous materials in their water supply. In addition, some additives used in disinfecting may have dangerous effects.

Oversupply of plant nutrients. As discussed in Chapter 8, certain nutrients are critical for plant growth, and if they (phosphorus, in particular) are oversupplied, growth of plankton can accelerate, resulting in eutrophication (rapid growth or oxidation that uses up all the dissolved oxygen; this can be lethal to marine life in a closed or restricted lake or reservoir) or in red tides (described later).

Oxygen-consuming products. Most organic waste products including human wastes oxidize when they reach the marine environment; if there is an insufficient mixing and oxygen supply, all the oxygen could be utilized, resulting in the death of most marine life.

Toxic chemicals and minerals. Many industrial and agricultural waste products are extremely poisonous to marine life. In some instances the long-term effects of what may be a relatively safe waste product is not known. A further discussion of the pesticides DDT, Aldrin, and Dieldrin is given below.

Sediments. The rivers of the United States discharge about 491 million tons of soil and sediment per year, or about 1.3 million tons per day into their estuaries and nearshore areas. This large amount of soil and sediment can be damaging to marine life such as oysters or clams as well as sometimes requiring dredging to keep waterways open. The disposal of the dredged material is again another pollution problem.

In the case of several of these pollutants it is their effect on the dissolved oxygen in restricted or poorly mixed bodies of water that causes the harm. The effect, however, on the total oxygen content of the ocean and its implications for humans have sometimes been exaggerated. For example, it has been suggested that if something decreased or eliminated the oxygen in the ocean that it would in turn drastically affect the atmospheric oxygen content. This is not correct since the production and consumption of oxygen in the ocean is essentially in a steady state, and the net exchange with the atmosphere is fairly small. It has been estimated that if all marine photosynthesis was to stop, the atmospheric concentration of oxygen would drop about 10 percent in 1 million years. Such a loss, although undesirable, could probably be tolerated by most species of life.

PESTICIDES

Among the more dangerous pesticides found in the marine environment are DDT, Aldrin, and Dieldrin. Most of the publicity has focused on DDT, and its use is restricted in many areas of the world. DDT is a fairly stable compound that tends to concentrate in organisms to much higher quantities than the normal background concentration. This process is known as *biological magnification* (Figure 13-3) and can be very effective. For instance, in an exceptional example oysters exposed continuously to waters containing only 0.1 ppb of DDT were able to concentrate up to 7.0 ppm of it in their tissue—an increase of 70,000 times. DDT can have

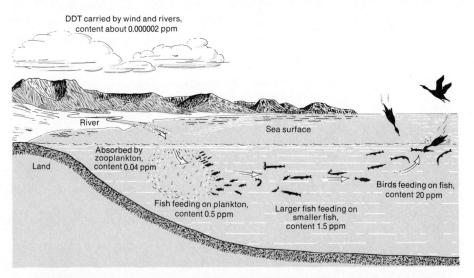

FIGURE 13-3 The process of biological magnification of DDT. At each level in the food chain the concentration of DDT is increased and subsequently passed on to the next level. Animals like humans and birds therefore can receive quantities of pesticides in concentrations millions of times higher than the amount originally introduced into the environment.

different effects on organisms. Its presence in some marine birds has been shown to result in the production of very thin egg shells, which has made reproduction difficult. Some phytoplankton show a significant reduction in photosynthesis rate when DDT is present in concentrations of about 10 ppb; other plankton suffer no adverse effects. Once the DDT reaches the ocean it remains initially in the upper mixed layers of the ocean but eventually is carried to deeper layers either by incorporation into organic matter or by sedimentation. These processes are not completely understood.

In recent years, with the decrease in the use of DDT, there was an increase in the use of the pesticides Aldrin and Dieldrin. These two pesticides are even more toxic than DDT and fortunately the Environmental Protection Agency (EPA) has tried to severely restrict or to stop their use. These actions, although challenged by various industries, have been moderately successful, but large amounts of these insecticides still remain in the environment. Both are more soluble in water than DDT although they also are carried to the ocean mainly by the atmosphere. Aldrin and Dieldrin, like DDT, tend to be concentrated by organisms in their tissue and are found in increased concentration that can exceed the background concentration by factors of thousands or more (Figure 13-3) as one looks at animals higher in the food chain.

The control of pollution by pesticides like DDT and others depends almost entirely on stopping their use. Prohibition would be difficult, however, since they are necessary in many countries to control insects carrying diseases such as malaria. Thus a dilemma exists since there are no real substitutes for these pesticides.

SYNTHETIC ORGANIC COMPOUNDS

The synthetic organic compounds category includes a wide and unfortunately increasing variety of chemical pollutants. It includes materials like those disposed of in the Love Canal area as well as products that were initially thought to be quite harmless, such as synthetic plastics, fibers, polymers, solvents, fertilizers, and the like. Many of these materials reach the ocean via river and atmospheric discharge or via shipping or other means of transport. Their fate and pathways, once in the ocean, are rarely known as is also their effect on marine animals and plants. Among the more dangerous are the chlorinated and halogenated hydrocarbons that are commonly used in items like flame retardants, fire extinguishers, solvents, and, until recently banned in the United States, aerosol propellants. These compounds are known to cause liver damage and are sometimes carcinogenic.

One of the most dangerous chemicals of this category is polychlorinated biphenyls, or PCB's, that until recently were used in making plastics, electrical insulation, fire retardants, and in heat exchangers. The toxic effects of PCB's were noted in the 1930s, but it became illegal to manufacture them only in 1979. Over 600,000 tons of this pollutant have entered the ocean and an estimated 50 to

TABLE 13-2 PCB Concentrations from
Organisms, Sediments, and Water
of the Atlantic Ocean

Source	PCB concentration (ppb)
Sea mammals	3,000
Sea birds	1,200
Mixed plankton	200
Finfish from upper ocean waters	50
Finfish from midwater depths	10
Bottom-living invertebrates	1
Deep-sea sediments	1
Seawater	0.001

Source: From Harvey, 1974.

80 percent of it may be in the North Atlantic. PCB's have been shown to reduce the growth rates of some phytoplankton even when present in amounts as small as 10 ppb. The reduced growth rate in turn could affect the phytoplankton species composition and influence the entire food chain. PCB's are fairly common in the environment and appear to show the influence of biological magnification (Table 13-2). Studies of PCB pollution in the Great Lakes have shown that some salmon and trout have PCB contents that are 100,000 to 1 million times that of the surrounding water. These concentrations can, in part, be passed on to those organisms that eat such fish.

The effects of PCB's on humans are varied, involving nausea and vomiting, abdominal pain, and jaundice. Long-term exposure can weaken ability to recover from other diseases and affect reproduction, including the possibility of having deformed children.

EXCESS NUTRIENTS

Nearshore waters may receive large amounts of industrial and human wastes that contain nutrients such as nitrogen and phosphorus. If there is an oversupply of these nutrients and the wastes are not well-circulated, the excess can lead to extremely heavy growth of algae and other phytoplankton. The growth, besides coloring the water and making boating or swimming unappealing, can also clog water intakes and filters. If the process continues, the decay of the large amount of organisms will deplete the available oxygen in the water and result in large fish kills. One such type of growth is called *red tide* and seems to have become far more common in recent years. For example, red tides used to occur along the Florida Gulf coast about once every 16 years but now occur almost every year. Red tides are caused by the rapid growth of certain dinoflagellate species that are

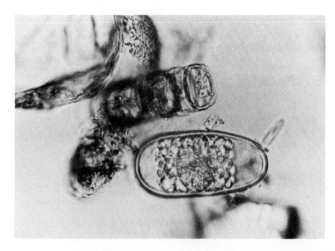

FIGURE 13-4 A cyst of *Gonyaulax ta-marensis* (center of picture), a dinoflagellate that can cause red tides. The organism is about 50 μ long; above it is a small diatom. (Photograph courtesy of Dr. Don M. Anderson.)

generally present in very small numbers but whose concentration can increase by factors of a million or more under certain conditions (Figure 13-4). These phytoplankton are generally not poisonous to fish, mussels, and clams that feed on them, but toxins can be produced within these animals, from the dinoflagellates, that can poison humans who eat them. The incidence of red tides may be increasing due to the increased supply of nutrients by pollution in areas such as the Florida Gulf coast.

HEAVY-METAL POLLUTION

Among the more dangerous pollutants are heavy metals, which are introduced into the marine environment by waste and sewage products. These heavy metals are usually present in extremely small concentrations, on the order of 1 ppb, and include elements like mercury, cadmium, silver, nickel, and lead. In certain situations, marine plants and animals can build up high concentrations of these metals without apparent harm to themselves, but if human beings consume these organisms, dangerous situations and even poisoning can result. One of the most recent instances of this was the ban on swordfish because of their mercury content.

One of the first examples of heavy-metal pollution was in Japan with the discovery of Minamata disease. This disease takes its name from a town on the west coast of Kyushu Island. One of the main industries of that town was a factory that produced certain chemical products including mercury compounds. As early as 1953, animals, including dogs and cats, were found dying of some sort of convulsive disease. These symptoms also occurred among the fishermen of the town, who were found to have intense damage to their nervous systems. Investigators looking into the problem did not receive much help from the company, which would not indicate what chemical agents were being used or what products were being introduced into the sea. Eventually, it was found that the mercury compounds discharged from the factory were being concentrated in the shellfish of the area, which in turn were eaten by the fishermen. By 1969 over 110 people had contracted this disease and over 45 people had died from it. Several unborn children were also found to have symptoms of this disease.

One of the questions about heavy metals as yet unanswered is, What amount of these heavy metals can humans tolerate, assuming that they eat a normal balanced diet? Another question is, What amounts do marine organisms need in their diet? Several heavy metals, especially copper and zinc, appear to be necessary in some amount for life in the ocean. As with DDT, however, there can be an amplification of heavy-metal content within the organism as one proceeds up the food chain (Figure 13-3). Lower members of the food chain such as plankton can concentrate heavy metals, and when the plankton are eaten by herbivores or carnivores higher in the food chain they pass on their heavy metal content. The chain eventually ends with predators like swordfish and tuna with concentrations high enough to cause concern about human consumption. The problem is complex because it is not clear how much mercury or zinc tuna may have had prior to our awareness of this problem and prior to the increased introduction of these metals into the marine environment. For example, the mercury content of museum specimens of tuna, some caught almost 100 years ago, was similar to the mercury content in recently caught fish (see report by G. E. Miller and others, 1972).

It is difficult to evaluate input of heavy metals since there are some entering the ocean at higher rates from natural activity than from humans. For example, again considering mercury, uncontrolled human input to the ocean is about 4,000 to 5,000 tons per year, but natural weathering of rocks and sediment from land supplies 5,000 tons per year, most of which is carried by rivers to the ocean. However, some of the highest values of mercury in the deep sea occur near the mid-ocean ridges, suggesting still another source (see, for example Figure 7-18). The important questions are, What and how much is being introduced and is it being consumed and retained by organisms that we eat? Again, this example points out the lack of data concerning the pathways certain elements take in the ocean. In the case of mercury, for example, very few samples have been analyzed for it and its oceanwide distribution is not known. Studies made along the coast of Europe show concentrations as low as 16 ppb, whereas concentrations along the oceanic ridges can be as high as 400 ppb, and even higher values occur near industrial areas. Although seawater values are generally very low, there is, as previously mentioned, a concentrating effect among organisms. For example, *Mercenaria mercenaria*, the common quahog, which is found in estuaries, may have as much as 0.1 to 0.4 ppm mercury. This value is near, but below, the limit (0.5 ppm) set by the United States for safe human consumption. Thus in theory, organisms with this amount of mercury are safe to eat. The questions remain, How much can one eat of such organisms? and What is the effect of such concentrations to human beings?

Ocean Dumping

The disposal of waste products at sea is another major form of pollution. Many countries routinely dump various wastes in shallow waters off their coasts, in many instances assuming that the ocean can easily absorb such ingredients. The annual dumping tonnage for the United States exceeds 45 million tons (Table 13-3), and much of this material is extremely detrimental to the environment. Certain areas off some coasts are set aside exclusively for dumping; one such locality exists about 24 km or 15 mi off New York City.

One of the dilemmas in using the ocean for dumping is that it is often the better of two alternatives—the other choice is land. In many cases the ocean is the final choice from economic considerations since land is generally considered as more valuable than the ocean, especially coastal land. Another argument for dumping of waste products at sea is that the material may quickly decompose and be oxidized, perhaps even resulting in some beneficial aspects to the marine environment. Some studies have shown, however, that microbial decay of organic matter in the ocean can proceed at a rate that is 10 to 100 times slower than at similar temperatures on land. This surprising discovery came from a unique and unplanned experiment when the research submersible *Alvin* sank in about 1,540 m (5,052 ft) of water. Fortunately, there was no loss of human life. Aboard the submersible when it sank were sandwiches, fruit, and thermos bottles filled with bouillon. The submersible was recovered about 10 months later and the food materials were found to be surprisingly well preserved. Studies about why this occurred are inconclusive but one important implication is obvious: The thought of using the deep sea as a dumping site for organic wastes from land may need some serious rethinking. The results from the *Alvin* "experiment" suggest that wastes dropped into the deep sea may remain there for very long periods of time.

In the United States, ocean dumping is regulated by the EPA. Recently Congress passed the Ocean Dumping Act, which will require all ocean dumping

TABLE 13-3   Ocean Dumping: Types and Amounts (in tons)

Waste	Atlantic	Gulf	Pacific	Total	Percentage of total
Dredge spoils	15,808,000	15,300,000	7,320,000	38,428,000	80
Industrial waste	3,013,200	696,000	981,300	4,690,500	10
Sewage sludge	4,477,000	0	0	4,477,000	9
Construction and demolition debris	574,000	0	0	574,000	
Solid waste	0	0	26,000	26,000	} 1
Explosives	15,200	0	0	15,200	
Total	23,887,400	15,996,000	8,327,300	48,210,700	100

Source: From Council of Environmental Quality, 1970, and Ketchum, 1973.

FIGURE 13-5 *Vulcanus*, a vessel especially equipped to burn waste products at sea. Material is burned at a temperature of over 1,100°C (over 2,000°F) at a rate of 25 tons per hour. Tests indicate that this may be an efficient method to dispose of some toxic substances. (Photograph courtesy of the U.S. Environmental Protection Agency.)

of sewage sludge be stopped no later than December 31, 1981. This act, if not modified, will require present ocean disposal to change to alternative methods such as land disposal. There are many who feel that land sites are even more valuable, and certainly more limited, than the ocean option and thus this act could create an even more critical pollution problem on land.

During the past few years several expeditions have found large amounts of plastic and polystyrene particles floating on the ocean surface. Concentrations as high as 12,000 particles per square kilometer have been detected in the Sargasso Sea off the east coast of the United States. Such large concentrations probably reflect increased production of plastics on land and subsequent dumping at sea. The plastic particles can serve as an area of attachment for small plants and animals. Many of the plastics contain PCB's, however, which are a dangerous pollutant.

There have been some innovative ways developed to handle waste product disposal in the ocean. One technique is to burn the material. This procedure is especially appealing for chlorinated hydrocarbons, which can be converted into carbon dioxide, hydrochloric acid, and water—substances easily absorbed by the ocean. A Dutch vessel called *Vulcanus* (Figure 13-5) has been especially equipped for such work.

Oil Pollution

Oil is one of the major sources of pollution in the ocean. It has been estimated that more than 8,000 spills occur yearly in the United States. Most of these are

TABLE 13-4 Input of Petroleum Hydrocarbons into the Ocean

Source	Million metric tons (per year)
Transportation	2.133
Tankers, dry docking, terminal operation, bilges, accidents	
Coastal refineries, municipal and industrial waste	0.8
Offshore oil productions	0.08
River and urban runoff	1.9
Atmospheric fallout	0.6
Natural seeps	0.6
Total	6.113

SOURCE: From U.S. National Academy of Sciences, 1975.

small accidents but they add up to almost 30 million tons of oil products per year. On a worldwide scale the numbers are more dramatic. As much as 2.2 billion tons of crude oil are used per year in the world, and about half is carried by tankers. There are about 6,000 tankers of which about 6 percent, or 360, are involved in collisions or groundings each year. About 0.1 percent of all oil transported by ships ends up in the ocean. Estimates of the total amount of hydrocarbons that enter the ocean vary, but generally (Table 13-4) are at least as high as 6 million tons (about 34.5 million bbl, or 1.9 billion gal). This loss comes from several sources, but transportation is the major polluter (Figure 13-6). Discharge at sea can occur because of normal leaking or accidents as well as the illegal practice of cleaning oil tanks at sea by flushing them out with seawater. The latter problem could be partially solved by using the load-on-top method that is described later in this section. Not included in these numbers are the major catastrophes such as the grounding of the *Torrey Canyon* in 1967, which lost over 700,000 bbl of crude oil—a barrel contains 42 gal of oil, a gal contains 3.785 liters (L). The *Torrey Canyon* was a large ship at that time but small by comparison with some more modern tankers (Figure 13-7). Considerable efforts were made to control the spread of its oil, but they were generally ineffective. Much of the oil eventually reached the beach where it caused severe environmental effects. It was estimated, however, that 90 percent of the marine animal deaths were caused by the detergent used to clean up the oil. The detergents cause the oil to form smaller drops, which are more easily spread in the marine environment and which tend to be more detrimental to marine life. In addition, the detergents themselves are a pollutant.

Another major oil spill occurred off Santa Barbara, California, in 1969. In this instance a high-pressure offshore oil well blew out resulting in over 700,000 gal of oil (a little less than 17,000 bbl) being introduced into the environment. Much of this oil drifted ashore coating beaches and covering many sea birds who dove unknowingly into the oil-covered water. Oil leaked into the area for over 300 days and caused several millions of dollars damage. There has been considerable debate about the environmental effects of this spill, which again is in large

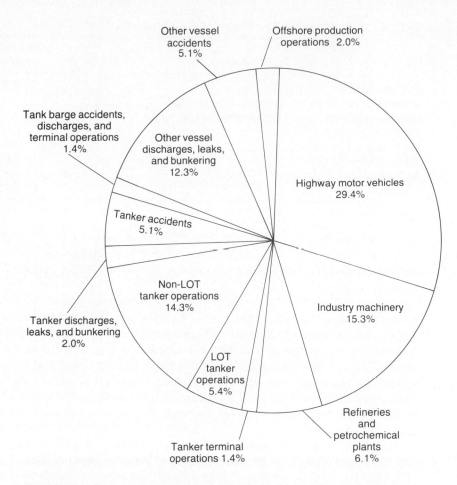

Other vessel
accidents
5.1%

Offshore production
operations 2.0%

Tank barge accidents,
discharges, and
terminal operations
1.4%

Other vessel
discharges, leaks,
and bunkering
12.3%

Highway motor vehicles
29.4%

Tanker accidents
5.1%

Non-LOT
tanker operations
14.3%

Industry machinery
15.3%

Tanker discharges,
leaks, and bunkering
2.0%

LOT
tanker
operations
5.4%

Refineries
and
petrochemical
plants
6.1%

Tanker terminal
operations 1.4%

FIGURE 13-6 Sources of petroleum pollution in the oceans. LOT means load on top (Figure 13-14). (Adapted from Parricelli and Keith, 1974.)

part due to lack of adequate background data. For example, What was the quantity and quality of life in this area before the spill? The offshore Santa Barbara area is especially vulnerable to oil pollution since natural seepage in the area may be as high as 50 to 75 bbl per day [or about 2,100 to 3,150 gal (about 8,000 to 12,000 L) per day]; this natural rate could, in as little time as 222 days, equal the amount spilled from the well. Natural oil and gas seeps are common occurrences in some areas (Figure 13-8) and contribute to a modest amount of oil in the environment (Table 13-4). Nevertheless, the presence of natural seeps does not condone the Santa Barbara spill but does show some of the problems in evaluating its impact.

A large spill occurred in August 1974 when the *Metula*, a Dutch supertanker, ran aground about 3 km (2 mi) north of Tierra del Fuego at the extreme end of South America (Figure 13-9). The tanker was carrying 1.58 million bbl of oil but

FIGURE 13-7 The Atlantic Richfield supertanker *Arco Anchorage*. This is one of the largest U.S. tankers and can carry 950,000 bbl of oil. (Photograph courtesy of Bethlehem Steel Corporation who built the vessel.)

FIGURE 13-8 Evidence of a naturally occurring oil seep from the Gulf of Mexico. (Photograph courtesy of Dr. Richard A. Geyer, College of Geosciences, Texas A & M University.)

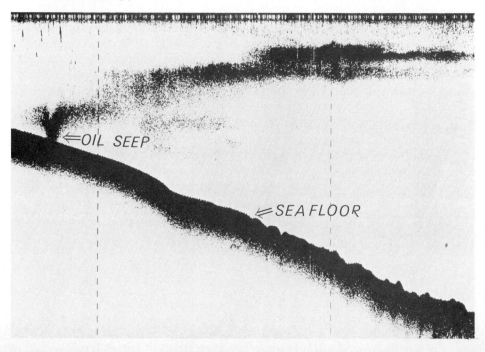

FIGURE 13-9 The 325 m (1066 ft) tanker *Metula* lying aground in the Strait of Magellan near the coast of Chile. A tug is trying to assist the ship. The *Metula* lost over 300,000 bbl of oil. (Photograph courtesy of U.S. Coast Guard.)

lost less than half of its cargo. About 25 percent of the oil reached Chile's shores and covered about 120 km or 75 mi of beach. Estimates are that at least 40,000 birds were killed and that it would take 10 years for the environment to return to normal conditions.

The more recent years have seen a dramatic increase in the size of some oil spills. In 1976 the Liberian-registered *Argo Merchant* went aground off Nantucket, Massachusetts, and spilled 7.6 million gal of oil (about 181,000 bbl) (Figure 13-10). This spill was the largest ever within U.S. waters. The record for the biggest tanker accident belongs to the *Amoco Cadiz* which broke up off the Brittany coast in March 1978. The vessel was carrying 68 million gal (257 million L) of oil, much of which reached the nearby coastal areas and caused considerable damage to nearshore shellfisheries as well as polluted much of the coastal region.

The record for the largest oil spill, however, belongs to the Mexican Ixtoc I blowout in the Gulf of Mexico, Bay of Campeche region. The well blew out on June 3, 1979 and continued spilling oil and gas until March 23, 1980 (Figure 13-11). In the course of the disaster an international debate developed between Mexico and the United States into whose waters some of the oil reached. The argument concerned responsibility for the accident, cleanup, and compensation for damages. NOAA sent a research team to the area aboard their vessel *Researcher* (Figure 1-3) but permission from Mexico for this effort did not come until near the end of the spill period. The exact amount of oil released into the environment may never be known, but during its worse times as much as 30,000 bbl of oil per day was thought to be escaping. The Mexicans tried several techniques to stop the flow, including plugging the hole with steel balls, drilling an adjacent relief well, and finally and successfully, capping it with a steel, sombrero-looking device.

Scientists studying the spill found it difficult to assess the actual damage that resulted from the spill. It clearly had an impact on the entire biological system; but the ocean, as one scientist said, showed a definite ability to clean

(a)

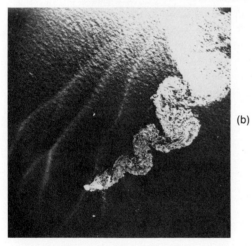

(b)

FIGURE 13-10 (a) A close view of the *Argo Merchant*. (Photograph courtesy of U.S. Coast Guard.) (b) An aerial photograph of the *Argo Merchant* taken from an elevation of 1,676 m (about 5,500 ft). The oil is clearly visible. (Photograph courtesy of National Aeronautics and Space Administration.)

FIGURE 13-11 A photograph of the Ixtoc I blowout. Photograph taken as part of a NOAA cruise to the region. (Courtesy of Dr. John Farrington.)

itself up. The potential harmful effects of this spill may have been modified by the favorable winds, offshore location of the well, and the fact that much of the oil was burned off at the surface.

One of the best studied oil spills occurred in 1969 when the oil barge *Florida* ran aground in Buzzards Bay, Massachusetts, off West Falmouth (Figure 13-12). The grounding resulted in the discharge of between 650 and 700 tons of No. 2 fuel oil into the coastal waters. Onshore winds and tides carried much of the material into very productive marshlands and killed many organisms. This particular area had been the subject of studies by scientists of the Woods Hole Oceanographic Institution before the spill occurred. They therefore had a unique opportunity to document the damage caused by the oil, and found that it retained its toxicity for several years! In some areas the oil penetrated as much as 60 cm or about 2 ft into the sediment. Shellfish in the area were especially affected by the oil. Total ecological damage in terms of local fishing and shellfish loss eventually exceeded $1 million. Visually, the area looked normal to a casual visitor within a few weeks of the disaster. Careful study showed, however, that the area had not reached its original condition even 6 years after the spill. It should be emphasized that the Falmouth spill was a relatively small one; what distinguished it was the fact that it was studied in detail and that these studies showed how long-lasting pollution effects can be.

The effects of oil spills can be of either an acute or a chronic nature. An acute effect results from a single spill or discharge. The major effect is the death of numerous organisms, and depending on the area and oceanographic conditions, the fauna and flora can eventually recover. It is estimated that about 3 to 4 percent of the total annual input of oil into the environment results from acute spills.

Chronic effects occur when the oil pollution happens either continuously or without sufficient time for the area to recover. Obviously, both kinds of effects are to be avoided but the impact of the chronic effect is more longlasting.

FIGURE 13-12 The oil barge *Florida* lying off Buzzards Bay, Massachusetts. Between 650 to 700 tons of fuel oil ended up in the coastal waters. (Photograph courtesy of *Falmouth Enterprise*.)

It should be noted that many elements can enter the environment by the burning or combustion of fossil fuels (coal, gas, and oil). For example, it is thought that more lead enters the ocean from the atmosphere by the burning of tetraethyl lead, which is used in gasoline as an antiknock agent, than is carried in by rivers. Numerous other dangerous elements, such as sulfur and mercury, can also enter the ocean in this manner.

The question of the risks of oil pollution is a highly emotional one, but some important facts should be considered. For example, according to a National Academy of Science report in 1975 entitled *Petroleum in the Marine Environment*, the amount of oil entering the total marine environment (Table 13-4) from offshore drilling (80,000 tons per year) is 7.5 times less than from natural seeps (600,000 tons per year) and 26 times less than that from transportation (2,133,000 tons per year). (These statistics do not include the recent Ixtoc I spill.) Most public concern, however, usually centers on offshore drilling rather than on oil pollution from tankers. A more realistic appraisal might be to ask what the best way is, ecologically, to obtain the needed oil. For example, the potential drilling along the east coast of the United States has drawn considerable opposition from people concerned about the environment. But, offshore drilling in the United States has been *relatively* safe. From 1964 to 1971 there were 16 major spills from 10,234 producing wells. The oil released in the 8 years was about 46,000 tons; this number, although it could be lowered by stronger government regulation, is less than twice that spilled by the *Argo Merchant* and only a small fraction of that spilled by the *Amoco Cadiz*. An offshore drilling program with buried pipelines (a relatively safe way of transporting oil) and inshore refineries would in many instances be ecologically several times safer than bringing oil in by tankers. The best alternative, of course, would be to reduce our use of oil, and that is happening; nevertheless, offshore drilling will still be necessary.

When oil enters the marine environment, regardless of the type of spill, several things happen. First, it should be appreciated that hydrocarbons are a very complex material composed of thousands of different organic compounds. In fact it appears that no one sample of oil has ever been completely analyzed. Once in the environment, some of the compounds dissolve into the water, some of the lighter and more volatile compounds evaporate into the atmosphere, some sink, some decay by biological weathering, some get into the food chain if eaten by organisms, and some may enter the bottom sediments (Figure 13-13). These processes are affected by the wind (aids in dispersal and mixing), by temperature (with colder temperature the oil forms lumps as well as degrades less), and by the original composition of the oil.

Hydrocarbons are very toxic to most forms of marine life. These effects can persist for many years, especially if they get into the bottom sediments and are slowly released over time (for example, when the bottom sediment is stirred by a storm). After a spill it could take a considerable period of time before an area might return to its natural state. The process in some instances may resemble a forest fire in the initial destruction and the time it takes for things to return to

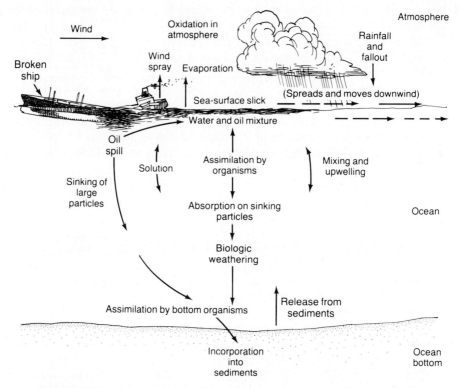

FIGURE 13-13 Some of the possible pathways that an oil spill may take. (From Ross, 1980.)

normal. Even small amounts of oil in the environment can affect biological processes such as feeding and reproduction. Oil spills have the potential for more damage when they occur in shallow water (where, of course, tankers go aground) since this is often the more biologically fertile area and has the better potential for oil to enter the bottom sediment.

One way to reduce oil pollution by tankers is by use of the load-on-top system. Tankers, once they discharge their oil, must add some water as ballast to their tanks; otherwise they will sit too high when at sea. When this water is added, it mixes with the oil clinging to the walls of the tank. After a few days and if the seas are not too rough, the oil and water mixture starts to separate (oil on top). On many ships the oil and water mixture is completely pumped into the ocean just prior to receiving a new cargo. If the water were to be pumped out from below, removing the relatively clean water, and the oil–water mixture were to be saved and collected in one tank, however, a great deal of pollution could be prevented (Figure 13-14).

Arriving at discharge port:
Full cargo. Clean ballast tank empty

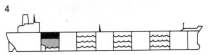

After discharging cargo and proceeding to sea:
Clean ballast tank full (clean seawater).
Cargo tanks partially full (dirty ballast).

After several days at sea:
Oil settles on top. Clean water pumped from
bottom; cleaning of empty tanks; tank
wash water collected in waste tank.

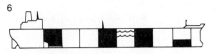

At sea:
Clean ballast for docking. Waste tank containing
waste and all residues for separation.

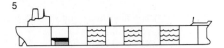

Arriving at load port:
Clean ballast for docking. Waste tank drained of
all clean water, leaving only collected residue.
Before loading, all clean water pumped into sea.

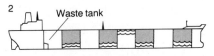

During loading cargo:
Waste tank loaded on top of residues.

▨ Clean seawater   ■ Crude Oil   ▨ Oil-contaminated
                                                 seawater

FIGURE 13-14   Diagram showing how the load-on-top technique works. (Adapted from *Oil Spills and Spills of Hazardous Substances,* Environmental Protection Agency, 1975.)

Pollution Resulting from Exploration and Exploitation of Marine Mineral Resources

The mineral resources of the ocean, besides the oil and gas mentioned in the previous section, also include sand and gravel on the continental shelf and manganese nodules in the deep sea (see Chapter 12 for more details).

Dredging of sand and gravel in shallow water obviously affects bottom-living creatures, both by the dredging operation itself and by the large amount of suspended matter introduced into the environment. At present, many states have regulations regarding nearshore dredging, but things may change as land sources of sand and gravel are exhausted (Figure 13-15).

The mining of manganese nodules from the deep sea (Figures 12-12 and 12-15) presents some interesting oceanographic and pollution problems. The water that comes to the surface (most nodules will be mined from relatively deep depths) with the nodules will be colder and usually of a different salinity than the surface

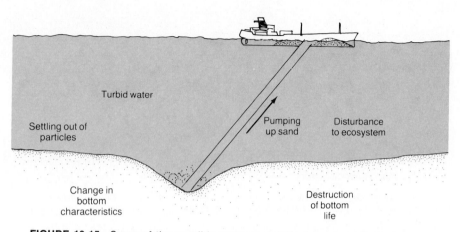

Turbid water

Settling out of
particles

Pumping
up sand

Disturbance
to ecosystem

Change in
bottom
characteristics

Destruction
of bottom
life

FIGURE 13-15 Some of the possible environmentally damaging effects that can occur from a sand dredging operation. (From Ross, 1980.)

waters. In addition, it will contain large amounts of mud from the bottom and probably have a higher nutrient content than the surface waters. The effects can vary. If the water from depth was to spread out on the surface, its relatively high content of nutrients could be beneficial for phytoplankton growth but its large suspended-matter content could reduce the transparency of the water and therefore reduce the water depth available for photosynthesis. The suspended material could also take years before it settled back to the bottom and in doing so could alter the chemistry of the water.

The effect of a manganese nodule mining operation on the bottom environment is considerably more difficult to ascertain. The characteristics and ecological parameters of the deep-water biological population are not well known, but they probably would be adversely affected by the mining operation. The falling suspended matter from the water carried to the surface could bury benthic organisms. The density of organisms in a manganese nodule area, however, is generally very low (many consider it to be a biological desert) and may have little importance in the overall biological system of the ocean. Alternatively, one must emphasize that little is known about this part of the ocean and the long-range effects of mining could have unanticipated results. On the other hand, the mining of copper, nickel, cobalt, and manganese from land mines is generally a very destructive operation and is probably several more times environmentally damaging than a marine operation.

Radioactive and Thermal Wastes

The input of radioactive material into the ocean from nuclear explosions does not appear to have created any major problems. Indeed, some of the radioactive isotopes have been very valuable for oceanographic research (see Chapter 7).

The treaty on the prohibition of emplacement of nuclear weapons on the seabed (which only refers to areas outside the territorial sea and contiguous zone—see Figure 15-1) and the general ban on atmospheric testing of nuclear bombs should keep this form of pollution low in the future. There are also fairly strict international controls on the dumping of radioactive waste from nuclear plants into the ocean; even so, the marine environment is occasionally suggested as a "final" dumping site for nuclear wastes. Presently, the artificial radioactive substances introduced into the ocean by atomic bomb blasts easily exceed those from land-based reactions or nuclear fuel processing plants. The estimated input from bombs, however, is only about 0.001 of the total natural radioactivity in the ocean. The danger is not in polluting the ocean but rather of a large input into a small, restricted area where it cannot be dispersed or diluted. With proper controls and regulations this possibility can be kept to a low probability. There has been much discussion in recent years about using the deep sea as a disposal site for large amounts of nuclear waste. This item is discussed further in Chapter 14.

A relatively new problem for the ocean, and especially coastal areas, is thermal pollution or thermal waste. Thermal pollution can result from the large volumes of water needed to cool electrical or nuclear power plants. These plants are generally situated near a source of water such as estuaries, lakes, or coastal areas. The water is often returned to the environment without prior cooling and can be as much as 10 to 15°C (18 to 27°F) higher in temperature than it was initially. This increase in temperature can be extremely harmful or fatal to most marine fauna, although there are possibilities for its being used beneficially. The heated effluents could be used in an aquaculture program to increase growth of oysters, lobsters, or other organisms. These effluents can sometimes be good places to find fish, which occasionally are attracted to these areas, perhaps because of food or the heat itself.

There are two main adverse biological results from thermal pollution: it affects the metabolic activities of the fauna and it decreases oxygen solubility. Both of these can eventually cause eutrophication. Another problem of using water for cooling is that generally a very large volume is needed. This involves complex pumping systems that in themselves may draw in marine life, especially plankton, often with physically damaging effects.

Summary

Pollution is an emotional and controversial subject often discussed in the media without adequate scientific basis. Some aspects are exaggerated; others, understated. There are many different kinds of pollution occurring in the ocean, especially in the nearshore zone. In many instances, adequate data concerning pollution input or prior conditions are not available. It is probable that some forms of marine pollution may become worse before controls are established. The inputs of some pollutants into the marine environment are immense. The curtailment or reduction of many forms of pollution will require international co-operation and large-scale financing. Many developing countries view pollution as a problem of the highly industrialized countries since they are the ones that cause most of it; hence such countries are unwilling to contribute to its solution.

Pollution can have three main effects in the marine environment:

1. It can directly destroy the organisms within the polluted area.
2. It can alter the physical and chemical properties of the environment, thus favoring or excluding specific organisms.
3. It can introduce substances, sometimes through biological magnification, that are dangerous to higher forms of life, such as human beings, but that are relatively harmless to lower forms of life.

There are numerous pathways that pollutants can take on their way to the ocean. These can include direct input into the ocean via rivers, sewer outfalls, runoff, ocean dumping, and from the atmosphere. Most of the entering pollutants come from human activity, but some also are a result of natural processes.

Probably the most critical aspect of marine pollution is the large quantity of pollutants that enters the coastal zone and estuaries. In many instances this quantity exceeds the capacity of these waters to cleanse themselves. Even with proper treatment and management, large amounts of potentially damaging pollutants can remain in the bottom sediments of the coastal zone and estuaries.

One of the most visible pollutants is that resulting from hydrocarbon activities including exploitation (drilling) and especially shipping. The institution of more rigid rules concerning the movement of hydrocarbons and the adoption of load-on-top techniques could reduce this toxic and widespread form of pollution.

The open ocean also has its share of detectable pollutants; however, its size and capacity make it less an immediate problem than our nearshore waters. As emphasized many times before, these nearshore areas are the localities of greatest marine use, both by people and by the organisms of the sea. One of the more reasonable solutions to the pollution problem is to recycle as many materials as we can, rather than to dispose of them so easily as we do now.

Suggested Further Readings

AHERN, W. R., *Oil and the Outer Coastal Shelf: The Georges Bank Case*. Cambridge, Mass.: Ballinger, 1973.

AMOS, A. F., C. GARSIDE, K. C. HAINES, and O. A. ROELS, "Effects of Surface Discharged Deep-Sea Mining Effluent," *Marine Technology Society Journal*, **6,** no. 4 (1972), pp. 40–45.

DUCE, R. A., "How Does Air Pollution Affect the Oceans?" *Maritimes*, **22,** no. 3 (1978), pp. 4–7.

HARVEY, G. R., "DDT and PCB in the Atlantic," *Oceanus*, **18,** no. 1 (1974), pp. 18–23.

PALMER, H. D., and G. M. GROSS, eds., *Ocean Dumping and Marine Pollution*. Stroudberg, Pa.: Dowden, Hutchinson, and Ross, 1979, 269 pp.

Petroleum in the Marine Environment. Washington, D.C.: National Academy of Sciences, 1975.

ROSS, D. A., *Opportunities and Uses of the Ocean*. New York: Springer-Verlag, 1980.

United Nations, Economic and Social Council, 51st Session, *The Sea: Prevention and Control of Marine Pollution*, 1971.

U.S., Environmental Protection Agency, *Oil Spills and Spills of Hazardous Substances*. Washington, D.C., 1975.

14

Innovative Uses of the Ocean

PROBABLY ONE OF THE more exciting questions in oceanography is what the future will bring. It seems clear, looking over the interesting developments of the past 10 years of marine science, that the next decade or two will also bring some startling discoveries as well as lead to some innovative and unique uses of the marine world. This chapter describes some of the immediate possible ocean uses and innovations; especially emphasized is the ocean as a source of energy. It should be stressed that before this book is even published some new development may occur that is not mentioned or that some of the innovative items discussed here will become commonplace.

Energy from the Ocean

Anybody who has ever walked along a beach or seen the ocean in a movie has to be impressed with it as a source of potential energy. In recent years, with the threat of oil shortages, there has been an increased exploration of the continental shelf and slope for new deposits of oil and gas. This turn to the ocean is reasonable since many of the potential oil and gas areas on land, especially in the United States and Canada, have already been explored and exploited (see Chapter 12). There are, however, other potential sources of energy from the ocean, for some of which there may be more enthusiasm than is justified. These ideas should nevertheless be explored since some may have the potential to succeed and reduce the world's dependence on oil and gas.

Energy from the ocean can come in several forms. Among the more obvious are waves, tides, and ocean currents, but other possibilities are use of the temperature differences between surface and deep waters, salinity differences, and the large volume of biological material in the ocean. It should be remembered that the ocean, which covers 72 percent of the world, receives a major portion of the incoming solar energy. More of this energy reaches the equatorial regions than the polar regions, producing a circulation that moves warm water and air toward the poles. This circulation also, directly or indirectly, causes the waves and currents of the ocean.

The power levels associated with the different areas and processes of the ocean can compare favorably with the world's power demands (Tables 14-1 and 14-2). To tap and utilize these sources of power, however, is often not very easy. Some of the more probable are discussed in the following pages.

TIDES

One source of energy considered since ancient times is tides. They have been used in the past in mills, but rarely in modern times. There have been numerous suggestions for building large, modern tidal energy plants but only two have come to fruition. The world's first major tidal power system, built on the Rance River in France, cost about $100 million. The Soviet Union has started an experimental tidal power station on the Barents Sea, about 80 km (50 mi) from Finland, and has announced its intention to build additional, larger tidal power plants in the future. Neither the Rance River plant (Figure 14-1) nor the Soviet one is a major source of power. Using tides for energy has some inherent difficulties, one of which is that tides do not flow continuously but change directions several times a day (Figure 9-30) and vary in strength over a two-week period (Figure 9-32). It is possible to alleviate the problem to some degree by storing water in a reservoir, as is done at the Rance power plant, at high tide and letting it out slowly to turn turbines on a more continuous basis.

One area where the tide range is sufficiently high for a tidal plant is the Bay of Fundy and the Passamaquoddy area of Maine. However, economics and politics have prevented development of this region. In spite of the enthusiasm for tidal plants they really can produce only a small fraction of the energy that a nuclear power plant could, and often would not be competitive on a cost basis. A tidal power plant can affect the ecology of the area since it interferes with or changes the tidal regime within an estuary. On the positive side, tidal plants have a fairly long life once built, and tides are essentially a free commodity although they have to be at least 5 m (16 ft) or so high for the plant to work.

WAVE ENERGY

Waves are another possible source of energy. During times of storms, wave energy can be awesome, moving large boulders, piers, or sea walls that weigh thousands of tons. When one thinks about using waves for energy, however, one should consider not the storm situation but the average state of the sea; the energy, even in small waves, can be immense. For example, a wave 3 m (almost 10 ft) high can transmit energy at the rate of about 100 kW per meter (3.3 ft) of its wave crest. Such power is essentially equivalent to that of a line of automobiles with their engines running at full power.

TABLE 14-1 Estimates of Power Levels in Natural Processes of the Planet Earth

Available sources of power	Total power in watts
Direct solar power	
Where sun hits atmosphere	10^{17}
At earth's surface	10^{16}
Photosynthesis (Stores sunlight in the form of chemical energy in fats, proteins, and carbohydrates, all of which are combustible.)	
Marine plants	10^{14}
Arable lands, forests	10^{13}
Bioconversion of waste materials	
Plant residues and manure (Can be converted by bacteria to gaseous fuels—hydrogen and methane—by storing them in airless containers at proper temperatures.)	10^{12}
Garbage, sewage, and dumps (Can be converted by the same process.)	10^{12}
Ocean thermal power	
Solar heat absorbed by ocean water (Can possibly be put to use by exploiting temperature differences between surface and depths, producing power to drive turbine.)	10^{13}
Steady surface-wind power, such as that from trade winds	10^{12}
Variable surface-wind power (in middle latitudes where winds are unsteady)	10^{12}
Hydroelectric power (from harnessing the kinetic energy of moving waters)	
Power in rainfall (Conceivably could be harnessed, but the world's total rainfall, even if the rain dropping on the oceans is included, would satisfy only 10% of the world's power demand.)	10^{12}
Flow of rivers (harnessable by traditional hydroelectric plants)	10^{11}
National evaporative exchanges between large bodies of water (The Mediterranean Sea and Red Sea are examples: evaporation is greater in them than in the ocean at large; therefore, there is a continual flow into them from the oceans to replace evaporated water. This flow can be harnessed just as in a millrace.)	10^{9}
Damming of evaporative sinks (By damming ocean openings to Red Sea and Mediterranean Sea, letting these seas evaporate until a drop of 100 m or more occurs, and then letting the ocean flow in, turning mill wheels, power might be obtained. It is not very practicable to build these dams, however.)	10^{11}
Tidal flow (This may be done at such places as the Bay of Fundy, where flow can be harnessed.)	10^{9}
Power of the great ocean currents, such as the Gulf Stream and Kuroshio Current (Theoretically these can be harnessed the way rivers are, with some sort of "water wheel.")	10^{8}

TABLE 14-1 (*continued*)

Available sources of power	Total power in watts
Ocean surface waves at the coastline (The power of waves is available at a potential total yield of 10^6 W/km of coastline.)	10^{10}
Geothermal power (This could be harnessed particularly at the "ring of fire" around the Pacific Ocean basin, so called because this is where tectonic plates merge and volcanoes erupt; the same happens along mid-ocean ridges.)	10^{10}

Present power demands	Total power in watts
Worldwide power demand for all needs of civilization	10^{13}
Human metabolism (Total power in terms of food needed to sustain present population level of 4 billion)	10^{11}

SOURCE: From von Arx, 1979.

TABLE 14-2 Some Energy Units

1 watt[a]	=	1 joule/s
1 calorie	=	4.184 joules
1 British thermal unit (Btu)	=	1,055 joules
1 Btu	=	252 calories
1 watt–hour (W-h)	=	3,600 joules
1 Kilowatt–hour (kWh)	=	3,413 Btu
1 Kilowatt (kW)	=	56.92 Btu/min
1 megawatt (MW)	=	1 million W
1 gallon of gasoline	=	125,000 Btu
1 barrel of oil	=	5,800,000 Btu
1 cubic foot of gas	=	1,031 Btu
1 ton lignite coal	=	20–40 million Btu
1 cord of wood	=	20 million Btu
Energy use in the United States per person	=	300 million Btu/y
Energy use in India per person	=	5 million Btu/y

[a] A 100 watt light bulb used for 10 hours is equal to 1 kilowatt hour.

Three basic techniques to extract energy from waves have been suggested by Richards, 1976.

1. Use the vertical rise and fall of the crests and troughs of successive waves to drive an air- or water-powered turbine.
2. Use the rolling motion of waves to move vanes or cams to turn turbines.
3. Converge waves into channels and concentrate their energy.

The first method has been used for several years in buoys to provide energy for lights or to make noise. Larger systems, up to several hundred feet in diameter,

FIGURE 14-1 The Rance power plant is a 750-m-long (2,460 ft) structure that has a reservoir of 184 million m³ of water. Tides in this region, which can have an amplitude up to 13.5 m (44 ft), drive turbines (below the surface) during both incoming (rising tide) and outgoing (falling tide) conditions. Energy production can be about 240 MW. (Photograph courtesy of French Engineering Bureau.)

have been proposed in which the waves could compress air to drive a turbine. Realistically, such systems would probably not generate any large amounts of power. In Japan a floating experimental device weighing 500 tons has been developed that will produce 125 kW power using the up-and-down movements of waves to drive turbines.

A more exciting development, using the third technique, has been used by Lockheed Corporation to develop a device called the *Dam Atoll*. This is a dome-shaped device that can be up to 76 m or 250 ft in diameter that sits just below the water surface (Figure 14-2). When the waves reach the device they enter an opening in the top and move through a central vortex, eventually turning a turbine that produces electrical energy. In 1980 Lockheed was awarded a $593,000 grant from the Department of Energy for tests on their system. According to the project designer 1 mi of waves could yield 64 MW of electricity and a group of 500 to 1,000 Dam Atolls could produce as much electricity as Hoover Dam.

A simpler device has been proposed by Isaacs and colleagues (1976) (Figure 14-3). Their device, called a *wave power pump*, consists of a vertical riser with a flapper valve and buoy float. It is loosely attached to the sea floor and can move about with wave motion. In operation the valve is closed for about half the wave

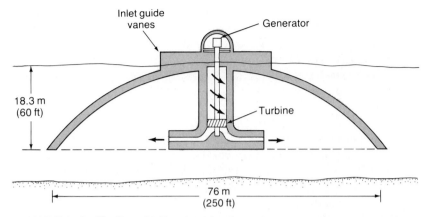

FIGURE 14-2 The Dam Atoll system developed by the Lockheed Corporation.

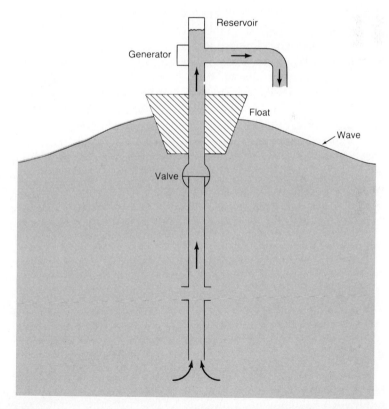

FIGURE 14-3 Schematic diagram of a wave power pump. (Adapted from Isaacs et al., 1976.)

cycle, which permits the water in the column to move upward with the float. When the float starts to descend because of the passage of the wave trough, the internal forces move the water inside the column even higher. As subsequent waves pass, the water continues to rise until it is high enough to produce power. Isaacs made a test of a 91 m (about 300 ft) device and was able to increase the pressure head of the waves by over 20 times. It is thought that a series of these pumps might provide a modest amount of energy. The device is very simple, has few moving parts, and no fuel is involved; it could be used locally, such as near a drilling platform for a small source of power.

Even with waves there can be some ecological effects. Obviously they are nonpolluting, but the interruption of the normal pattern of waves may cause some environmental and ecological problems since this will reduce or change the patterns of breaking waves, the longshore drift, and the movement of sand. Such effects have been discussed in Chapter 11.

ENERGY FROM OCEAN CURRENTS

Another potential source of energy might be some of the current systems of the ocean—in particular, the Gulf Stream or the Kuroshio Current that flows off Japan. The amount of possible energy in some currents is impressive; the Gulf Stream, off Florida, for example, has as much as 30 million m^3 (39 million yards^3) of water moving at almost 3 kn. It has been suggested that devices such as underwater windmills or turbines might be put into the flow of these currents and the moving water could then drive turbines and produce energy (Figure 14-4). Even more imaginative ideas include a series of parachutes attached to a continuous cable. These parachutes would open and catch the current when flowing with it but would close when coming back against it (the movement of the cable could be used to generate electricity). Obviously all of these devices would have to be extremely large and strong if they were to operate and remain in a major current—no such devices are presently in operation.

ENERGY FROM THERMAL DIFFERENCES

One of the more interesting and perhaps viable sources of obtaining energy from the ocean is to use the differences in the temperature of surface and deep waters. The incoming solar radiation heats the surface waters of the ocean, making them considerably higher in temperature than deeper waters. Between these two areas is a thermocline of rapid temperature change (Figures 9-7 and 9-18). The technique that seems to have the best chance of immediate success is one that is called OTEC, or Ocean Thermal Energy Conversion. The system is very simple: Warm water is drawn in at the surface and colder water is drawn from depth. A working fluid, such as ammonia, comes in contact with the warmer seawater and evaporates; the gas can then be used to drive the turbine of an electrical generator.

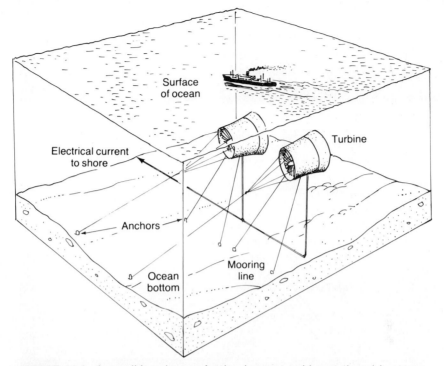

FIGURE 14-4 A possible scheme of using buoyant turbines tethered in strong currents.

This is a typical technique used in conventional electricity generation. The colder seawater, which is pumped up from depth through a large diameter tube, is then used to condense the gas back to its liquid state. The liquid (ammonia) is then again put in contact with the warm seawater and the cycle begins again (Figure 14-5). The system is a closed one; the heating and cooling take place via heat exchangers and can operate 24 hours a day. Some more complex systems have been developed but this seems to be the most promising one.

The OTEC concept is actually not a new one, having been first proposed in the 1880s by a French physicist named Jacques d'Arsonval. A small plant following this idea was actually built in Cuba in 1930 but lasted only a few weeks before being damaged by heavy seas. The concept has become much more popular in recent years because of the need for alternative sources of energy, and several models (Figures 14-6, 14-7, and 1-1) and working systems have been proposed.

It is generally thought that a temperature difference (from surface to 1,000 m or 3,281 ft) of about 20°C (about 36°F) should be available for an OTEC system to work. Obviously this restricts areas where the devices can be built since there are not that many areas where such a temperature difference exists within a relatively close distance to land.

An OTEC system is only about 2 percent efficient. In addition a 100 MW plant has only about one-tenth the capacity of a modern nuclear power plant or fossil fuel plant. Thus it would be important to build more, rather than larger

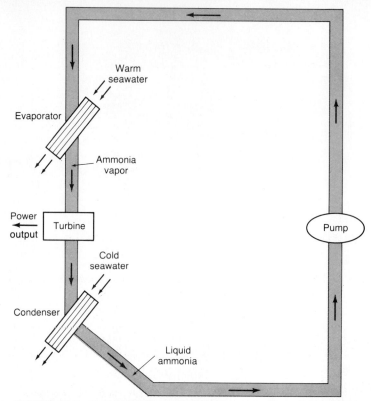

FIGURE 14-5 Basic aspects of the OTEC system. (From Ross, 1980.)

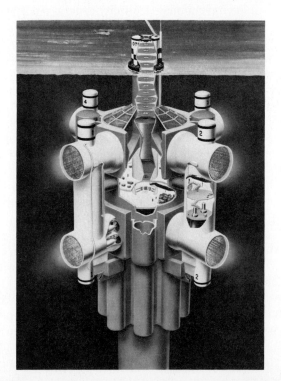

FIGURE 14-6 The Lockheed Ocean Thermal Energy Conversion system showing the major components. See Figure 14-7 for details of operation. (Courtesy of Lockheed Missiles and Space Company, Inc.)

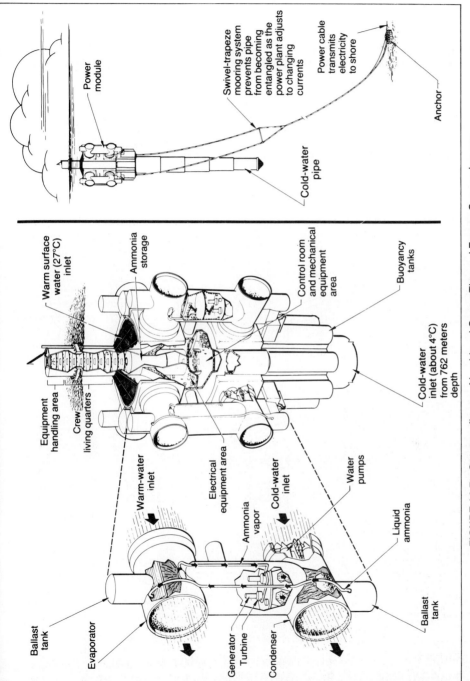

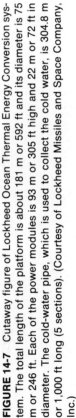

FIGURE 14-7 Cutaway figure of Lockheed Ocean Thermal Energy Conversion system. The total length of the platform is about 181 m or 592 ft and its diameter is 75 m or 246 ft. Each of the power modules is 93 m or 305 ft high and 22 m or 72 ft in diameter. The cold-water pipe, which is used to collect the cold water, is 304.8 m or 1,000 ft long (5 sections). (Courtesy of Lockheed Missiles and Space Company, Inc.)

ones, and it would be necessary to have many of these devices scattered over parts of the ocean. On the other hand, however, they cannot be too far from land because of the problems of transporting the energy, unless it is used to produce hydrogen or ammonia. For the United States it has been suggested that the plants cannot be much more than 160 km or 100 mi offshore and probably would have to be located in areas such as the Gulf of Mexico or off Hawaii (Figure 14-8).

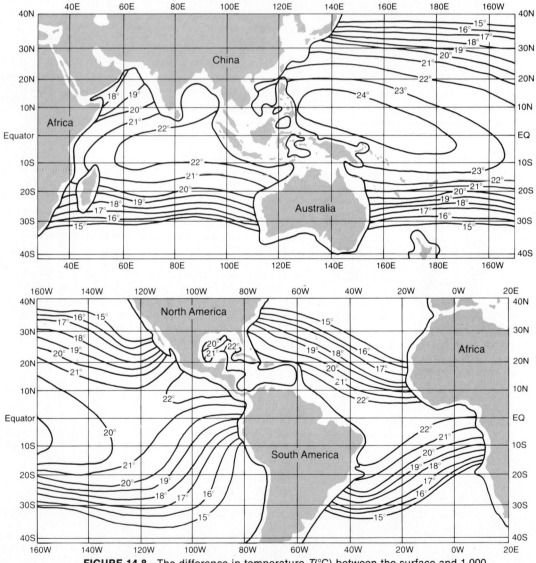

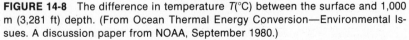

FIGURE 14-8   The difference in temperature $T(°C)$ between the surface and 1,000 m (3,281 ft) depth. (From Ocean Thermal Energy Conversion—Environmental Issues. A discussion paper from NOAA, September 1980.)

There still are some who question whether OTEC can ever work effectively since there are problems associated with the system; corrosion can damage and reduce the efficiency of heat exchangers, and marine organisms can foul the pipes and other components. The actual transmission of energy can also be a limiting factor. If the system is not a floating one, it would have to be anchored to the sea floor, which can cause other problems.

In 1979 a test OTEC system (called *mini-OTEC*) was successfully operated for over 125 hours off Hawaii. The device, built without federal funds, produced 50 kW of power, with 40 kW being used to operate the system and 10 kW being surplus and available for other uses. The device had a 609 m (2,000 ft) long pipe that was 56.1 cm (22.1 in.) in diameter. A more advanced system called OTEC-1 is presently being tested that has a 1 MW (1,000 kW) capacity. One difficulty with any OTEC system is the vast quantities of water it needs. A commercial 250 MW system would need 1,416 m^3 (50,000 ft^3) of water per second, which is equivalent to the flow of a large river like the Missouri.

There are also some legal questions associated with OTEC. For example, Who, if anybody, "owns" the temperature difference of the ocean water outside the territory of the coastal state? Also, the system itself will change the surface temperatures of the ocean, and this can, as we have already seen in Chapter 10, Figure 10-7, possibly cause some dramatic effects on climate.

The OTEC devices should be more valuable to tropical islands (such as Hawaii or Puerto Rico) or less-developed countries situated near the equator that are highly dependent on imported oil. Countries such as Japan, France, Sweden, and Holland, in addition to the United States, are also developing OTEC plans. Two U.S. laws were passed in 1980 that will affect OTEC development: One mandates that OTEC plants be producing 10,000 MW energy by 1999; the other establishes licensing regulations and provides some loan guarantees for OTEC plants.

ARTIFICIAL UPWELLING

One interesting sidelight of the OTEC scheme is that it can be used as an artificial upwelling system whereby nutrient-rich deeper water is brought to the surface where it can be used in an aquaculture or mariculture operation. If this technique could be incorporated into an OTEC system, then a biological product could also be produced and could reduce the operating costs of the energy-generating aspect. The nutrient-rich deeper water could be used to grow algae that, in turn, could be used to grow other organisms higher in the food chain. The cold water could also help moderate some of the high temperatures of the tropical regions where most OTEC systems would be located (Figure 14-9). This moderation could be especially effective near islands that are situated in major wind systems with large amounts of humid air. The cold water could then be used to recover moisture from the atmosphere by passing the cold seawater through a series of condensers placed in the flow of the moisture-laden winds. As the winds were cooled some

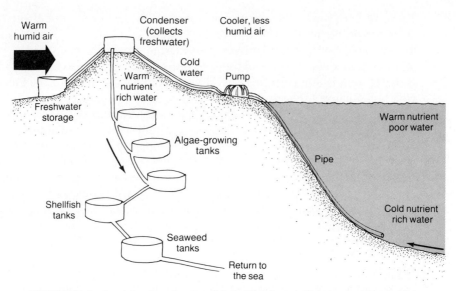

FIGURE 14-9 A schematic of a possible artificial upwelling mariculture system.

of the moisture would condense out and could be recovered and used for drinking water. Another possibility is that the cold water from an artificial upwelling system could be used to cool a nuclear power plant or some similar system.

ENERGY FROM MARINE BIOMASS

It should be clear from Chapter 12 and elsewhere that the ocean presents a vast potential for the growing of marine organisms. One suggestion is that certain organisms, plants and algae in particular, can be used as a source of energy. Under the right conditions, these organisms can grow very rapidly and after harvesting and treatment be converted into natural gas and other potential products.

One of the more commonly discussed species is the giant California kelp (*Macrocystis*). This alga can grow as much as a foot a day and is presently harvested as a source of various biological products. Natural gas and other products such as fertilizer or livestock feed can also be obtained from the physical, chemical, and biological breakdown of the kelp. Kelp or perhaps other types of plants could be incorporated into an artificial upwelling system or an OTEC system that would increase the nutrient supply to the plant and in theory help its growth. A model of such a system is shown in Figure 14-10.

Some have suggested that a marine farm could be developed in relatively shallow water that would produce food, fuel, and other products. Kelp, or a similar plant, could be harvested by cutting off the upper portions on a fairly regular basis; other organisms such as microscopic algae also could be used. It

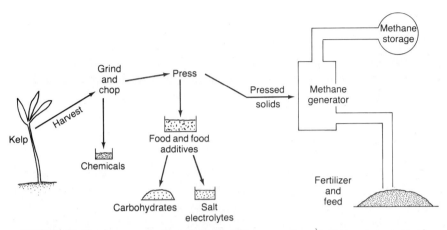

FIGURE 14-10 A conceptual model of a kelp processing plant showing some of the possible byproducts. (Adapted from Flowers and Bryce, 1977.)

might even be possible to develop these "farms" in ponds or small lakes using local waste products as a source of nutrients.

The concept of taking solar energy and the products that result via photosynthesis is fairly reasonable since photosynthesis itself is such a major source of energy on the earth. It is easy to get too optimistic about solar energy, which is renewable and essentially inexhaustible as a form of energy, but it should be realized that the efficiency of going from solar energy to a biological product and then harvesting and converting to energy is really very small. Thus although this process may work in some situations, it is not evident if these processes can ever lead to a major large energy supply.

OTHER SOURCES OF ENERGY

Some other ideas have been suggested for developing energy from the ocean; these include damming up large bodies of water, especially those partially restricted by land, in regions having a high evaporation rate. An example could be the southern end of the Red Sea, the Straits of Hormuz in the Persian Gulf, or the Strait of Gibraltar. The concept is simple: The high evaporation rate can reduce the water height on one side of the dam whereas the water on the other side, if it is connected to the open ocean, will remain at its original elevation. After a period of time, perhaps a year or so, there will be an elevation difference with the oceanic side being high and the other side being low. This difference could then be used as a hydrostatic head to drive power. This idea, although interesting, probably will not come to fruition because it would involve serious disruptions of ocean commerce.

An interesting plan to modify a small sea has been proposed by the government of Israel. The Dead Sea, the saltiest body of water in the world, situated more than 400 m (1,312 ft) below sea level, is drying out. It has been suggested that a canal be built connecting the Mediterranean to the Dead Sea. The water moving down to the Dead Sea could provide hydroelectric power as well as recharge the Dead Sea and maintain it as a source of potash. The program is controversial since it has ecological and political (Jordan also borders on the Dead Sea) problems and is still being discussed.

Freshwater from the Sea

Besides energy and food, the coming years will see other major problems on the earth. For example, many countries and regions will have continuing shortages of freshwater. This situation is especially common in many less-developed or underdeveloped countries. Two of the obvious places to look for freshwater are the ocean and ice caps or glaciers. As we have seen in Table 7-1, the ocean contains about 97.2 percent of the world's water supply and the glaciers and ice caps about 2.15 percent. These are extremely large percentages when you realize that all the water in freshwater lakes, rivers, and streams totals only about 0.01 percent. On the other hand, however, it is not that easy to go from saltwater to drinkable freshwater. But often the water needed is not principally for drinking but rather for agricultural and industrial use, in which case the water does not have to be as pure as for drinking. Desalination of seawater, although it involves a fairly complex technology, is really a simple process.

There are three principal types of desalination techniques: membrane, distillation, and crystallization or freezing. In the membrane process a thin sheet or membrane is used to separate saltwater from freshwater; membranes can selectively allow some substances to pass through it. Distillation techniques involve heating saltwater until some boils or vaporizes; the salts remain behind. The water vapor can then be cooled and condensed into freshwater. In freezing salt water the salts are left behind and the crystals are freshwater, which can then be melted. There presently are over 1,500 desalination plants in the world that each produce over 25,000 gal of freshwater per day; 350 are in the United States.

One of the alternatives to using desalinated or freshwater for agriculture is to develop agricultural products that can grow in salty or brackish waters. Some recent experiments have suggested this can be a reasonable possibility. Forms of barley have been developed that can grow in seawater, and some other crops have been developed that can grow in brackish waters.

Another innovative source of freshwater could be icebergs. They have received a lot of publicity—the idea being to tow larger icebergs (the best ones come from Antarctica) by some mechanism to the areas where water is most needed. A moderately large iceberg can contain over 1 million acre-feet of water, which could supply up to 10 million people and be worth $20 million. Such numbers compare very favorably with desalination techniques. One suggestion that has received considerable press has been a Saudi Arabian plan; the original idea probably came from Professor John Isaacs of the Scripps Institution of Oceanography who suggested it many decades ago. In 1977 a conference was held in Iowa called the First International Conference in Iceberg Utilization. The meeting was sponsored by a Saudi Arabian prince and was attended by over 200 scientists and representatives from over 18 different countries.

The movement of an iceberg from Antarctica involves all kinds of technical, environmental, and legal problems many of which have never before been considered. On the other hand, icebergs contain a fairly good quality freshwater, in some instances perhaps even better than typical drinking water. Other possible use areas besides Saudi Arabia could be California, Australia, and New Zealand; the latter two are much closer to Antarctica.

The mechanisms for moving such a large piece of ice and to prevent its melting are immense. The trip itself could take over a year. Tow speeds of icebergs, regardless of the technique of propulsion, would be less than 1.6 km per hour or 1 mi per hour. As an iceberg moved through the water it would produce interesting biological effects, would cool the water, and indeed could even affect weather, producing perhaps fog or rain. If and when the iceberg arrived at its destination, mechanisms would have to be developed for breaking it down and melting it. The Saudi Arabian who sponsored the conference was considering moving a 100 million-ton iceberg from Antarctica to Saudi Arabia, having it wrapped in cloth and plastic and towed by several ships. It was estimated that the trip would take 8 months and cost $100 million. It was also assumed that the mile-long iceberg would lose only about 20 percent of its mass in travel. If these calculations were correct, the water that could result would be cheaper than that of desalination. There are those, of course, who have a hard time visualizing towing an iceberg across the equator and who suspect that by the time one reached the equator, one would be towing nothing but a piece of rope. Nevertheless, it is an interesting idea.

Disposal of Nuclear Wastes in the Deep Sea

The ocean has always been an area where people have dumped their waste products. The present technological and economic development of the world has led to a new and unique waste disposal problem: the disposal and management of high-level waste products from the world's civilian and military nuclear power plants. The volume of this very toxic material continues to increase; some has an extremely long lifetime—up to 100,000 years or so. Basically, three types of options for disposal exist: (1) transportation into space, (2) transmutation or elimination of the dangerous components by nuclear processes into acceptable compounds, and (3) safe storage within the earth's environment. It appears at this time that the technology for the first two options is not dependable or adequate.

Shooting nuclear products into space may use as much energy as they themselves produce, and the risk of failure is fairly high. At present no process exists whereby the nuclear components can be changed into safe products. Thus the option of storage within the earth seems the most promising at this moment.

There are several possibilities within this option; one is to store nuclear waste on land in strong, secured, guarded containers. These containers might have to be maintained for a period of time of up to tens to hundreds of thousands of years; an extensive period of time considering that our recorded history is only about four thousand years.

A second possibility is to dispose of the material in subsurface rock or salt formations. This has, for many years, been a favorite option but recent investigations have shown that the rocks and salt are not as stable as anticipated and that there can be dangerous reactions and movements.

A third possibility is to put the material in canisters into the major ice sheets of the world. However, as was seen in Chapter 10, climatic conditions can change, and the possibility of these sheets' melting or moving is very high. The fourth possibility, and the one that is receiving more and more consideration, is to store the material below the deep-sea floor of the ocean.

None of the above possibilities is really acceptable; they all have environmental risks. The deep-sea option does have some advantages. For example, the deep sea is one of the least valuable pieces of real estate on the earth and probably has no significant fishing or oil and gas use. It could be mined for manganese nodules but the deep-sea region is so immense that numerous sites are available. The second and perhaps more important reason favoring the deep sea is that it is an extremely stable part of the ocean. This should not be confused with previous suggestions that radioactive material be placed in deep-sea trenches. Trenches are very unstable areas and zones of subduction with a high incidence of faulting and earthquake activity. Thus there is a good possibility of storage containers there being damaged and radioactive materials escaping into the environment.

The deep-sea portions being considered are the central parts of the major ocean plates (Figure 6-11), which have very slow sedimentation rates in the order of a centimeter (about 0.4 in.) or so per 1,000 years, low currents, and appear to be little influenced by major climatic or geologic conditions on earth. Several important questions still remain, in particular, Can the sediments themselves produce an efficient barrier to any radioactive material that may escape? And, How will the material be placed into the sea floor if, indeed, it is done? The anticipation is that holes will be drilled into the sea floor and canisters be emplaced (Figure 14-11).

Another question, and perhaps an equally important one, is, What are the legal aspects of placing radioactive material in the sea floor? The Law of the Sea Treaty presently being negotiated, although covering many aspects of the ocean, really does not address nuclear waste aspects. If marine sediments can act as an effective barrier to any radionuclei, the deep-sea option might become very appealing.

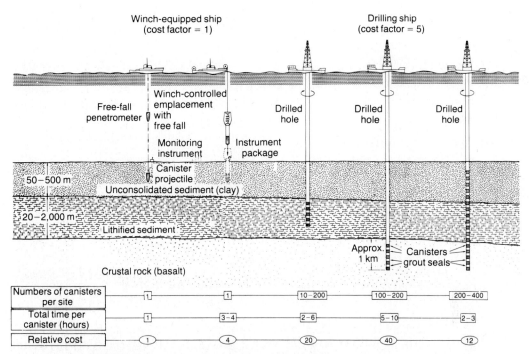

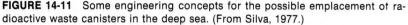

FIGURE 14-11 Some engineering concepts for the possible emplacement of radioactive waste canisters in the deep sea. (From Silva, 1977.)

Innovative Uses of Marine Organisms

It should be obvious that there are considerable differences that exist between the plants and animals of the ocean and those on land. Some of these differences present some interesting challenges and opportunities. In the ocean, for example, fishing is similar to hunting on land, which is generally inefficient. If mechanisms or techniques were developed whereby one could herd or track fish, then the catch per unit effort would certainly increase. The mechanism of using sound or other fish to attract fish has been attempted with moderate success.

Another technique is to develop procedures whereby certain fish, like salmon, return to specific localities. The principle here is that salmon, which breed in freshwater, are released early in their life to the ocean where they grow and fatten and eventually return to freshwater to breed. The salmon follow certain odors to their freshwater spawning grounds, and if they are exposed to a distinct chemical at the proper time in their growth, they will return to that particular area where the chemical is being released. In other words, they become imprinted with the smell of that chemical and will return to such waters, where they can be captured.

(a)   (b)

FIGURE 14-12 (a) A fish ladder at Newport, Oregon, used by the Weyerhaeuser Corporation. When the salmon return they will be attracted to this area because of the chemistry of the water. Note the bay in the background. (b) Coho salmon being harvested. The man is standing at the top of the fish ladder shown in (a).

Weyerhaeuser Corporation has developed a hatchery system that first uses waste heat to help salmon gain a year of growth. In this system salmon eggs are hatched, fed, and grown to 10 to 13 cm or about 4 to 5 in. in 6 months versus the 18 months or so that it would take outside or in a cold-water hatchery. The fish are kept for various periods of time, during which they are imprinted with the odors and location of a specific area to which they will return as adults (Figure 14-12). One of the advantages of having fish return to where they can be harvested is that they can be immediately caught and processed with no problems or delays associated with handling or transportation. An interesting aspect of this procedure is that when the fish are released they become public property, and the company has no control over them in the 2 to 5 years or so that they may be in the ocean; any fisherman may catch them.

Some marine animals appear to be quite intelligent—in particular, whales and porpoises. Most people have seen the tricks that these animals can perform in captivity. It has been suggested that perhaps they can be used to assist in the herding of fish or recovering devices lost on the sea floor. Porpoises and whales have already been trained to dive to depths of more than 500 m (1,640 ft) to return such items.

KRILL

Krill may be one of the more important biological products of the ocean. It is a group of crustaceans that has as many as 85 different species, the most common one is called *Euphausia superba* and averages about 2 to 4 cm (0.8 to 1.6 in.) in length (Figure 8-31). Krill is one of the more important foods for whales; however, with the reduction of the whale population, the krill stock seems to have increased considerably. Estimates of the amount of krill presently available in the ocean range between 500 million to 5 billion tons. It is found mainly in the Antarctic regions and some feel that it could be harvested at a rate of about 150 million tons a year (or almost 3 times the present total fish catch). Catches can be immense sometimes reaching as high as 60 tons an hour. One of the problems with krill is processing, as the organism deteriorates very quickly. It does contain, however, about 20 percent protein and a high amount of amino acids, and could be a valuable food.

The anticipation of the development and size of the krill fishery has been considerable and has been mentioned quite often in the press. In large part this is due to the leveling off of the world's fish harvest in the last few years to between 60 and 70 million tons per year.

Krill is an extremely important organism in the Antarctic food web; besides whales, seals, birds, squid, fish, and penguins feed upon it in one way or another. One thing that gives krill such a potential for high catches is that they tend to congregate in large swarms and therefore are relatively easy to catch. Actual fishing of krill has been done by the Japanese since the early 1960s and several products are available, including krill meat and krill protein concentrate. It is not evident that krill will ever really be so dramatic a product from the ocean as anticipated. The economics of catching the animals so far from the place where they are sold and the difficulty of preserving them and selling the product have increased the costs.

In spite of the growth of the krill fishery, there are still a lot of aspects that are unknown. For example, How much will it affect the other animals in the ecosystem? Ironically, people, who are the main killers of whales by hunting them, may also be affecting their growth by catching their prey.

Another important use for the animals in the sea is as a source of new and valuable drugs. In the past, several important drugs have been obtained from the ocean—in particular, seaweed as a source of iodine. The systematic search for drugs from marine organisms, however, is a relatively new phenomenon. It is based on the fact that many marine organisms are toxic; therefore, they have the potential for use in drugs since poisons usually indicate strong potential physiological activity. The diversity of marine life also makes it an especially attractive place to look for organisms since most land species have already been examined.

Many types of algae and seaweed are already used in cosmetic, drug, and food industries. The brown algae is used to produce alginates, and red algae, to produce agar and carrageenan. Calcium alginate can reduce bleeding and is used in some dental operations. Agar can be used for the growth of bacteria in cultures and other things. It has also been suggested that marine organisms be used to extract certain elements from seawater since some marine animals concentrate trace elements from the ocean into their skeletons or tissues by factors thousands of times the concentration of the element in seawater.

Other Innovative Ideas

SATELLITES

Satellite technology holds great promise for oceanography. The ability of these devices to constantly orbit the earth and take pictures or make measurements, or to remain in one place and watch one area constantly provides an excellent opportunity to observe many oceanographic phenomena.

One device, the Earth Resource Technology Satellite (*ERTS*) now called *LANDSAT*, revolves around the earth 18 times each day taking pictures every 103 minutes in an almost polar orbit of about 497 mi or 800 km in height. Each strip is contiguous to that of the previous day, with a small overlap. Therefore, every 18 days the satellite passes over every location on the earth at essentially the same time with similar lighting. A more recent satellite called *SEASAT* was the first satellite designed for oceanography but unfortunately lasted only 89 days before failing, but it still gave some extremely valuable information about the ocean including data on its waves, surface slopes, and temperature.

Pictures from satellites can be used for such things as measuring or monitoring dumping and oil pollution, looking for minerals on land, studying structural relationships, locating schools of fish, or detecting and measuring coastal zone erosion (Figures 3-17, 6-13, and 6-14). Areas of biological productivity and major schools of fish can sometimes be noted by differences in water color or by detecting films of fish oil or algae. A major advantage of using satellites is that they cover a large area in a very short time. A research vessel can make only one measurement at a time and usually travels no faster than 12 kn.

MARINE ARCHAEOLOGY

The possibility of finding a sunken wreck or buried treasure has attracted many people to the sea, and it is common to read of such discoveries by amateur skin divers or professional treasure hunters. For example, a very active group, at least judging by reports in the press, has been exploring for the *Titanic* which was lost in 1912 some 556 km or 300 mi off Newfoundland. This supposedly unsinkable vessel sank on its maiden voyage after a collision with an iceberg. A 6-week expedition, using magnetometers (to detect large pieces of iron) and echo sounders, apparently detected several possible areas where the vessel might be situated; but no photographs or reliable evidence of such a find has yet been presented.

There is another type of person, the professional marine archaeologist, who rather than looking for treasure is interested in how man lived in ancient times. The proper use of underwater archaeological techniques requires the same amount of skill and effort as does land work. The modern marine archaeologist still uses hearsay or old sea tales to find prospective sites, but more often depends on modern techniques such as side-scan sonar (Figure 3-32), echo sounding, metal detectors (magnetometers), and even submersibles. Often a site is excavated using a system of grids on the sea floor for reference and mapping (Figure 14-13). Every

(a)

(c)

(b)

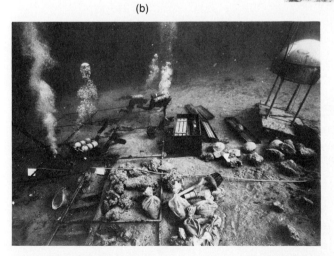

FIGURE 14-13 A series of photographs of divers associated with the Institute of Nautical Archaeology (Texas A & M) working on an eleventh century nautical site off Turkey. (a) Expedition barge anchored above the site. (b) Divers working on the site. Note the grid that marks the area. (c) Divers carrying small sections of the hull to the surface. (Photographs courtesy of Institute of Nautical Archaeology, Texas A & M University.)

item collected is photographed while still on the sea floor. Excavation can be by an air lift, which sucks up bottom sediment; the sediment can then be screened to detect artifacts—large items are uncovered but remain in place by this technique. One of the world's leaders in the field of underwater archaeology is Dr. George Bass who is president of the Institute of Nautical Archaeology at Texas A & M University (Figure 14-13).

One unique problem with underwater artifacts is their preservation once exposed to the air. Wood, for example, can quickly dry out and be damaged. In many instances the wood may have been destroyed or made very fragile by worms. For other materials, especially some metals, chemical reactions can cause considerable damage, often because seawater is a good electrolyte. Gold, fortunately, is not very active in an electrolyte solution, and gold coins often survive marine burial fairly well. Another problem is diving limitations. Extended work by free divers beyond depths of 45 m (about 150 ft) can be very dangerous. Working from diving bells or submersibles can increase the range and time spent on the bottom. Still another problem concerning underwater archaeological sites is the fact that there are few laws covering such regions and once one is found, they are fair game for anybody.

An exciting, recent underwater archaeological event concerned the attempted raising of the U.S.S. *Monitor*. The vessel, launched in 1862, was an ironclad vessel built for use in the Civil War. It fought only one battle before

FIGURE 14-14 Drawing of rescue operations being made aboard the U.S.S. *Monitor* on December 31, 1862. The vessel sank during a storm off the North Carolina coast while being towed by the U.S.S. *Rhode Island* (background). (Photograph courtesy of Naval History Division, Department of the Navy.)

being lost while under tow (Figure 14-14). It was discovered in 1973 by scientists aboard the Duke University research vessel *Eastward* in waters about 64 m (210 ft) deep off Cape Hatteras (Figure 14-15). After a series of investigations, it was decided that it would be too risky to attempt to salvage the vessel. Dives and underwater photographs showed the ship's remains to be very fragile and thin and that raising the vessel would probably destroy it. The site has been designated a marine sanctuary area and underwater operations are under strict control.

UNDERWATER HABITATS—OFFSHORE ISLANDS OR PLATFORMS

The vast size of the ocean suggests that it may be a suitable place for people to live and work. Underwater habitats have been placed at relatively shallow depths (usually less than 183 m or 600 ft) on the sea floor and women and men have lived and worked there for varying lengths of time.

With the probable development of OTEC and the continued exploration for oil and gas, offshore habitats will become more and more commonplace in the

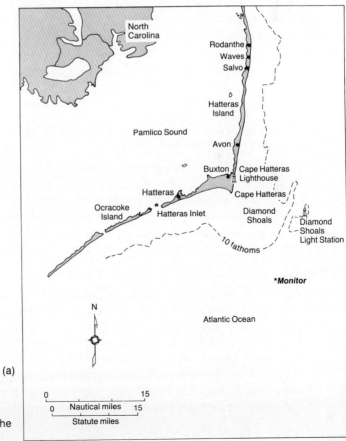

(a)

FIGURE 14-15 (a) Location of the U.S.S. *Monitor*.

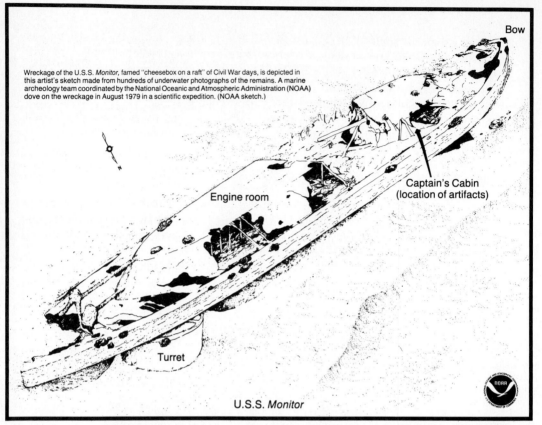

Wreckage of the U.S.S. *Monitor*, famed "cheesebox on a raft" of Civil War days, is depicted in this artist's sketch made from hundreds of underwater photographs of the remains. A marine archeology team coordinated by the National Oceanic and Atmospheric Administration (NOAA) dove on the wreckage in August 1979 in a scientific expedition. (NOAA sketch.)

Bow

Captain's Cabin
(location of artifacts)

Engine room

Turret

U.S.S. *Monitor*

FIGURE 14-15 (b) An artist's sketch of the wreckage of the U.S.S. *Monitor*, based on numerous underwater photographs. Much of the data was obtained during a series of dives coordinated by NOAA. (Photographs courtesy of National Oceanic and Atmospheric Administration.)

future. Often one sees reference to these places as habitats for individuals to live in when space on land is used up. This is probably an unreasonable use since it will be very expensive and inconvenient to live offshore. Nevertheless, offshore platforms could be used for nuclear power stations, landing strips, fish processing factories, recreational facilities, or still unthought of ideas (see chapter-opening photograph).

Summary

The future uses of the ocean are limited only by one's imagination. Innovative uses of oceanographic processes and characteristics have been proposed for new sources of energy. These include using tides, waves, ocean currents, marine biomass, and energy derived from the temperature differences between surface and deep waters. This last idea, often called OTEC (Ocean Thermal Energy Conversion), does appear to offer much promise. One small pilot plant has already been built and several larger ones should be operational within the coming years. However, because of the need for considerable differences in surface and deep-water temperature, the device will probably be limited to tropical or near-tropical areas. An OTEC device could be combined with an artificial upwelling system that would reduce the cost of the electricity produced as well as produce a biologically useful byproduct.

Obtaining freshwater from the ocean is another important use; one especially innovative idea is to tow icebergs to areas of possible need. The concept, although interesting, has not yet been shown to be feasible.

The ocean, in recent years, has been suggested as a site for the disposal of nuclear wastes. As unappealing as this may sound there really are few options for the safe storage of the large volumes of nuclear waste material.

The biological resources of the ocean might be increased by innovative approaches. These can include methods whereby fish are trained or imprinted to return to certain areas where they can be caught. Other possibilities include the harvest of new organisms, such as krill, and new methods of aquaculture.

Satellite monitoring of the ocean is a technique that holds considerable promise for the future. These devices could be used to detect areas of biological productivity, coastal changes, and pollution input and effects.

Marine archaeology involves not just the amateurs looking for lost treasure but also the professional trying to learn about ancient history. Several recent expeditions have obtained much data about ancient ships and civilizations.

Suggested Further Readings

AUBURN, F. M., "Deep Sea Archaeology and the Law," *International Journal of Nautical Archaeology and Underwater Exploration*, **2.1** (1973), pp. 159–62.

BASCOM, W., "The Disposal of Waste in the Ocean," *Scientific American*, **231**, no. 2 (1974), pp. 16–25.

FAULKNER, D. J., "The Search for Drugs from the Sea," *Oceanus*, **22**, no. 2 (1979), pp. 44–50.

FRALICK, R. A., and J. H. RYTHER, "Uses and Cultivation of Seaweeds," *Oceanus*, **19**, no. 4 (1976), pp. 32–39.

FREY, D. A., "Deepwater Archaeology," *Sea Frontiers*, (July/August, 1979), pp. 194–203.

MACLEISCH, WM., ed., "Harvesting The Sea," *Oceanus*, **22**, no. 2 (1979), 72 pp.

MACLEISCH, WM., ed., "High-Level Nuclear Wastes in the Seabed?" *Oceanus*, **20**, no. 1 (1977), 67 pp.

MACLEISCH, WM., ed., "Marine Biomedicine," *Oceanus*, **19**, no. 2 (1976), 60 pp.

MACLEISCH, WM., ed., "Ocean Energy," *Oceanus*, **22**, no. 4 (1979/80), 68 pp.

RICHARDS, A. F., "Extracting Energy from the Oceans: A Review," *Marine Technology Society Journal,* **10** (1976), pp. 5–21.

ROSS, D. A., *Opportunities and Uses of the Ocean*. New York: Springer-Verlag, 1980, 320 pp.

VON ARX, W. S., "Prospects: A Social Context for Natural Science," *Oceanus*, **22**, no. 4 (1979), pp. 3–11.

15

The Law of the Sea

FOR CENTURIES THE OCEAN, except for a narrow strip near the shore, was considered to belong to nobody but rather to be available for everyone's use. In recent years new uses of the ocean, such as mining and fishing its resources or dumping waste products into it, have encouraged many countries to declare extensive territorial claims to the sea. It is very probable that within a few years at least 37 million n mi^2 of the ocean floor will be under the jurisdiction of the adjacent state. This is an area equivalent to about 87 percent of the land area of the earth.

Before describing the present legal regime of the ocean, a brief history is given of how the law of the sea has developed.

Early History of the Law of the Sea

FROM EARLY HISTORICAL TIME TO THE TRUMAN PROCLAMATION

Certain freedoms and regulations of the sea have been understood almost since the time that people began to sail the ocean. The first significant legal statement, however, concerning the law of the sea is usually credited to a Dutch lawyer, Hugo Grotius, who in 1609 published his *Mare Liberum*, or concept of the freedom of the seas. Basically, Grotius said that nations are essentially free to use the sea for whatever purpose they wish as long as this use does not interfere with another nation's use of the sea. By this time in history it was also realized that a coastal state (country that has a shoreline) should have some control over the ocean immediately adjacent to its coast. This led to the definition of a zone called the *territorial sea* in which the coastal nation had complete sovereignty. This zone was established as being as wide as the distance from shore that a cannon could fire, which then was about 3 mi (about 5.5 km). The 3-mi-wide territorial sea has never been officially ratified by an international agreement.

The area outside the territorial sea, called the *high sea*, was considered as *res nullius*, or belonging to no one. In general, international law protected the rights of all countries to use all parts of the high seas in essentially whatever manner or purpose they chose. This concept was not really challenged prior to World War II since the interests of the countries that were using the sea, at least during peace time, tended to coincide.

After World War II, an increasing interest in the ocean developed in part because of some actions by the United States and in part because of a new awareness that some of the resources of the sea, such as fish, were not inexhaustible and that some minerals on or under the sea floor might be extremely valuable.

The first major challenge to the 3-mi territorial sea concept was by U.S. President Harry S Truman in 1945 in what is now called the Truman Proclamation on the Continental Shelf. In the proclamation he established a national policy with respect to the natural resources of the subsoil and seabed of the continental shelf—a much wider area than the 3 mi territorial sea. The proclamation stated that the United States felt

> that the exercise of jurisdiction over the natural resources of the subsoil and seabed of the continental shelf by the contiguous nation is reasonable and just, since the effectiveness of measures to utilize or conserve these resources would be contingent upon cooperation and protection from the shore, since the continental shelf may be regarded as an extension of the landmass of the coastal nation and thus naturally appurtenant to it, since these resources frequently form a seaward extension of a pool or deposit lying within the territory, and since self-protection compels the coastal nation to keep close watch over activities off its shores which are of the nature necessary for utilization of these resources. [Thus] the Government of the United States regards the natural resources of the subsoil and seabed of the continental shelf beneath the high seas but contiguous to the coasts of the United States as appertaining to the United States, subject to its jurisdiction and control.[1]

One of the motivations for such an action at that time was the "long range worldwide need for new sources of petroleum and other minerals."[2] Thus, in 1945 some leaders had already anticipated the energy problems of the 1970s.

Essentially, what President Truman did was to claim ownership for the United States of all seabed resources on or under its entire continental shelf and suggest that some sort of international agreement be reached concerning regulation of fisheries and conservation. These proclamations, of course, caused considerable controversy and had some unanticipated results. Lost in much of the discussions was the fact that the United States was not really changing its 3-mi territorial sea position. But, in claiming ownership of part of the sea floor, it obviously was influencing some of the basic freedoms of the ocean.

---

[1] U.S., *Policy of the United States with Respect to the Natural Resources of the Subsoil and Seabed of the Continental Shelf,* Proclamation No. 2667, 1945.
[2] Ibid.

The Truman Proclamation also caused some problems between U.S. coastal states and the federal government about offshore mineral claims. Prior to 1945 the individual states of the United States believed that they had title to the lands immediately off their coasts. In fact, oil exploration was made in the 1880s off California without any federal control and, initially, without even any state control. Since the Truman proclamation claimed federal ownership of seabed mineral resources, a series of legal battles started between some states and the federal government. In 1947, the Supreme Court decided (*United States* v. *California*) that the federal government had full dominion and power over the lands, minerals, and other things lying seaward of the ordinary low-water mark on the coast of California and outside the inland waters. Subsequent cases between the United States and Texas and the *United States* v. *Louisiana* also reaffirmed the federal government's ownership of the submerged land off these states.

The U.S. Congress, which is more sensitive to states' interests, changed the Court's actions by two congressional acts passed in 1953. The Submerged Lands Act gave the states title to the lands and resources out to a distance of 3 geographic miles in the Atlantic and Pacific Oceans and out to 3 leagues or 10.5 statute miles in the Gulf of Mexico. Within this zone, however, the federal government retained its rights concerning navigation, commerce, national defense, and international affairs. This act defined the inner continental shelf as extending from the low-water line to the end of the 3-mi territorial sea. The outer continental shelf is the rest of the continental shelf seaward of the territorial waters. The second act, the Outer Continental Shelf Lands Act, gave control of the seabed and subsoil of the outer continental shelf to the federal government.

Later several Gulf states challenged the width of their title, which was based in part on historical claims when they joined the Union. Another question was whether their claim should start from offshore islands or from the coast. Florida (west coast only) and Texas won boundaries that extend 9 mi (about 16.6 km) from the coast, whereas Louisiana, Mississippi, and Alabama received only 3 mi (about 5.5 km). The courts also ruled against states' starting their claims other than from mean lower low water.

The 1958 and 1960 Geneva Conventions on the Law of the Sea

The Truman Proclamation of 1945 actually did not define the limits of the continental shelf although a White House press statement released on the same day as the proclamation said generally that submerged land that is contiguous to the continent and covered by no more than 100 fathoms (600 ft, or 183 m) of water, is considered as the continental shelf. The same press release said that the Truman Proclamation does not abridge the right of free and unimpeded navigation of waters of the character of high seas above the shelf, nor does it extend the present limits of the territorial waters of the United States. Nevertheless, many saw this as a challenge to the existing regime, and disclaimers in a press release were not very effective.

The Truman Proclamation was treated with suspicion by some countries, by indifference by others, and as an opportunity to adjust supposed injustices by still other countries. The last group included several South American countries that had few offshore resources (and a narrow shelf) and were anxious to protect what they had. Chile, Ecuador, and Peru in 1952 extended their jurisdiction and sovereignty over the sea, seabed, and subsoil out to a distance of 200 mi (370 km) from their coastline; in other words, they were declaring a 200-mi territorial sea. Their move differed considerably from the Truman Proclamation in that they included the sea surface and water as well as the ocean bottom and thus dramatically extended their jurisdiction. Peru based its claim on its important guano industry (bird droppings used for fertilizer) on islands off its coast.

The unilateral extension of the territorial sea by these and eventually other countries clearly indicated then that an international convention concerning the sea was necessary. This occurred with the 1958 and 1960 Geneva Conferences on the Law of the Sea. These two conferences, which included representatives from 86 countries, resulted in the following four basic conventions that were eventually approved by many of the participants:

1. Convention on the Territorial Sea and the Contiguous Zone.
2. Convention on the Continental Shelf.
3. Convention on the High Seas.
4. Convention on Fishing and the Conservation of the Living Resources of the High Seas.

CONVENTION ON THE TERRITORIAL SEA AND CONTIGUOUS ZONE

The Convention on the Territorial Sea and Contiguous Zone established the point that "the sovereignty of the state extends beyond its land territory and its internal waters to a belt of sea adjacent to its coast, described as a territorial sea."[3] The landward baseline for measuring the breadth of a territorial sea was defined as "the low-water line along the coasts as marked on large-scale charts officially recognized by the coastal state."[4] Unfortunately, the convention did not give a distance for the width of the territorial sea. Thus it avoided one of the key issues for having the conference.

The contiguous zone was defined as an area of high seas contiguous to a coastal state's territorial sea and according to the convention, it "may not extend beyond 12 mi (22.2 km) from the baseline from which the breadth of the territorial

---

[3] 1958 Convention on the Territorial Sea and the Contiguous Zone—Convention on the Law of the Sea Adopted by U.N. Conference at Geneva, 1958.
[4] Ibid.

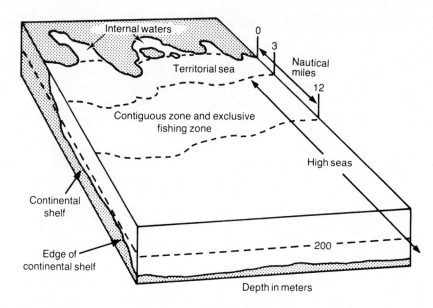

FIGURE 15-1 General divisions of the ocean after the 1958 Geneva Convention.

sea is measured."[5] Within this zone the coastal state can control fishing by foreign states. Thus although the breadth of the territorial sea was not actually defined in this convention, it can be argued that because of the 12-mi limit to the contiguous zone, the territorial sea cannot be more than 12 mi wide (that is, a contiguous zone of 0 mi in width). For many states a 12-mi territorial sea covered only a portion of their continental shelf.

The coastal state within its internal waters and territorial sea has essentially the same judicial authority as it has in its land territory. It can, if it wants, exclude foreign and alien ships, except for the right of foreign ships to innocent passage within the territorial sea. Passage is considered "innocent" if it does not harm the peace, good order, or security of the coastal state.

After the Geneva conferences, the countries of the world who agreed to the conventions generally divided their waters into three or four main categories (Figure 15-1).

1. *Internal Waters:* This includes rivers, lakes, and all waters landward of the inner limit of territorial seas (low-water line) as well as estuaries, harbors, and bays between the coast and offshore islands. In these waters the coastal state has complete sovereignty, except that if these waters had once been territorial or high seas in which case the right of innocent passage or other historical rights may be allowed.

2. *Territorial Seas:* This is a zone of undefined width, although most states then claimed about 12 mi. The rights of the coastal state are similar to those listed under internal

---

[5] Ibid.

496

waters except that foreign states have the right of innocent passage, innocent overflight, and entry under distress conditions.

3. *Contiguous Zone and Exclusive Fishing Zone:* This zone does not extend beyond 12 mi from where the territorial sea starts but, except for fishing and a few other restrictions, the contiguous zone is part of the high sea.

4. *High Seas:* The high seas are defined as all waters not included within territorial or internal waters. Within the high seas almost complete freedom exists.

Most states in the late 1950s and early 1960s accepted a 9-mi exclusive fishery contiguous zone combined with a 3-mi territorial sea, for a total of 12 mi. But some countries still claimed a 200-mi width to their territorial sea.

CONVENTION ON THE CONTINENTAL SHELF

The Convention on the Continental Shelf defined the shelf as "the seabed and subsoil of the submarine areas adjacent to the coast but outside the area of the territorial sea to a depth of 200 meters, or beyond that limit to which the superadjacent waters admits of the exploitation of the natural resources of the said areas (and) to the seabed and subsoil of similar submarine areas adjacent to the coasts of islands."[6] This clause is one of the more critical ones of the four conventions in that it states that the coastal state could claim that part of the continental shelf beyond 200 m (656 ft) that it is capable of mining. This is commonly referred to as the so-called *exploitability clause.* This convention also was a failure in part since it did not actually define where the continental shelf ends.

The Convention on the Continental Shelf gave the coastal state sovereign rights over the continental shelf "for the purpose of exploring and exploiting natural resources."[7] Natural resources consisted

> of the mineral and other nonliving resources of the seabed and subsoil together with living organisms belonging to sedentary species, that is to say, organisms which, at the harvestable stage, are immobile on or under the seabed and are unable to move except in constant physical contact with the seabed or subsoil. [It also stated that] the exploitation of its natural resources must not result in any unjustifiable interference with navigation, fishing, or the conservation of the living resources of the sea, nor result in any interference with fundamental oceanographic or other scientific research carried out with the intention of open publication.[8]

The question of scientific research is of course an important one for oceanography. On this point the convention stated that "the consent of the coastal state shall be obtained in respect of any research concerning the continental shelf

---

[6] 1958 Convention on the Continental Shelf—Convention on the Law of the Sea Adopted by the U.N. Conference at Geneva, 1958.
[7] Ibid.
[8] Ibid.

and undertaken there. Nevertheless, the coastal state shall not normally withhold its consent if the request is submitted by a qualified institution with a view to purely scientific research into the physical or biological characteristics of the continental shelf, subject to the proviso that the coastal state shall have the right, if it so desires, to participate or to be represented in the research and that in any event the results shall be published.''[9] This consent requirement concerning freedom of scientific research was later to become an obstacle to some research programs. This point will be discussed in a later section.

CONVENTION ON THE HIGH SEAS

The Convention on the High Seas defined the term "high seas" as "all parts of the sea that are not included in the territorial sea or in the internal waters of a state.''[10] It stated that the high seas are open to all countries and they can be freely used for things like navigation, fishing, and overflight.

CONVENTION ON FISHING AND CONSERVATION OF LIVING RESOURCES OF THE HIGH SEAS

The Convention on Fishing and Conservation of Living Resources of the High Seas mainly considered the biological resources of the ocean and stated that all countries have the right for their nationals to fish on the high seas, subject to past treaties and to articles adopted at the conference. It also said that the different countries should cooperate in conserving the living resources of the high seas.

SHORTCOMINGS OF THE 1958 AND 1960 GENEVA CONVENTIONS

The results of the 1958 and 1960 Conventions led in later years to controversy on several important aspects, especially on three points

1. The lack of definition of the width of the territorial sea.
2. The 200-m outer limit of the continental shelf.
3. The exploitability of marine mineral resources.

The failure to define the width of the territorial sea encouraged some coastal states to establish their own width, which in some instances resulted in claims of widths on the order of hundreds of miles. States that did not ratify the conventions, of course, were not bound by it.

The second controversy was that the convention's definition of the conti-

---

[9] Ibid.
[10] 1958 Convention on the High Seas—Convention on the Law of the Sea Adopted by the U.N. Conference at Geneva, 1958.

nental shelf usually did not coincide with the geological or oceanographic definition of the shelf. The use of the 200-m (656-ft) depth contour is an arbitrary one, and in most instances the edge of the continental shelf is shallower and in fewer instances, deeper. A shelf defined by depth was not acceptable to countries with very narrow shelves, such as Peru and Chile, nor to countries that have shelves that extend for hundreds of miles but parts of which are more than 200 m deep.

The third controversy resulting from the conventions was a result of the ambiguity caused by the exploitability aspect included in the definition of the continental shelf, which said that the shelf extended to "a depth of 200 meters or, beyond that limit, to where the depth of the superadjacent waters admits of the exploitation of the natural resources of the said areas."[11] The exploitability concept left the ownership of the sea floor open to numerous interpretations including that an entire ocean could be divided among the countries bordering it (Figure 15-2) or that the United States or any other country that had the capability could mine its shelf, continental slope, continental rise, ocean basin, ocean ridge, and perhaps right across and up onto the other side of the ocean. The United States has accepted this exploitability concept in that it has leased areas for oil exploration and exploitation in water depths greater than 200 m off California and over 100 mi off the coast in the Gulf of Mexico. The exploitability clause was not so important to the conference delegates in 1958 and 1960 as it became in the 1970s and 1980s when the promise of exotic mineral and biological resources from the ocean caused many countries to show considerable interest in exploiting them.

In one respect, these conferences were important because it was one of the last times that the high seas would be considered as an area of unqualified freedom. Following the conference, over 100 nations eventually indicated that they have some jurisdiction over the minerals in the areas adjacent to their coasts. These claims have taken many forms, including individual and unilateral declarations, domestic legislation, treaties, offshore concessions, or ratification of the 1958 Convention on the Continental Shelf.

One irony to these conferences was that the United States, perhaps realizing the Pandora's box it had opened by the 1945 Truman Proclamation, argued strongly for a 3-mi territorial sea but was not successful. In the second conference, held in 1960, the United States supported a 6-mi-wide territorial sea but it failed to carry by one vote.

EVENTS LEADING TO THE THIRD LAW OF THE SEA CONFERENCE

The years immediately following the 1958 conference were essentially a transition period until the conventions were ratified by the various governments (22 were needed) in the early 1960s. During this time, however, marine technology made

---

[11] 1958 Convention on the Continental Shelf—Convention on the Law of the Sea Adopted by the U.N. Conference at Geneva, 1958.

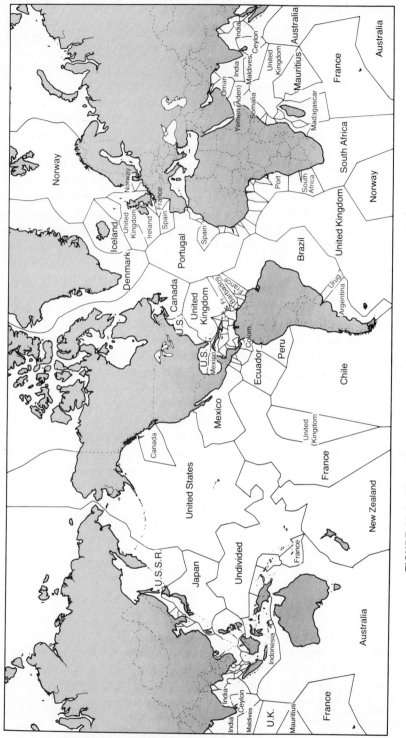

FIGURE 15-2 Illustration showing how the sea floor could look if it was divided along lines equidistant from the closest points of adjacent or opposite coastal states in 1967. This could have been a division of the sea floor based on the exploitability concept. (Adapted from Christy and Herfindahl, 1967. Base chart is H.O. 1262F.)

considerable advances so that portions of the seabed beyond 200 m actually became available for mineral exploration or, in other words, the "exploitability clause" became more realistic. Among these technological advances were the capability for offshore drilling for oil (Figure 12-1) and the developing technology to mine and refine the manganese nodules on the sea floor (see Chapter 12). There were also considerable expansion and mechanization of fishing fleets, allowing them to stay at sea for months and process their catch using large support ships. Thus the legal vacuum created by the 1958 conference on the ownership of resources beyond the 200-m depth was starting to become critical.

Although the United States and other maritime nations were concerned about the extension of claims into the ocean, it was a representative of the small country of Malta who provided a dramatic impetus toward a reconsideration of ocean policy. In 1967, Malta proposed that the U.N. General Assembly consider the development of a treaty that would reserve the seabed and ocean floor beyond the limits of national jurisdiction (the territorial sea) and their resources in the interest of humankind. This proposal and a subsequent speech by the then Maltese Ambassador, Dr. Arvid Pardo, noted the rapid marine technological advances by the developed countries that could cause the ocean floor and its resources to be subjected to further national claims and appropriations. He suggested that the ocean floor outside of national jurisdiction be reserved for peaceful uses and that its resources become the *common heritage of mankind*. The financial resources derived from such uses could be used to further the development of less-developed countries. This proposal, which was somewhat controversial, did obtain almost unanimous support. An Ad Hoc Committee initially consisting of 35 nations was formed in 1967 to study the peaceful uses of the seabed and ocean floor beyond the limits of national jurisdiction.

The U.N. Ad Hoc Seabed Committee met several times in 1968 and 1969 without reaching any major agreement. Finally in December 1969, after considerable debate and opposition by the United States and the Soviet Union, the United Nations adopted a resolution (65 to 12 with 30 abstentions) to convene

A conference on the law of the sea to review the regimes of the high seas, the continental shelf, the territorial sea and contiguous zone, fishing and conservation of the living resources of the high seas, particularly in order to arrive at a clear, precise, and internationally accepted definition of the seabed and ocean floor which lies beyond national jurisdiction, in the light of the international regime to be established for that area.[12]

The Seabed Committee was also expanded at that time to include a total of 86 members (later the People's Republic of China and four other countries joined for a total of 91 members). The U.S. reason for objecting was apparently to avoid a broad-scale conference but rather to try to reach an international consensus by

---

[12] U.N. Document A/2574A (1969).

unilateral declarations of policy. The U.S. Congress at that time was also not enthusiastic for a new convention since it felt that the Geneva conferences were sufficient concerning the U.S. ownership of resources on the continental margin. In effect, the Congress was supporting the exploitability clause, which was not surprising since the United States was the most marine technologically advanced country at that time.

The conference was to start in 1973 with the Seabed Committee serving as the Preparatory Committee for the Conference. The committee was to be responsible for the agenda, location, and date of the meeting. In March 1971, at its 45th meeting, the Seabed Committee divided itself into three subcommittees in preparation for the conference. Subcommittee I was to be concerned with the seabed area beyond national jurisdiction and what would be the powers and functions of any international regime that administered this area. Subcommittee II was to prepare a list of issues for the conference to consider including width of the territorial sea, regime of the high seas, international straits, and fisheries. Subcommittee III was concerned with protection of the marine environment and scientific research.

These three committees were to consider principles and arguments and present draft treaties for consideration by the conference. Many countries eventually submitted draft articles on various subjects, and it was hoped that this exchange of ideas would lead to compromises that would be satisfactory to all or most of the participants. This did not turn out to be the case; in fact, not one single major draft treaty was agreed upon before the conference started. Actually, the large ranges of interests and desires among the participants in the conference indicated that agreement would be difficult. In addition, the Third Law of the Sea Conference was to have almost more than twice the number of participants than the earlier two conferences. More important is that many of the new countries were developing nations who wanted a change in the old freedom-of-the-sea concept. Simply said, these developing countries felt that the old systems allowed the few developed countries to dominate the ocean and its resources. There was also considerable interest among these states in establishing an economic resource zone, perhaps as much as 200 nautical miles wide, where the coastal state had almost if not all control over the marine resources.

The Basic Issues of the Third Law of the Sea Conference

The important issues of the Third Law of the Sea Conference are as follows:

1. A definition of the width of the territorial sea and any other adjacent zones and the degree of coastal state control in these zones.
2. Ownership of the resources of the water, seabed, and subsoil. This item has two main parts: the resources within the area under the coastal state's jurisdiction (and thus the question of the extent of this jurisdiction) and the resources in the area beyond national jurisdiction or the deep or high seas.

3. Right of overflight and navigation through what are now international straits but which could be included within expanded territorial seas or an economic resource zone.

4. Management of living resources in the ocean, especially those species that migrate and those in areas where many countries have traditionally fished for them (such as Georges Bank) but which may now come under a single country's jurisdiction.

5. Protection and reduction of pollution in the ocean.

6. Freedom of scientific research in the ocean.

7. A regime for control or management of the high seas.

The conferences started in 1974 and have continued into the 1980s (see Table 15-1).

The U.S. position on these issues has varied with time but basically it will accept extended coastal state jurisdiction over resources if certain other considerations are met. These include, among other things, the prevention of unreasonable interference with other uses of the sea such as navigation rights, mechanisms for the mining of marine resources and sharing of the derived revenues with other nations, and a compulsory method of settling disputes.

The United States has also indicated its willingness to adopt a 12-mi (22.2 km) territorial sea if there is an international guarantee of free and unimpeded passage through international straits. This is an extremely important point since there are 116 straits that have a high-sea passageway under a 3-mi territorial sea (since they are wider than 6 mi—that is, 3 mi for each bordering country) that would become entirely territorial seas if there was a change from a 3- to a 12-mi territorial sea (since they are less than 24 mi wide). These include some very important straits such as Gibraltar, Malacca, and Bab el Mandeb (southern end of the Red Sea). If these straits were to become territorial seas with no guarantee of unimpeded passage, countries would have only the right of innocent passage through them. This would require submarines to travel on the surface through them and would not allow military planes to fly over them. This could severely limit the U.S. (or Soviet) military position around the world. The strait issue is also important to oil-importing countries, such as Japan, whose maritime trade could be seriously affected by unreasonable transit regulations by coastal states through whose straits their ships had to pass.

The probable new divisions of the sea floor are a 200-nautical mile exclusive economic zone (EEZ) and a territorial sea of 12 mi width. In certain situations the coastal state could also control marine resources and scientific research out to an amazing distance of 350 n mi (648 km) from their coast. With the establishment of just a 200-n mi economic zone, 37,745,000 n mi^2 (about 128,000,000 km^2) will come under the jurisdiction of coastal states. These numbers represent a considerable portion of the ocean or of the land (about 36 percent of the ocean). It should be emphasized that each of these depths or distances generally has no real significant oceanographic or geologic meaning associated with them. Many countries have continental shelves shallower than 200 m and several are deeper.

TABLE 15-1   Chronology of the Law of the Sea Conferences

1958	First U.N. Conference on Law of the Sea (LOS)—New York Adoption of the four underlying conventions
1960	Second U.N. Conference on LOS (UNCLOS)—New York Failed to agree on width of territorial sea
1970	U.N. General Assembly Declaration of the common heritage of mankind Resolutions to start UNCLOS III
1973	Third U.N. Conference on LOS
1973	First Session—New York Organizational meeting, list of issues and subjects to be negotiated
1974	Second Session—Caracas Transition from U.N. Seabed Committee to a 150-member conference Organization of proposals into working papers
1975	Third Session—Geneva Distribution of Informal Single Negotiating Texts covering all subjects before the conference
1976	Fourth Session—New York Revised Single Negotiating Text issued Deep seabed mining, area of chief disagreement
1976	Fifth Session—New York Issuance of Revised Text on dispute settlement Concentration of deep seabed mining agreement and economic zone delineation
1977	Sixth Session—New York Production of Informal Composite Negotiating Text (ICNT)
1978	Seventh Session—Geneva Establishment of seven negotiating groups (NG) to deal with "hard core" issues
1978	Resumed Seventh Session—New York Frustration with pace of progress; concern over possibility of unilateral legislation
1979	Eighth Session—Geneva Revision of ICNT Remaining concern over deep seabed mining, marine scientific research, continental shelf, and delimitation of offshore boundaries between adjacent or opposite nations
1979	Resumed Eighth Session—New York Setting of deadlines for adoption of convention
1980	Ninth Session—New York Continuing consultations
1980	Resumed Ninth Session—Geneva Development of draft convention
1981	Tenth Session—New York U.S. Review
1981	Resumed Tenth Session—Geneva
1982	Eleventh Session—New York

SOURCE: Adapted from Peterson, 1980.

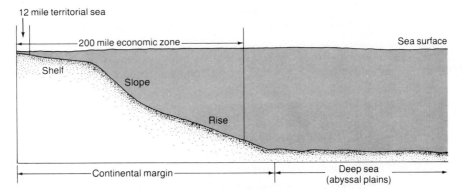

FIGURE 15-3 The probable divisions of the sea floor that will arise from the Third Law of the Sea Conference. In cases where the 200-mi economic zone does not extend to the edge of the continental margin, additional land may be claimed. Compare this figure with that of Figure 15-1, which shows the division of the sea floor following the 1958 (First) Law of the Sea Conference.

Likewise, a distance criterion of 200 n mi would, for some countries such as Argentina and the Soviet Union, not extend over all parts of their shelves. Ironically the United States will acquire the largest amount (2,222,000 n mi^2, about 7,500,000 km^2); actually most of the overall winners in acquiring sea floor are developed countries. The new divisions of the sea floor are shown in Figure 15-3.

To understand the position of developing countries, many of whom argued for the EEZ, it should be remembered that most of these countries lack a basic knowledge of their offshore geology and biology and thus know little of its resource potential. In many instances, this is due to an absence of marine scientific expertise within the country. Likewise, developing countries generally realize that total freedom of the seas offers the most advantages to those countries that are sufficiently technologically advanced to use and exploit the resources of the ocean. Therefore, based on these two points, it is logical that developing countries would wish to have an extensive economic resource zone over which they would have considerable control. It should also be noted that developing countries generally put a higher priority on achieving economic growth by using marine or other resources than in being concerned about environmental quality and pollution of the sea. Developing countries generally view pollution as a problem of the developed countries in that they caused much of it by their own industrial development and that they (the developing countries) are the only ones who can afford to deal with this problem.

The 1974 conference met in Caracas, Venezuela, and did not come to any major conclusions other than to meet in Geneva in 1975 and then return to Caracas to sign a treaty. The 1975 meeting in Geneva, however, also ended without any conventions being drafted. This pattern has continued until the ninth session in

1980 with small compromises having been made each year. At the end of the first part of the ninth session a text thought to be close to the final one has emerged. Nevertheless, there were still controversies concerning the mining of manganese nodules (see Chapter 12) and the international regime called the *Enterprise* that would control many activities in the deep sea. In the summer of 1980 a congressional act was passed that would allow U.S. companies to start mining nodules by January 1, 1988, if there was no treaty (the treaty, if signed, would take precedence over the act). A major concern about these meetings is that without a Law of the Sea Treaty or the strong possibility of one, various countries will continue to make unilateral actions and initiate further claims on the ocean.[13] Whether unilaterally or by convention, it is clear that a large portion of the ocean will be added in some exclusive manner to the coastal states. Ironically, this division of the ocean will make Dr. Pardo's original statement concerning the ocean as the common heritage of mankind a hollow dream. This is because most, if not almost all, of the oil and gas and biological resources that might be found in the ocean will be found within the proposed 200-mi economic zone. The remaining part of the ocean is, with the exception of manganese nodules and some exotic mineral deposits associated with oceanic ridges (Figure 7-18), essentially devoid of known significant resources. The manganese nodules can be mined for their copper, manganese, nickel, and cobalt content, but there are several factors that limit the amount of nodules and thus their financial return that could be mined in a given year (see Chapter 13, pages 410–418). Only if the countries of the world were to share the resources within their own 200-mi zone off their coast would the common heritage concept have the meaning for which Dr. Pardo had hoped.

The Law of the Sea Conference took a dramatic turn in early 1981 when the United States indicated that it wanted to make a thorough review of the Draft Treaty. The new administration was especially concerned about how the treaty would impact on U.S. deep-sea mining interests. A possible result of the review could be that the United States withdraws from further participation in the conference. However, another session is planned for 1982.

Freedom of Scientific Research Issue

Freedom of scientific research is an especially complex aspect in the Law of the Sea negotiations. It is extremely important for developing basic oceanographic knowledge as well as for learning about marine resources. The concept of freedom of scientific research in the ocean had been treated as an almost universally accepted fact by marine scientists essentially since the study of the ocean began. The only exception to this freedom was in a country's inland waters and in its 3-mi (5.5 km) territorial sea, where the country has complete sovereignty, but

---

[13] For example, the U.S. passed the Fishery Conservation and Management Act that restricted foreign fishing in a 200-n mi zone off the United States starting in March 1977.

where it could give permission for scientific research. In later years, as countries were extending their marine claims, the effect on scientific research was not initially felt. When countries had restrictions on working in their territorial waters, these could usually be alleviated or removed by contacting a scientific colleague in that country, inviting him or her to come on the cruise, and hoping in this way to gain permission to make the study. Another reason for marine scientists' not worrying about restrictions in the 1950s and 1960s was that so many areas of the ocean still had not been studied, so if permission was not given, the scientific program could be easily modified to avoid entering those areas and to work in other equally interesting but more hospitable areas. In the past decade, however, there has been an increase in the number of restrictions on scientific investigations. Recently, the State Department has taken a more active role in helping to obtain scientific permission. The State Department can be of little help, however, in obtaining permission to work in an area in which there is a disputed claim. For example, the State Department at present does not recognize territorial seas greater than 3 mi in width or contiguous zones farther than 12 mi (22.2 km) off the coast. Thus if one wishes to work in the waters off Peru, which claims 200 mi (370 km) off their coast, the State Department generally would not be able to help in obtaining permission.

If one considers the position of a foreign country with little oceanographic capabilities, there are several reasons why it might have concern about foreign countries' or scientists' working within their territorial waters or exclusive economic zone. Some states or administrators do not understand or appreciate the difference between basic and applied scientific research and thus feel that basic marine science, which attempts mainly to understand the principles and processes of the ocean, could also have a detrimental effect on that nation's control over their own resources. In some instances, they feel that because they do not have the proper scientific expertise to evaluate the scientific findings, they could very easily be exploited. Another concern of some coastal states is that they think that marine research can be a form of espionage and thus could jeopardize their national security. Many scientists and administrators in the United States appreciate these problems and try to overcome them by helping the developing countries acquire a marine science capability of their own that would give them an understanding of marine research and allow their meaningful participation in marine research programs.

During the preparations for the Law of the Sea Conference, there was much discussion about freedom of scientific research. The principle result of the conference on science has been to establish a consent regime (the coastal state must give permission) for research within their 200-mi zone. Most agree that this permission or consent will often be difficult to obtain, therefore there has been considerable concern about the ability to do scientific research within this zone. The problem is a critical one because for an oceanographer to understand the important processes of the ocean, it is imperative that one be able to study the boundary conditions where the land and water meet, both along the coastline and on the continental shelf. This region is the source of sediments for the deep sea, where waves and currents end, where most upwelling of ocean water occurs, and where most pollutants and other chemicals first get into the ocean. Indeed, many ocean phenomena are global in scale and are not affected by, nor do they respect, political boundaries. If oceanographers can only study the deep sea or that area outside the 200-n mi (370 km) limit, it could become a relatively sterile and esoteric science.

The magnitude of the area that could be eliminated from scientific research can be seen in Figure 15-4, which shows what part of the ocean would be affected if the entire 200 nautical mile economic resource zone was restricted from scientific research. The 200-n mi zone includes about 35.8 percent of the ocean, and if the region around Antarctica was included, the amount would be even greater. Some countries have even suggested regulating research in the deep sea and having it be controlled by an international authority, but this view probably will not be included in the treaty.

In 1971 when the Seabed Committee divided itself into three subcommittees,

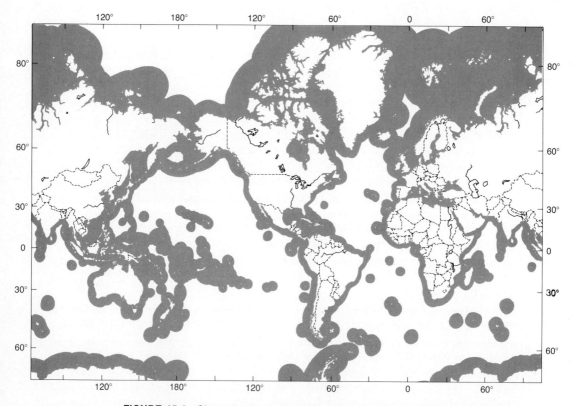

FIGURE 15-4 Chart showing area that would fall within a 200-nmi economic resource zone. Note that the Mercator projection of this figure distorts polar regions and that Australia is really 3 times the size of Greenland.

FIGURE 15-5 Visit of research vessel *Knorr* (of Woods Hole Oceanographic Institution) to the United Nations. The vessel and some of its work was shown to delegates to the Law of the Sea Conference. (Photograph courtesy of Woods Hole Oceanographic Institution.)

Subcommittee III was established to deal with the preservation of the marine environment (including prevention of pollution) and the freedom of scientific research. Several marine scientists have met with this subcommittee and other delegates to the Law of the Sea Conference, both formally and informally, to discuss the mutual advantages of marine science research. In one instance, a research vessel was even brought to the United Nations to demonstrate its use in marine science (Figure 15-5). Dr. John Knauss, Provost of the Graduate School of Oceanography of the University of Rhode Island, made a statement to a Preparatory Meeting of the Law of the Sea Conference in which he tried to show the importance of science to society and said "all the technological and economic advances of modern times have been based on the findings of science. A strong case can be made for the thesis that if it were not for the advancements in technology generated by science, there would probably be no need for a new Law of the Sea Conference."[14]

It is difficult to predict the eventual effects of the restrictions on the freedom of marine scientific research. There can be some compensatory advantages, perhaps, in encouraging oceanographers to have cooperative research programs with scientists from underdeveloped countries and to develop research objectives that will have much more use and value for humankind in the future. It is also probably true that those countries that put severe restrictions on basic marine scientific research in their waters will probably only be hurting themselves in the long run.

Dr. Phillip Handler, when he was president of the National Academy of Sciences, commented on the freedom of scientific research to one of the subcommittees of the Seabed Committee. He said that

> "regulations would not mean the end of oceanography. Research will continue, but the more it is subject to controls, the greater the danger that it will become second class research. Progress will be slower, research will be more expensive and new avenues of inquiry will not be pursued. By directing science toward short term goals, we may lose the unique scholar driven only by his own curiosity, who has been responsible for so many of the great scientific discoveries of the past. I submit that mankind cannot afford such a loss."[15]

Summary

The ocean has, until very recently, been considered as belonging to no one country and available for everyone's use. The one exception to this was a narrow strip (the territorial sea) immediately adjacent to the land. The Truman Proclamation in 1945, which claimed the natural resources of the U.S. continental shelf, may have been the action that started other countries making claims and extensions of their territorial seas.

These events led to the First and Second Law of the Sea Conferences held in Geneva in 1958 and 1960. Four conventions were adopted: Convention on the Territorial Sea and the Contiguous Zone; Convention on the Continental Shelf; Convention on the High Seas; and Convention on Fishing and the Conservation of the Living Resources of the High Sea. These conventions, although codifying many aspects concerning the use of the ocean, actually were somewhat of a failure in that they failed to define the width of the territorial sea and gave an arbitrary

---

[14] Statement by Professor John A. Knauss before Subcommittee III of the Committee on the Peaceful Uses of the Seabed and the Ocean Floor Beyond the Limits of National Jurisdiction; August 3, 1972.

[15] Statement made by Dr. Philip Handler in 1973 to Subcommittee III of the Committee on the Peaceful Uses of the Seabed and the Ocean Floor Beyond the Limits of National Jurisdiction.

definition to the edge of the continental shelf (200-m depth) and to where it can be exploited.

In the following years countries continued to extend their claims seaward in large part as a result of gaps in the conventions, the development of marine technology, and the awareness of marine resources, and also simply for political reasons. These actions and a desire by some to consider the resources of the ocean as a common heritage of humankind to be shared by all, led to the Third Law of the Sea Conference that started in 1973 and has continued into the 1980s. This conference has considered almost all aspects of the ocean and in many instances has resulted in a confrontation between developed and developing countries. Results include a 12-n mi territorial sea and a 200-n mi exclusive economic zone (EEZ). In the EEZ the adjacent coastal state will have control over many activities including exploiting its resources and marine science. An international regime, details of which are unresolved, will probably control or manage the exploitation of the resources of the deep-sea region, which are mainly manganese nodules.

The desire to share the resources of the ocean following the common heritage concept has been lost since essentially all resources (oil and gas and biological) lie within the EEZ and will be claimed by the adjacent state. The control over marine science within the EEZ could have a damaging effect in restricting the scientific understanding of the ocean.

Suggested Further Readings

ANDERSON, E. V., "The Law of the Sea Battle—An Overview," *Marine Technology Society Journal,* **8,** no. 6 (1974), pp. 4–14.

BROWN, S., and L. L. FABIAN, "National Interests Vs. International Needs," *Marine Technology Society Journal,* **8,** no. 6 (1974), pp. 29–39.

CADWALADER, G., "Freedom of Science in the Ocean," *Science,* **182** (1973), pp. 15–20.

HOLLICK, A. L., "The Clash of U.S. Interests: How U.S. Policy Evolved," *Marine Technology Society Journal,* **8,** no. 6 (1974), pp. 15–28.

KNAUSS, J. A., "Marine Science and the 1974 Law of the Sea Conference—Science Faces a Difficult Future in Changing Law of the Sea," *Science,* **184** (1974), pp. 1335–41.

PETERSON, S., "The Common Heritage of Mankind? Regulating the Uses of the Ocean," *Environment,* **22,** no. 1 (1980), pp. 6–11.

ROSS, D. A., "Threat to the Freedom of Scientific Research in the Deep Sea?" *Oceanus,* **21,** no. 1 (1978), pp. 69–71.

SWING, J. T., "Law of the Sea Revisited," *Oceans Magazine,* Sept. 1979, pp. 58–59.

WOOSTER, W. S., ed., *Freedom of Oceanic Research.* New York: Crane, Russak, 1973.

WOOSTER, W. S., "The Ocean and Man," *Scientific American,* **221,** no. 3 (1969), pp. 218–34.

References

ABEL, R. B., "Careers in Oceanography," *Sea Technology,* **19** (1978), pp. 25–27.

ALBERS, J. P., M. D. CARTER, A. L. CLARK, A. B. COURY, and S. P. SCHWEINFURTH, *Summary Petroleum and Selected Mineral Statistics for 120 Countries, Including Offshore Area.* U.S. Geological Survey Professional Paper 817, Washington, D.C.: U.S. Government Printing Office, 1973.

ALEXANDER, L. M., "Indices of National Interest in the Oceans," *Ocean Development and International Law Journal,* **1** (1973), pp. 21–49.

ANDERSON, E. V., "The Law of the Sea Battle—An Overview," *Marine Technology Society Journal,* **8** (1974), pp. 4–14.

ARRHENIUS, G., "Pelagic Sediments," in *The Sea,* Vol. 3, ed. M. N. Hill. New York: Interscience, 1963.

ATKINS, W. R. G., "The Phosphate Content of Sea Water in Relation to the Growth of the Algal Plankton, Part III," *Journal Marine Biological Association,* **14** (1926), pp. 447–67.

ATKINS, W. R. G., "The Seasonal Variation in the Copper Content of Sea Water," *Journal Marine Biological Association,* **31** (1953), pp. 493–94.

BARAZANGI, M., and J. DORMAN, "World Seismicity Map Compiled From ESSA Coast and Geodetic Survey Epicenter Data, 1961–1967," *Seismological Society of America Bulletin,* **59** (1969), pp. 369–80.

BARTH, T. F. W., *Theoretical Petrology.* New York: John Wiley, 1952.

BENDER, E., "A National Oceanic Satellite System," *Sea Technology,* **21** (1980), pp. 27–31.

BERNER, R. A., *Principles of Chemical Sedimentology,* International Series in the Earth and Planetary Sciences. New York: McGraw-Hill, 1971.

BERTEAUX, H. O., and N. P. FOFONOFF, "Oceanographic Buoys Gather Data From Surface to Sea Floor," *Oceanology International* (July–August, 1967), pp. 39–42.

BIGELOW, H. B., *Oceanography: Its Scope, Problems, and Economic Importance.* Boston: Houghton Mifflin, 1931.

BISCHOFF, J. L., and F. T. MANHEIM, "Economic Potential of the Red Sea Heavy Metal Deposits," in *Hot Brines and Recent Heavy Metal Deposits in the Red Sea,* ed. E. T. Degens and D. A. Ross. New York: Springer-Verlag, 1969.

BISKO, D., J. R. DUNN, and W. A. WALLACE, "Planning for Non-Renewable Resources in Urban-Suburban Environs," Rensselaer Polytechnic Institute, Report for the U.S. Bureau of Mines, 1969.

BLUMER, M., "Scientific Aspects of the Oil Spill Problem," *Environmental Affairs,* **1,** no. 1 (1971), pp. 64–73.

BLUMER, M., H. L. SANDERS, J. F. GRASSLE, and G. R. HAMPSON, "A Small Oil Spill," *Environment,* **13,** no. 2 (1971), pp. 2–12.

BONATTI, E., T. KRAEMER, and H. RYDELL, "Classification and Genesis of Submarine Iron-Manganese Deposits," in *Ferromanganese Deposits on the Ocean Floor,* ed. D. R. Horn. New York: Arden House Harriman, 1972.

BORN, G. H., J. A. DUNNE, and D. B. LAME, "Seasat Mission Overview," *Science,* **204,** (1979), pp. 1405–06.

BOSTROM, K., and M. N. A. PETERSON, "Precipitates from Hydrothermal Exhalations on the East Pacific Rise," *Economic Geology,* **61,** (1966), pp. 1258–65.

BOURNE, D. W., and B. C. HEEZEN, "A Wandering Enteropneust from the Abyssal Pacific, and the Distribution of 'Spiral' Tracks on the Sea Floor," *Science,* **150** (1965), pp. 60–63.

BOWIN, C. O., R. BERNSTEIN, E. D. UNGAR, and J. R. MADIGAN, "A Shipboard Oceanographic Data Processing and Control System," *Institute of Electrical and Electronic Engineers Transactions GE-5,* no. 2 (1967), pp. 41–50.

BROECKER, W. S., "Radio Isotopes and Large-Scale Oceanic Mixing," in *The Sea,* **Vol. 2,** ed. M. N. Hill. New York: Interscience, 1963.

BROECKER, W. S., *Chemical Oceanography.* New York: Harcourt Brace Jovanovich, Inc., 1974.

BROECKER, W. S., R. D. GERARD, M. EWING, and B. C. HEEZEN, "Geochemistry and Physics of Ocean Circulation," in *Oceanography,* ed. Mary Sears. American Association Advancement of Science Publication No. 67, 1961.

BROWN, S., and L. L. FABIAN, "National Interests Vs. International Needs," *Marine Technology Society Journal,* **8,** no. 6 (1974), pp. 29–39.

BRUJEWICZ, S. W., *Tikhii Ocean (Pacific Ocean).* Moscow Publishing House, 1966.

BUDYKO, M. I., *Climate and Life.* New York: Academic Press, 1974.

BURKE, W. T., "Marine Science Research and International Law," Occasional Paper No. 8, Law of the Sea Institute, University of Rhode Island, 1970.

CADWALADER, G., "Freedom of Science in the Ocean," *Science,* **182** (1973), pp. 15–20.

CARPENTER, E. J., S. J. ANDERSON, G. R. HARVEY, H. P. MIKLAS, and B. B. PECK, "Polystyrene Spherules in Coastal Waters," *Science,* **178** (1972), pp. 749–50.

CARPENTER, E. J., and K. L. SMITH, JR., "Plastics on the Sargasso Sea Surface," *Science,* **175** (1972), pp. 1240–41.

CHARLIER, R. H., "Other Ocean Resources," in *Ocean Yearbook,* **1,** ed. E. M. Borgese, and N. Ginsburg. Chicago: University of Chicago Press, 1978, pp. 160–210.

CHEEK, C. H., "Law of the Sea: Effects of Varying Coastal States Having Controls on Marine Research," *Ocean Development and International Law Journal,* **1,** no. 2 (1973), pp. 209–19.

CHRISTY, F. T., JR., and H. HERFINDAHL, "A Hypothetical Division of the Sea Floor: A Chart Prepared for the Law of the Sea Institute," University of Rhode Island, 1967.

CLARKE, G. L., "Dynamics of Production in a Marine Area," *Ecological Monographs,* **16** (1946), pp. 323–35.

COHEN, M. R., and I. E. DRABKIN, *A Source Book in Greek Science.* New York: McGraw-Hill, 1948.

COX, A., *Plate Tectonics and Geomagnetic Reversals.* San Francisco: W. H. Freeman and Company, Publishers, 1973.

COX, R. A., "The Physical Properties of Sea Water," in *Chemical Oceanography,* ed. J. P. Riley and G. Skirrow. New York: Academic Press, 1965.

COX, R. A., and N. D. SMITH, "The Specific Heat of Sea Water," *Proceedings Royal Society London A.,* **151** (1959), pp. 51–62.

CRONAN, D. S., "Geochemistry of Some Manganese Nodules and Associated Pelagic Deposits." Ph.D. dissertation, Imperial College, University of London, 1967.

CRONAN, D. S., "Composition of Atlantic Manganese Nodules," *Nature, Physical Science,* **235,** no. 61 (1972), pp. 171–72.

CRONAN, D. S., and J. S. TOOMS, "The Geochemistry of Manganese Nodules and Associated Pelagic Deposits from the Pacific and Indian Ocean," *Deep-Sea Research,* **16** (1969), pp. 335–61.

CURRAY, J. R., "Late Quaternary Sea Level; A Discussion," *Bulletin Geological Society of America,* **72** (1961), pp. 1707–12.

CURRAY, J. R., "Late Quaternary History, Continental Shelves of the United States," in *The Quaternary of the United States,* ed. H. E. Wright, Jr. and D. G. Frey. Princeton, N.J.: Princeton University Press, 1965.

DEACON, M., *Scientists and the Sea, 1650–1900: A Study of Marine Science.* London: Academic Press, 1971.

DEGENS, E. T., and D. A. ROSS, *Hot Brines and Recent Heavy Metal Deposits in the Red Sea.* New York: Springer-Verlag, 1969.

DEUSER, W. G., and E. T. DEGENS, "$^{18}O/^{16}O$ and $^{13}C/^{12}C$ Ratios in Fossil Foraminifera and Pteropods from the Area of the Hot Brine Deeps of the Central Red Sea," in *Hot Brines and Recent Heavy Metal Deposits in the Red Sea,* ed. E. T. Degens and D. A. Ross. New York: Springer-Verlag, 1969.

DIETRICH, G., *General Oceanography.* New York: John Wiley, 1963.

DIETZ, R. S., "Continent and Ocean Basin Evolution by Spreading of the Sea Floor," *Nature,* **190** (1961), pp. 854–57.

DIETZ, R. S., and J. C. HOLDEN, "Reconstruction of Pangaea: Breakup and Dispersion of Continents, Permian to Present," *Journal of Geophysical Research,* **75** (1970), pp. 4939–56.

DILL, R. F., "Contemporary Submarine Erosion in Scripps Submarine Canyon," Unpublished Ph.D. dissertation, University of California, San Diego, 1964.

DU TOIT, A. L., *Our Wandering Continents, and Hypothesis of Continental Drifting.* Edinburgh, England: Oliver and Boyd, 1937.

DUOMANI, G. A., "Exploiting the Resources of the Seabed," in *Science, Technology, and American Diplomacy:* Report Prepared for the Subcommittee on National Security Policy and Scientific Developments of the Committee on Foreign Affairs, U.S. House of Representatives. Washington: U.S. Government Printing Office, 1971.

EDMOND, J. M., "GEOSECS Is Like the Yankees: Everybody Hates It and It Always Wins . . . K. K. Turekian (1978)," *Oceanus,* **23,** no. 1 (1980), pp. 33–39.

Effective Use of the Sea. Report of the Panel on Oceanography of the President's Science Advisory Committee, June 1966.

EMERY, K. O., "Atlantic Continental Shelf and Slope of the United States, Geologic Background," *U.S. Geological Survey Professional Paper 529-A,* 1966, pp. 1–23.

EMERY, K. O., "Geological Methods for Locating Mineral Deposits on the Ocean Floor," *Transactions 2nd Marine Technological Society Conference, June 27–29, 1966,* 1966, pp. 24–43.

EMERY, K. O., "The Atlantic Continental Margin of the United States During the Past 70 Million Years," *Geological Association of Canada,* Special Paper 4, Geology of the Atlantic Region, 1967, pp. 53–70.

EMERY, K. O., "Continental Rises and Oil Potential," *Oil and Gas Journal,* **67,** no. 19 (1969), pp. 231–43.

EMERY, K. O., and D. A. ROSS, "Topography and Sediments of a Small Area of the Continental Slope South of Martha's Vineyard," *Deep-Sea Research,* **15** (1968), pp. 415–22.

EMILIANI, C., "Pleistocene Temperatures," *Journal Geology,* **63** (1955), pp. 538–78.

FAIRBRIDGE, R. W., B. C. HEEZEN, T. ICHIYE, and M. THARP, "Indian Ocean," in *Encyclo-*

pedia of Oceanography, ed. R. W. Fairbridge. New York: Reinhold, 1966.

FAIRBRIDGE, R. W., "Trenches and Related Deep Sea Troughs," in *Encyclopedia of Oceanography,* ed. R. W. Fairbridge. New York: Reinhold, 1966.

FAIRBRIDGE, R. W., J. L. REID, JR., E. OLAUSSEN, and M. N. A. PETERSON, "Pacific Ocean," in *Encyclopedia of Oceanography,* ed. R. W. Fairbridge. New York: Reinhold, 1966.

FISHER, R. L., and H. H. HESS, "Trenches," in *The Sea,* Vol. 3, ed. M. N. Hill. New York: Interscience, 1963.

FLEMING, R. H., "The Control of Diatom Population by Grazing," *Journal du Conseil Internat. p. l'Explor de la Mer,* **14,** no. 2 (1939), pp. 210–27.

FLEMING, R. H., "The Composition of Plankton and Units for Reporting Population and Production," *Proceedings of Sixth Pacific Science Congress California, 1939,* **3** (1940), pp. 535–40.

FLOWERS, A., and A. J. BRYCE, "Energy from Marine Biomass," *Sea Technology,* **18** (Oct. 1977), pp. 18–21.

FOLGER, D. W., and B. C. HEEZEN, "Trans-Atlantic Sediment Transport by Wind," (abstract), Annual Meeting of the Geological Society of America, New Orleans, 1967.

FROSCH, R. A., C. D. HOLLISTER, and D. A. DEESE, "Radioactive Waste Disposal in the Oceans," in *Ocean Yearbook,* **1,** ed. E. M. Borgese, and N. Ginsburg, Chicago: University of Chicago Press, 1978, pp. 340–49.

FUGLISTER, F. C., "Temperature and Salinity Profiles and Data from the International Geophysical Year of 1957–1958," *Atlantic Ocean Atlas,* The Woods Hole Oceanographic Institution Atlas Series, **1,** 1960.

FYE, P. M., A. E. MAXWELL, K. O. EMERY, and B. H. KETCHUM, "Ocean Science and Marine Resources," in *Uses of the Seas,* ed. E. A. Gullion, Englewood Cliffs, N.J.: American Assembly, 1968.

GARRELS, R. M., and F. T. MACKENZIE, *Evolution of Sedimentary Rocks.* New York: W. W. Norton & Company, Inc., 1971.

GOLDBERG, E. D., "The Oceans as a Chemical System," in *The Sea,* Vol. 2, ed. M. N. Hill. New York: Interscience, 1963.

GOLDBERG, E. D., and G. O. ARRHENIUS, "Chemistry of Pacific Pelagic Sediments," *Geochimica et Cosmochimica Acta,* **13** (1958), pp. 153–212.

GOULD, H. R., "Some Quantitative Aspects of Lake Meade Turbidity Currents," *Society of Economic Paleontology and Mineralogy Special Publication No. 2,* 1951, pp. 34–52.

GRIFFIN, W. L., "Law of Ocean Space," in *Encyclopedia of Marine Resources,* ed. F. E. Firth. New York: Van Nostrand Reinhold, 1969.

HAMILTON, E. L., "Thickness and Consolidation of Deep-Sea Sediments," *Bulletin Geological Society of America,* **70** (1959), pp. 1399–1424.

HARGROVE, J. L., "New Concepts in the Law of the Sea," *Ocean Development and International Law Journal,* **1,** no. 1 (1973), pp. 5–12.

HARTLINE, B. K., "POLYMODE: Exploring the Undersea Weather," *Science,* **205** (1979), pp. 571–73.

HARTLINE, B. K., "Coastal Upwelling: Physical Factors Feed Fish," *Science,* **208** (1980), pp. 38–40.

HARVEY, H. W., "On the Production of Living Matter in the Sea off Plymouth," *Journal Marine Biological Association,* **29** (1950), pp. 97–137.

HEATH, G. R., "Deep-Sea Manganese Nodules," *Oceanus,* **21,** no. 1 (1978), pp. 60–68.

HEDBERG, H. D., "The National–International Jurisdictional Boundary on the Ocean Floor," *Ocean Management,* **1** (1973), pp. 83–118.

HEDBERG, H. D., J. D. MOODY, and R. M. HEDBERG, "Petroleum Prospects of the Deep Offshore," *The American Association of Petroleum Geologists Bulletin,* **63,** no. 3 (1979), pp. 286–300.

HEEZEN, B. C., "The Deep Sea Floor," in *Continental Drift,* International Geophysics Series, ed. S. K. Runcorn. New York: Academic Press, 1962.

HEEZEN, B. C., and M. EWING, "Turbidity Currents and Submarine Slumps, and the 1929 Grand Banks Earthquake," *American Journal of Science,* **250** (1952), pp. 849–73.

HEEZEN, B. C., and M. EWING, "The Mid-Oceanic Ridge," in *The Sea,* Vol. 3, ed. M. N. Hill. New York: Interscience, 1963.

HEEZEN, B. C., and C. HOLLISTER, "Deep-Sea Current Evidence from Abyssal Sediments," *Marine Geology,* **1** (1964), pp. 141–74.

HEEZEN, B. C., and C. HOLLISTER, *The Face of the Deep.* New York: Oxford University Press, 1971.

HEEZEN, B. C., and P. J. FOX, "Mid-Oceanic Ridge," in *Encyclopedia of Oceanography,* ed. R. W. Fairbridge, New York: Reinhold, 1966.

HEEZEN, B. C., M. THARP, and M. EWING, "The Floors of the Ocean, North Atlantic," Geological Society of America Special Paper no. 65, 1959.

HEIRTZLER, J. R., G. O. DICKSON, E. M. HERRON, W. C. PITMAN, III, and X. LE PICHON, "Marine Magnetic Anomalies, Geomagnetic Field Reversals, and Motions of the Ocean Floor and Continents," *Journal of Geophysical Research,* **73** (1968), pp. 2119–36.

HERSEY, J. B., ed., *Deep-Sea Photography.* Baltimore: The Johns Hopkins Press, 1967.

HERSEY, J. B., and R. H. BACKUS, "Sound Scattering by Marine Organisms," in *The Sea,* Vol. 1, ed. M. N. Hill. New York: Interscience, 1962.

HESS, H. H., "History of Ocean Basins," in *Petrologic Studies: A Volume to Honor A. F. Buddington.* New York: Geological Society of America, 1962.

HESSLER, R. R., J. D. ISAACS, and E. L. MILLS, "Giant Amphipod from the Abyssal Pacific Ocean," *Science,* **175** (1972), pp. 636–37.

HESSLER, R. R., and H. L. SANDERS, "Faunal Diversity in the Deep-Sea," *Deep-Sea Research,* **14** (1967), pp. 65–78.

HOLLICK, A. L., "The Clash of U.S. Interests: How U.S. Policy Evolved," *Marine Technology Society Journal,* **8,** no. 6 (1974), pp. 15–28.

HOLMES, A., "A Revised Geological Time Scale," *Edinburgh Geological Society,* Transaction 17, Part 3, 1960, p. 204.

HOLT, S., "Marine Fisheries," in *Ocean Yearbook,* **1,** eds. E. M. Borgese, and N. Ginsburg. Chicago: University of Chicago Press, 1978, pp. 38–83.

HOOD, D. W., "Chemical Oceanography," *Oceanographic and Marine Biology Annual Review,* **1** (1963), pp. 129–55.

HOOD, D. W., "Seawater: Chemistry," in *Encyclopedia of Oceanography,* ed. R. W. Fairbridge. New York: Reinhold, 1966.

HORN, D. R., B. M. HORN, and M. N. DELACH, *Ferromanganese Deposits of the North Pacific.* Technical Report no. 1, NSF GX-33616-IDOE, 1972.

HORN, D. R., B. M. HORN, and M. N. DELACH, "Distribution of Ferromanganese Deposits in the World Ocean," in *Ferromanganese Deposits on the Ocean Floor,* ed. D. R. Horn. New York: Arden House and Lamont–Doherty Geological Observatory, 1972.

HORN, D. R., B. M. HORN, and M. N. DELACH, *Factors Which Control the Distribution of Ferromanganese Nodules and Proposed Research Vessel's Track North Pacific.* Technical Report no. 8, NSF GX-33616-IDOE, 1973.

HUGHES, J. T., J. J. SULLIVAN, and R. SHLESER, "Enhancement of Lobster Growth," *Science,* **177** (1972), pp. 1110–11.

HUNKINS, K. L., "Drifting Ice Stations," in *Encyclopedia of Oceanography,* ed. R. W. Fairbridge. New York: Reinhold, 1966.

HURD, B., and B. PASSERO, *Oceans of the World: The Last Frontier—An Annotated Introductory Bibliography on the Law of the Sea.* M.I.T. Sea Grant Report No. 74-17, 1974.

International Boundary Study: Limits in the Seas—Theoretical Areal Allocations of Seabed to Coastal States. The Geographer, Office of the Geographer, Bureau of Intelligence and Research.

ISAACS, J. D., D. CASTEL, and G. L. WICK, "Utilization of the Energy in Ocean Waves," *Ocean Engineering,* **3** (1976), pp. 175–87.

JANNASCH, H. W., K. EIMHJELLEN, C. O. WIRSEN, and A. FARMANFARMIAN, "Microbial Degradation of Organic Matter in the Deep Sea," *Science,* **171** (1971), pp. 672–75.

JENKINS, P. M., "Oxygen Production by the Diatom *Coscinodiscus excentricus* in Relation to Submarine Illumination in the English Channel," *Journal of Marine Biological Association,* **22** (1937), pp. 301–43.

JERLOV, N. G., "Optical Studies of Ocean Waters," *Reports of the Swedish Deep Sea Expedition, Physics and Chemistry,* **3,** no. 1 (1951), pp. 1–59.

KALLIE, K., and H. WATTENBERG, "Uber den Kupfergehalf des Ozeanwassers," *Naturwissenschaften,* **26** (1938), pp. 630–31.

KERR, R. A., "Oceanography: A Closer Look at Gulf Stream Rings," *Science,* **198** (1977), pp. 387–89, 430.

KERR, R. A., "Tidal Waves: New Method Suggested to Improve Prediction," *Science,* **200** (1978), pp. 521–22.

KETCHUM, B., "A Realistic Look at Ocean Pollution," *Marine Technology Society Journal,* **7,** no. 7 (1973), pp. 8–15.

KETCHUM, B., ed., *The Water's Edge.* Cambridge, Mass.: M.I.T. Press, 1972.

KING, C. A. M., "Ocean Waves," in *Encyclopedia of Oceanography,* ed. R. W. Fairbridge. New York: Reinhold, 1966.

KLIEN, D. H., and E. D. GOLDBERG, "Mercury in the Marine Environment," *Environmental Science and Technology,* **4,** no. 9 (1970), pp. 765–68.

KNAUSS, J. A., "Developing the Freedom of Scientific Research Issue of the Law of the Sea Conference," *Ocean Development and International Law Journal,* **1,** no. 1 (1973), pp. 93–120.

KNAUSS, J. A., "Marine Science and the 1974 Law of the Sea Conference—Science Faces a Difficult Future in Changing Law of the Sea," *Science,* **184** (1974), pp. 1335–41.

KNIGHT, H. G., "United States Ocean Policy—Perspective 1974," *Notre Dame Lawyer,* **49,** no. 2 (1973), pp. 241–75.

KOMINZ, M., "What Causes the Ice Ages?" *Maritimes,* **23,** no. 1 (1979), pp. 12–15.

KRAUSKOPF, K. B., "Factors Controlling the Concentrations of Thirteen Rare Metals in Sea Water," *Geochimica et Cosmochimica Acta,* **9** (1956), pp. 1–33.

KRAUSKOPF, K., and A. BEISER, *Fundamentals of Physical Science.* New York: McGraw-Hill, 1966.

KUENEN, P. H., and C. I. MIGLIORINI, "Turbidity Currents as a Cause of Graded Bedding," *Journal of Geology,* **58** (1950), pp. 91–129.

KULLENBURG, B., "The Piston Core Sampler," *Sv. Hydr-Biol. Komm. skr., 3, ser.,* Hydro. Bed. 1, Boteborg, Sweden, 1947.

LADD, H. S., E. INGERSON, R. C. TOWNSEND, M. RUSSELL, and H. K. STEPHENSON, "Drilling on Eniwetok Atoll, Marshall Islands," *Bulletin of the American Association of Petroleum Geologists,* **37** (1953), pp. 2257–80.

LAEVESTU, T., and T. G. THOMPSON, "Soluble Iron in Coastal Waters," *Journal Marine Research,* **16** (1958), pp. 192–98.

LA FOND, E. C., "Fixed Platforms," in *Encyclopedia of Oceanography,* ed. R. W. Fairbridge. New York: Reinhold, 1966.

LAMAR, D. L., and P. M. MERIFIELD, "Cambrian Fossils and the Origin of the Earth–Moon System," *Bulletin Geological Society of America,* **78** (1967), pp. 1359–68.

LANDSBERG, H. H., "Low-cost, Abundant Energy: Paradise Lost?" *Science,* **184** (1974), pp. 247–53.

LEOPOLD, L. P., and K. S. DAVIS, *Water.* New York: Time Incorporated, 1966.

LISITZIN, A. P., *Sedimentation in the World Ocean.* Tulsa, Oklahoma: Society of Economic Paleontologists and Mineralogists Special Publication no. 17, 1972.

LUDWIGSON, J., "Ferment in the Fleet," *Mosaic,* **11,** no. 2 (1980), pp. 2–11.

Lyman, J., "Chemical Considerations," *Physical and Chemical Properties of Sea Water*. National Academy of Science, National Research Council Publication No. 600, 1958.

Madsen, F. J., "Abyssal Zone," in *Encyclopedia of Oceanography*, ed. R. W. Fairbridge. New York: Reinhold, 1966.

Manheim, F. T., *Mineral Resources Off the Northeastern Coast of the United States*. U.S. Geological Survey Circular 669, Washington, D.C.: U.S. Government Printing Office, 1972.

Mason, R. G., and A. D. Raff, "Magnetic Survey off the West Coast of North America, 32° N Latitude to 42° N Latitude," *Bulletin Geological Society of America*, **72** (1961), pp. 1259–66.

Matthews, D. J., "Tables of the Velocity of Sound in Pure Water and Sea Water for Use in Echo-Sounding and Sound-Ranging," *London Hydrographic Department, Admiralty* (2nd ed.), 1939.

McCaslin, J. C., "Offshore Oil Production Soars," *Oil and Gas Journal*, **72,** no. 18 (1974), pp. 136–42.

McCoy, F. W., Jr., R. P. Von Herzen, D. M. Owen, and P. R. Boutin, "Deep-Sea Corehead Camera Photography and Piston Coring," Woods Hole Oceanographic Institution Reference No. 69–19, 1969.

McGuinness, W. T., "Acoustics" (Underwater), in *Encyclopedia of Oceanography*, ed. R. W. Fairbridge. New York: Reinhold, 1966.

McKelvey, V. E., "Seabed Minerals and the Law of the Sea," *Science*, **209** (1980), pp. 464–72.

McKelvey, V. E., and F. H. Wang, *World Subsea Mineral Resources*, U.S. Geological Survey Miscellaneous Geological Investigation Map I-632. Washington, D.C.: U.S. Government Printing Office, 1970.

Meade, R. H., "Landward Transport of Bottom Sediments in Estuaries of the Atlantic Coastal Plain," *Journal of Sedimentary Petrology*, **39** (1969), pp. 222–34.

Menard, H. W., *Marine Geology of the Pacific*. New York: McGraw-Hill, 1964.

Menard, H. W., and S. M. Smith, "Hypsometry of Ocean Basin Provinces," *Journal of Geophysical Research*, **71** (1966), pp. 4305–25.

Mero, J. L., "Manganese Nodules (Deep-Sea)," in *Encyclopedia of Oceanography*, ed. R. W. Fairbridge. New York: Reinhold, 1966.

Mero, J. L., *The Mineral Resources of the Sea*. New York: Elsevier, 1965.

Mero, J. L., "Ocean-Floor Manganese Nodules," *Economic Geology*, **57** (1962), pp. 747–67.

Metz, W. D., "Ocean Temperature Gradients: Solar Power from the Sea," *Science*, **180** (1973), pp. 1266–67.

Miller, A. R., D. C. Densmore, E. T. Degens, J. C. Hathaway, F. T. Manheim, P. F. McFarlin, R. Pocklington, and A. Jokela, "Hot Brines and Recent Iron Deposits in Deeps of the Red Sea," *Geochimica et Cosmochimica Acta*, **30** (1966), pp. 341–59.

Miller, D. J., "Giant Waves in Lituya Bay, Alaska," *U.S. Geological Survey Professional Paper*, No. 354-C, 1960, pp. 51–86.

Miller, G. E., P. M. Grant, R. Kishore, F. J. Steindruger, F. S. Rowland, and V. P. Gunn, "Mercury Concentrations in Museum Specimens of Tuna and Swordfish," *Science*, **175** (1972), pp. 1121–22.

Miller, S. L., "The Origin of Life," in *The Sea*, Vol. 3, ed. M. N. Hill. New York: Interscience, 1963.

Milliman, J. D., "Carbonate Sedimentation on Hogsty Reef, A Bahamian Atoll," *Journal of Sedimentary Petrology*, **37** (1967), pp. 658–76.

Milliman, J. D., and K. O. Emery, "Sea Levels During the Past 35,000 Years," *Science*, **162** (1968), pp. 1121–23.

Moody, J. D., "Petroleum Demands of Future Decades," *American Association of Pe-

troleum Geologists Bulletin, **54,** no. 12 (1970), pp. 2239–45.

MOORE, D. G., "The Free-Corer; Sampling Without Wire and Winch," *Journal of Sedimentary Petrology,* **31** (1961), pp. 672–80.

MOORE, T. C., and G. R. HEATH, "Abyssal Hills in the Central Equatorial Pacific: Detailed Structure of the Sea Floor and Subbottom Reflectors," *Marine Geology,* **5** (1967), pp. 161–79.

MORGAN, J. W., "Rises, Trenches, Great Faults, and Crustal Blocks," *Journal Geophysical Research,* **73** (1968), pp. 1959–82.

MURRAY, J., and A. F. RENARD, "Deep Sea Deposits, Scientific Results of the Exploration Voyage of H.M.S. *Challenger, 1872–1876,*" *Challenger Reports.* London: Longsman, 1891.

Ocean Dumping—A National Policy. A Report to the President Prepared by the Council on Environmental Quality, Washington, D.C.: U.S. Government Printing Office, 1970.

Ocean Margin Drilling—A Technical Memorandum. Office of Technology Assessment, Washington, D.C., 1980.

Ocean Yearbook, **1.** BORGESE, E. M., and N. GINSBURG, eds., Chicago: University of Chicago Press, 1978, 890 pp.

Oceanographic Vessels of the World, International Geophysical Year, World Data Center for Oceanography and the National Oceanographic Data Center, U.S. Navy Oceanographic Office, Washington, D.C. 1963.

"Oil and Gas Resources: Did U.S.G.S. Gush Too High?" *Science,* **185** (1974), pp. 127–29.

OPDYKE, N. D., B. GLASS, J. D. HAYS, and J. FOSTER, "Paleomagnetic Study of Antarctic Deep Sea Cores," *Science,* **154** (1966), pp. 349–57.

OSTENSO, N. A., "Arctic Ocean," in *Encyclopedia of Oceanography,* ed R. W. Fairbridge. New York: Reinhold, 1966.

OTHMER, D. F., and O. A. ROELS, "Power, Fresh Water, and Food from Cold, Deep Seawater," *Marine Technological Society Journal,* **8** (August 1974), pp. 39–43.

PADELFORD, N. J., "Prospects for a New Regime of the Seas: International Political Considerations," Presented at the 7th Annual Conference of the Marine Technology Society. Washington, D.C.: Report No. MITSG 72–5, 1971.

PATTULLO, J. B., "Tides," in *Encyclopedia of Oceanography,* ed. R. W. Fairbridge. New York: Reinhold, 1966.

PETERSON, M. N. A., "Scientific Goals and Achievements (Deep-Sea Drilling Program)," *Ocean Industry Magazine,* May 1969, pp. 62–66.

PETERSON, S., "The Common Heritage of Mankind?" *Environment,* **22,** no. 1 (1980), pp. 6–11.

Petroleum in the Marine Environment. Washington, D.C.: National Academy of Sciences, 1975.

PITMAN, W. C., III, E. M. HERRON, and J. R. HEIRTZLER, "Magnetic Anomalies in the Pacific and Sea Floor Spreading," *Journal of Geophysical Research,* **73** (1968), pp. 2069–85.

PRATT, R. M., "Great Meteor Seamount," *Deep-Sea Research,* **10** (1963), pp. 17–25.

PRITCHARD, D. M., "Observations of Circulation in Coastal Plain Estuaries," in *Estuaries,* ed. G. H. Lauff. American Association for the Advancement of Science Publication Number 83, 1967, pp. 37–44.

PROSPERO, J. M., and F. F. KOCZY, "Radionuclides in Oceans and Sediments," in *Encyclopedia of Oceanography,* ed. R. W. Fairbridge. New York: Reinhold, 1966.

QUASIM, S. Z., "Development of Marine Science Capabilities in Different Regions of the World," in Bologna Conference Report Held by the Johns Hopkins University (October 15–19, 1973).

RAITT, R. W., R. L. FISHER, and R. G. MASON, "Tonga Trench," in *The Crust of the Earth.* Geological Society of America, Special Paper 62, 1955, pp. 237–54.

RAKESTRAW, M. N., D. P. RUDD, and M. DOLE, "Isotopic Composition of Oxygen in Air Dissolved in Pacific Ocean Water as a

Function of Depth," *Journal of the American Chemical Society,* **73** (1951), p. 2976.

RAMPINO, M. R., S. SELF, and R. W. FAIRBRIDGE, "Can Rapid Climatic Change Cause Volcanic Eruptions?" *Science,* **206** (1979), pp. 826–28.

RAYMONT, J. E. G., *Plankton and Productivity in the Oceans.* New York: Macmillan, 1963.

REDFIELD, A. C., "On the Proportions of Organic Derivatives in Sea Water and Their Relation to the Composition of Plankton," *James Johnstone Memorial Volume,* 1934, pp. 177–92.

REDFIELD, A. G., "Ontogeny of a Salt Marsh Estuary," *Science,* **147** (1965), pp. 50–55.

REDFIELD, A. C., B. H. KETCHUM, and A. F. RICHARDS, "The Influence of Organisms on the Composition of Sea-Water," in *The Sea,* Vol. 2, ed. M. N. Hill. New York: Interscience, 1963.

RICHARDS, A. F., "Chemical Oceanography, General," in *Encyclopedia of Oceanography,* ed. R. W. Fairbridge. New York: Reinhold, 1966.

RICHARDSON, P. L., "Benjamin Franklin and Timothy Folger's First Printed Chart of the Gulf Stream," *Science,* **207** (1980), pp. 643–45.

RILEY, G. A., "Factors Controlling Phytoplankton Populations on Georges Bank," *Journal of Marine Research,* **6** (1946), pp. 54–78.

RILEY, J. P., "Historial Introduction," in *Chemical Oceanography,* ed. J. P. Riley and G. Skirrow. New York: Academic Press, 1965.

ROSS, D. A., and G. G. SHOR, JR., "Reflection Profiles Across the Middle America Trench," *Journal Geophysical Research,* **70** (1965), pp. 5551–71.

ROSS, D. A., "Current Action in a Submarine Canyon," *Nature,* **218** (1968), pp. 1242–45.

RUNCORN, S. K., "Changes in the Earth's Moment of Inertia," *Nature,* **204** (1964), pp. 823–25.

RUSSELL, F. S., "Hydrographical and Biological Conditions in the North Sea as Indicated by Plankton Organisms," *Journal du Conseil Internat. p. l'Explor de la Mer,* **14,** no. 2 (1939), pp. 171–92.

RYTHER, J. H., "Photosynthesis and Fish Production in the Sea," *Science,* **166** (1969), pp. 72–76.

RYTHER, J. H., and C. S. YENTSCH, "Primary Production of Continental Shelf Waters Off New York," *Limnology and Oceanography,* **3** (1958), pp. 327–35.

SACHS, P. L., and S. O. RAYMOND, "A New Unattached Sediment Sampler," *Journal of Marine Research,* **23** (1965), pp. 44–53.

SANDERS, J. E., K. O. EMERY, and E. UCHUPI, "Microtopography of the Continental Shelf by Side-Scanning Sonar," *Bulletin Geological Society of America,* **80** (1969), pp. 561–72.

SCHLEE, S., *The Edge of an Unfamiliar World: A History of Oceanography.* New York: Dutton, 1973.

SHEPARD, F. P., *Submarine Geology* (2nd ed.). New York: Harper and Row, Pub., 1963.

SHEPARD, F. P., *Submarine Geology* (3rd ed.). New York: Harper and Row, Pub., 1973, 517 pp.

SHEPARD, F. P., J. R. CURRAY, D. L. INMAN, E. A. MURRAN, E. L. WINTERER, and R. F. DILL, "Submarine Geology by Diving Saucer," *Science,* **145** (1964), pp. 1042–46.

SHEPARD, F. P., and G. EINSELE, "Sedimentation in San Diego Trough and Contributing Submarine Canyons," *Sedimentology,* **1,** no. 2 (1962), pp. 81–133.

SHEPARD, F. P., G. A. MACDONALD, and D. C. COX, "The Tsunami of April 1, 1946," *Bulletin Scripps Institution of Oceanography, University of California,* **5** (1950), pp. 391–455.

SHIGLEY, C. M., "Seawater as Raw Material," *Ocean Industry,* **3,** no. 11 (1968), pp. 43–46.

SILVA, A. J., "Physical Processes in Deep-Sea Clays," *Oceanus,* **20** (1977), p. 31–40.

SPAETH, M. G., and BERKMAN, S. C., "The Tsunami of March 28, 1964, as Recorded at Tide Stations," *Coast and Geodetic Survey Technical Bulletin,* **1**, no. 33, 1967.

SPIESS, F. N., J. D. MUDIE, and C. D. LOWENSTEIN, "Deeply Towed Marine Geophysical Observational System," *Transactions American Geophysical Union,* **48** (1967), p. 133.

STEVENSON, J. R., "Some Likely Outcomes From the Next Law of the Sea Conference," J. Seward Johnson Lectures in Marine Policy Held at the Woods Hole Oceanographic Institution, 1973.

STOMMEL, H., "The Anatomy of the Atlantic," *Scientific American,* **192** (Jan. 1965), pp. 30–35.

STOMMEL, H., "A Survey of Ocean Current Theory," *Deep-Sea Research,* **4** (1957), pp. 149–84.

STRONG, A. E., and R. J. DERYCKE, "Ocean Current Monitoring Employing a New Satellite Sensing Technique," *Science,* **182** (1973), pp. 482–84.

SVERDRUP, H. U., M. W. JOHNSON, and R. H. FLEMING, *The Oceans, Their Physics, Chemistry, and General Biology.* Englewood Cliffs, N.J.: Prentice-Hall, Inc., 1942.

TAIT, R. V., and R. S. DeSANTO, *Elements of Marine Ecology.* New York: Springer-Verlag, 1972.

TERZAGHI, K., "Varieties of Submarine Slope Failures," *Proceedings 8th Texas Conference on Soil Mechanics and Foundation Engineering Special Publication 29.* Austin, Texas: Bureau of Engineering Research, University of Texas, 1956.

THACHER, P. S., and N. MEITH-AVCIN, "The Oceans: Health and Prognosis," in *Ocean Yearbook,* **1**, ed. E. M. Borgese, and N. Ginsburg. Chicago: University of Chicago Press, 1978, pp. 293–339.

"The Limits of National Jurisdiction and the Common Heritage Concept: A Panel Discussion." Woods Hole Oceanographic Institution, 1973.

THOMPSON, T. G., "The Physical Properties of Sea Water," *Physics of the Earth,* Oceanography National Research Council, Bulletin 85, **5** (1932), pp. 63–94.

Tsunami Warning System in the Pacific. Intergovernmental Oceanographic Commission, Paris, France.

TUCKER, M. J., "Sideways Looking Sonar for Marine Geology," *Geo-Marine Technology,* **5** (1966), pp. 332–38.

UCHUPI, E., "The Continental Margin South of Cape Hatteras, North Carolina: Shallow Structure," *Southeastern Geology,* **8** (1967), pp. 155–77.

UCHUPI, E., "Maps Showing Relation of Land and Submarine Topography, Nova Scotia to Florida," *United States Geological Survey,* Miscellaneous Geological Investigations, Map I-451, 1965.

UCHUPI, E., "Slumping of the Continental Margin Southeast of Long Island, New York," *Deep-Sea Research,* **14** (1967), pp. 635–39.

United Nations Educational Scientific and Cultural Organization, Conference on the Law of the Sea *Scientific Considerations Relating to the Continental Shelf,* GUILCHER, A., P. H. LIEMEN, F. P. SHEPARD, and V. R. ZENKOVITCH, (13/2), 1957.

VACQUIER, V., A. D. RAFF, and R. E. WARREN, "Horizontal Displacements in the Floor of the Northeastern Pacific Ocean," *Geological Society of America Bulletin,* **72** (1961), pp. 1251–58.

VENING MEINESZ, F. A., "Gravity Expeditions at Sea," Waltman, Delft, Vol. 1, 1932.

VON ARX, W. S., *An Introduction to Physical Oceanography.* Reading, Mass.: Addison-Wesley, 1962.

WARREN, B., "Oceanic Circulation," in *Encyclopedia of Oceanography,* ed. R. W. Fairbridge. New York: Reinhold, 1966.

WATTENBERG, H., "Uber die Titrationsalkalinitat und den Kalziumkarbonatgehalt des Meerwassers, Deusche Atlantische Exped. *Meteor* 1925–1927," Wiss. erg., Bd., 8, 2 Teil, 1933, pp. 122–231.

WEEKS, L. G., "The Gas, Oil, and Sulfur Potentials of the Sea," *Ocean Industry,* June 1968, pp. 43–51.

WEGENER, A., "Die Entstehung der Kontinente," *Geolica Rundschau,* **3** (1912), pp. 276–92.

WELLS, J. W., "Coral Growth and Geochronometry," *Nature,* **197** (1963), pp. 948–50.

WEST, S., "DSDP: 10 Years After," *Science News,* **113,** no. 25 (1978), pp. 408–10.

WETHERILL, G. W., and C. L. DRAKE, "The Earth and Planetary Science," *Science,* **209** (1979), pp. 96–104.

WHITMORE, F. C., JR., K. O. EMERY, H. B. COOKE, and D. J. P. SWIFT, "Elephant Teeth from the Atlantic Continental Shelf," *Science,* **156** (1967), pp. 1477–81.

WIEGEL, R. L., "Waves, Tides, Currents, and Beaches: Glossary of Terms and List of Standard Symbols," *Council on Wave Research,* University of California, 1953.

WILKERSON, J. W., "Airborne Oceanography," *Geo-Marine Technology,* **5** (1966), pp. 287–93.

WILSON, R. D., P. H. MONAGHAN, A. OSANIK, L. C. PRICE, and M. A. ROGERS, "Natural Marine Oil Seepage," *Science,* **184** (1974), pp. 857–65.

WOOSTER, W. S., "Scientific Aspects of Maritime Sovereignty Claims," *Ocean Development and International Law Journal,* **1,** no. 1 (1973), pp. 13–20.

WORZEL, J. L., "Extensive Deep-Sea Subbottom Reflectors Identified as White Ash," *Proceedings of the National Academy of Sciences,* **45,** no. 3 (1959), pp. 349–55.

WRIGHT, L. D., and J. M. COLEMAN, "Variations in Morphology of Major River Deltas as Functions of Ocean Wave and River Discharge Regimes," *American Association of Petroleum Geologists,* **57,** no. 2 (1973), pp. 370–98.

WURSTER, C. F., "Aldrin and Dieldrin," *Environment,* **38** (1971), pp. 33–45.

WUST, G., W. BROGMUS, and E. N. NOODT, "Die zonale Verteilung von salzgehalt, Niederschlag, Verfunstung, Temperatur und Diche an der Oberflache der Ozeans," *Kieler Meeresforschung,* **10** (1954), pp. 137–61.

ZARUDSKI, E. F. K., "Swordfish Rams the *Alvin*," *Oceanus,* **4,** no. 4 (1967), pp. 14–18.

Glossary

ABYSSAL   Referring to that part of the ocean between a depth of about 2,000 to 6,000 m.

ABYSSAL HILLS   Small irregular hills, rising to a height of 30 to 1,000 m, that cover large areas of the ocean floor. They are especially common in the Pacific Ocean.

ABYSSAL PLAIN   A very flat portion of the ocean floor underlain by sediments. The slope of this feature is less than 1:1,000.

ACIDIC SOLUTION   A liquid whose hydrogen ion concentration is greater than its hydroxyl ion concentration or whose pH is less than 7.0.

ALBEDO   The ratio of the solar radiation reflected by a body or surface to the amount incident or received.

ALKALINE SOLUTION   A liquid whose hydroxyl ion concentration is greater than its hydrogen ion concentration or whose pH is greater than 7.0. Seawater is slightly alkaline, having a pH between 7.5 and 8.4.

ANADROMOUS   Animals that migrate from the sea to freshwater (usually a river) to spawn. (Salmon are an example.)

ANAEROBIC   A condition in which oxygen is absent. The Black Sea is an example.

ANION   A negatively charged ion. Examples are chloride, Cl^-, and oxygen, O^{2-}.

APHOTIC ZONE   That part of the ocean where not enough light is present for photosynthesis by plants.

AQUACULTURE   Farming of the ocean, whereby organisms, such as fish, algae, and shellfish, are grown under controlled conditions. At present this technique is used only in nearshore areas.

ASTHENOSPHERE   The upper portions of the earth's mantle. This region can move similarly to plastic flow, and magma is believed to form here.

ATOLL   A circular-shaped coral reef surrounding a lagoon.

ATOM   The smallest component of an element that has all the properties of the element. An atom consists of protons, electrons, and neutrons.

ATOMIC WEIGHT   The relative weight of an atom on the basis that the most abundant isotope of carbon (^{12}C) has a weight of 12. The atomic weight is essentially equal to the number of protons in the atom.

AUTHIGENIC DEPOSIT   Deposits that formed in place before the sediment was buried. They usually precipitated directly from the seawater.

AUTOTROPH   An organism which produces its own food from inorganic material using light or chemical energy.

AUTOTROPHIC BACTERIA   Bacteria that produce their own food from inorganic compounds.

BARRIER REEF   A reef mainly composed of coral that parallels land but is separated from it by a deep lagoon.

BASALT   An igneous rock, commonly found on the sea floor, composed mainly of feldspar and pyroxene minerals. Basalt rocks are thought to underlie most of the ocean basin.

BATHYAL   Referring to that part of the ocean between a depth of about 200 to 2,000 m.

BATHYMETRY   The measurement of the depth of the ocean.

BATHYTHERMOGRAPH   An instrument used to measure temperature in the ocean.

BEACHES   Unconsolidated sediments (generally sand or gravel) that cover parts or all of the shore.

BENTHIC OR BENTHONIC   The area of the ocean bottom inhabited by marine organisms.

BENTHOS   Organisms that live on or in close contact (such as certain fish) with the ocean bottom.

BIOLUMINESCENCE   The production of light by living organisms as a result of a chemical reaction.

BIOMASS   The amount of living organisms in grams per unit area or unit volume.

BIOSPHERE   A collective term for the area of habitat of the organisms of the earth.

BIOTOPE   An area where the principal habitat conditions and the living forms adapted to the conditions are uniform.

BOILING POINT   The temperature at which a liquid starts to boil. For pure water this temperature is 100°C or 212°F at normal pressure.

CALORIE   A measure or unit of heat generally defined as the amount of heat needed to raise 1 gram of water by 1°C (abbreviated cal).

CARBON-14   A radioactive isotope that can be used for dating. This isotope is especially useful in dating material that was once alive since all living matter contains carbon. The half-life of carbon-14 is 5,560 years.

CATADROMOUS   Animals that migrate from rivers to the sea to breed. (Some eels are examples.)

CATASTROPHIC WAVES   Large waves, resulting from intense storms or submarine slumping, that can cause immense damage and loss of life.

CATION   A positively charged ion. Examples are hydrogen, H^+, and sodium, Na^+.

CENTRIFUGAL FORCE   A force due to rotation that causes motion away or out from the rotating object.

CHLOROPHYLL   A group of green pigments found in plants that are essential for photosynthesis.

COLLIGATIVE PROPERTY   Those properties that vary with the number of chemical elements in the solution and not with their composition. In seawater boiling point and osmotic pressure increase with increasing salinity, and freezing point and vapor pressure decrease.

COLLOIDAL PARTICLES   Very small particles, usually smaller than 0.00024 mm.

COMMUNITY   An integrated group of organisms inhabiting a common area. These organisms may be dependent on each other or possibly on the environment. The community may be defined by their habitat or by the composition of the organisms.

COMPENSATION DEPTH   The depth at which the oxygen produced by a plant during photosynthesis equals the amount the plant needs for respiration (during a 24-hour period).

CONSERVATIVE ELEMENTS   Elements in seawater whose ratio to other conservative elements remains constant. Examples are chlorine, sodium, and magnesium.

CONTINENTAL DRIFT   The concept that continents can drift or move about on the surface of the earth.

CONTINENTAL MARGIN   That portion of the ocean adjacent to the continent and separating it from the deep sea. The continental margin includes the continental shelf, continental slope, and continental rise.

CONTINENTAL RISE   An area of gentle slope (usually less than half a degree or 1:100) at the base of the continental slope.

CONTINENTAL SHELF The shallow part of the sea floor immediately adjacent to the continent. It generally has a smooth seaward slope and terminates seaward at an abrupt change in slope that begins the continental slope.

CONTINENTAL SLOPE A declivity, averaging about 4°C that extends from the seaward edge of the continental shelf down to the continental rise or deep-sea floor.

CONTINUOUS SEISMIC PROFILE A record produced by using a high-energy acoustic device that shows thickness and structure of the upper layers of the crust.

CONVECTION CURRENTS Motion within a fluid due to differences in density or temperature.

CORIOLIS FORCE An apparent force due to the earth's rotation. This force causes moving objects to turn to the right in the Northern Hemisphere and to the left in the Southern Hemisphere.

COVALENT BOND The bond or linkage between two atoms in a molecule formed by the sharing of electrons.

DEAD RECKONING A type of navigation that mainly uses the speed and direction of the ship to estimate position.

DECIBAR A measure of pressure equal to one-tenth normal atmospheric pressure and approximately equal to the pressure change of 1 m depth in seawater.

DEEP-SCATTERING LAYER A sound-reflecting layer caused by the presence of certain organisms in the water. The layer, or layers, which may be 100 m thick, usually rises toward the surface at night and descends when the sun rises.

DEEP SEA DRILLING PROJECT A large-scale scientific project whose main aim was to drill numerous deep holes into the sediments on the ocean floor (see JOIDES).

DENSITY Mass per unit volume.

DESALINATION A variety of processes whereby the salts are removed from seawater resulting in water that can be used for human consumption.

DETRITAL DEPOSITS Sedimentary deposits resulting from the erosion and weathering of rocks.

DIAGENESIS Chemical and physical changes that sediments experience after their deposition.

DIURNAL Occurring daily. Referring to tides, one low and one high tide within one lunar day (about 24 hours and 50 minutes).

DIVERGENCE The flow of water in different directions away from a particular area or zone; often associated with areas of upwelling.

EARTHQUAKE A sudden motion of the earth caused by faulting or volcanic activity. Earthquakes can occur in the near-surface rocks or down to as deep as 700 km below the surface. The actual area of the earthquake is called the focus; the point on the earth's surface above the focus is called the epicenter.

ECHO SOUNDING A method of determining the depth of the ocean by measuring the time interval between the emission of an acoustic signal and its return or echo from the sea floor. The returning signal is usually printed to give a visual picture of the topography of the sea floor. The instrument used in this method is called an echo sounder.

ECOLOGY The study of the interactions of organisms with their physical, chemical, and biological environments.

ECOSYSTEM An ecological unit including the environment and the organisms, each interacting with the other.

ELECTRICAL CONDUCTIVITY A measure of the ability of a material to conduct electricity.

EPICENTER See earthquake.

EROSION The physical and chemical breakdown of a rock and the movement of these broken or dissolved particles from one place to another.

ESTUARY A semienclosed coastal body of water having a free connection with the open sea and within which seawater is diluted by freshwater derived from land drainage.

EUPHOTIC ZONE   That part of the water that receives sufficient sunlight for plants to be able to photosynthesize.

FATHOM   A common unit measure of depth equal to 6 ft (1.83 m).

FAULT   A fracture of rock along which the opposite sides have been relatively displaced.

FETCH   The distance over the sea surface that the wind blows in the area where waves are generated.

FISH PROTEIN CONCENTRATE   An odorless, tasteless protein concentrate that can be made from almost any kind of fish. This product is thought by many to be a partial solution to some of the world's food problems.

FJORD   A glaciated valley or trough partially covered by the sea.

FOCUS   See earthquake.

FOLD   A bend or curvature in rock or sediment layers.

FOOD CHAIN   A complex system that involves many different organisms, each of which is the food for an organism higher up in the chain or sequence.

FRACTIONATION   The separation or division into different components that occurs with some isotopes in the marine environment.

FRACTURE ZONE   A large linear and irregular area of the sea floor characterized by ridges and seamounts. These features are commonly associated with the median ridge common to most ocean basins.

FREEZING POINT   The temperature at which a liquid freezes or solidifies at normal pressure. For freshwater this temperature is 0°C, or 32°F; for normal seawater (salinity about 35 ‰) the freezing temperature is −1.9°C, or about 28°F.

FREQUENCY   The number of cycles or events in a given period of time.

GABBRO   An igneous rock having a relatively high amount of manganese and a low amount of silica, compared to granites.

GEOMORPHOLOGY   The study of the shape of the earth's surface and the processes that control and modify these features.

GEOSTROPHIC CURRENT   A current in which the horizontal pressure gradient is balanced by the Coriolis force. These currents can be calculated by careful measurement of temperature and salinity at closely spaced localities.

GLACIATION   The change or alteration of land or ocean floor caused by the movement of a glacier over it.

GRANITE   A coarse-grained igneous rock consisting mainly of quartz and alkali feldspar.

GRAVIMETER   A device used to measure differences in the earth's gravitational field.

GRAVITY   The force of attraction that causes objects on earth to fall toward the center of the earth. The universal law of gravitation as first given by Newton states that every particle in the universe attracts every other particle with a force that is proportional to the product of their masses and inversely proportional to the square of the distances between the particles.

GRAZING   The feeding by zooplankton upon phytoplankton.

GROSS PRODUCTION   The amount of organic matter photosynthesized by plants over a certain period and within a certain area or volume.

GUANO   The accumulated excrement of sea birds that is often used as a fertilizer.

GUYOT   A flat-topped seamount.

GYRE   A large, essentially closed, circulation system.

HADAL   The deepest parts of the ocean; that part below a depth of 6,000 m.

HALF-LIFE   The time it takes for one-half the atoms of a radioactive isotope to decay into another isotope. The half-life is different for each radioactive element.

HALOCLINE   A zone, usually 50 to 100 m below the surface and extending to perhaps 1,000 m, where the salinity changes rapidly. The salinity change is greater in the halocline than in the water above or below it.

HEAT CAPACITY   A ratio of the amount of heat

absorbed or released by an object to the change in temperature of the object. In other words, how much heat is necessary to raise the temperature of the object.

HERBIVORES   Organisms that eat plants.

HETEROTROPH   An organism that requires preformed organic compounds for food and energy.

HETEROTROPHIC BACTERIA   Bacteria that use organic material, produced by other organisms, for their food.

HOLOCENE   See Pleistocene.

HOLOPLANKTON   Organisms that spend their entire life as plankton.

HYDROCARBON   Organic compounds composed of carbon and hydrogen. Petroleum and natural gas are hydrocarbons.

HYDROLOGICAL CYCLE   The system whereby the water is removed from the ocean by evaporation into the atmosphere and eventually returns to the ocean either directly as precipitation or indirectly by rivers.

HYDROTHERMAL DEPOSITS   Deposits resulting from high-temperature water, either by altering existing rocks or by forming their own precipitates.

IGNEOUS ROCKS   Rocks formed by the solidification of molten magma. Magma is composed of numerous minerals (mainly silicates) and gases derived from the earth's crust and mantle and is in a molten state.

INTERFACE   A surface between two media that have a discontinuity in some property, such as their sound velocities. The major interfaces in the ocean are the water–atmosphere, water–biosphere, and water–sediment.

ION   An atom or group of atoms having an electrical charge. Most of the atoms in seawater are in the ionic form. An ion having a positive charge is called a cation; one having a negative charge is called an anion.

IPOD   See JOIDES.

ISOTOPE   Different forms of the same element, differing mainly in their atomic weights.

JOIDES   Joint Oceanographic Institutions

Deep Earth Sampling—a program that initially drilled six holes into the sediments on the continental shelf and slope off eastern Florida. The second phase of the program, The Deep Sea Drilling Program (DSDP), has resulted in over 800 holes drilled into the Atlantic, Pacific, and Indian Oceans and several marginal seas. The third phase of this program, IPOD (International Phase of Ocean Drilling), will drill extremely deep holes in selected portions of the ocean basin and the continental margin.

KILOMETER   A metric measure of distance equal to 1,000 m, 0.62 statute mi, or 0.54 n mi and abbreviated km.

KNOT   A unit of velocity equal to 1 n mi (6,080 ft) per hour. It is approximately equal to 50 cm per second or 1.69 ft per second and is abbreviated kn.

LAGOON   A shallow pond or lake separated from the sea by a shallow bar or bank.

LANGMUIR CIRCULATION   A cellular type of surface water circulation having alternating vortices with its axes in the direction of the wind.

LATENT HEAT   The amount of heat absorbed or released by a substance during a change of state, under conditions of constant temperature and pressure. The latent heat of evaporation is the amount of heat necessary to go from the liquid to the gaseous state.

LAVA   Liquid rock that comes to the surface, usually from a volcano or a fissure.

LEEWARD   The direction toward which the wind is blowing.

LITHOSPHERE   The solid, outer portion of the earth that includes its crust.

LITTORAL   The benthic zone between high tide out to a depth of about 200 m.

LONGSHORE CURRENTS   Currents in the nearshore region that run essentially parallel to the coast.

MAGMA   Molten rock material derived from the earth's crust and mantle. When it is extruded and flows on the earth's surface (above or below water), it is called lava.

MAGNETIC ANOMALY   A departure from the regularity of the earth's magnetic pattern. The departure may be due to the concentration of magnetic minerals or to a buried (or exposed) igneous rock.

MAGNETOMETER   An instrument used to measure the direction and intensity of the earth's magnetic field.

MEROPLANKTON   Animals that spend only a portion of their life as plankton, ultimately becoming nekton or benthos.

METABOLISM   The process that includes the formation of protoplasm by organisms from food or photosynthesis, the eventual breakdown of the protoplasm, and the release of waste products and energy.

METAMORPHIC ROCKS   Rocks whose composition or general characteristics changed as the result of large changes in temperature, pressure, or chemical environment. (An example is limestone changed into marble.)

MICROTEMPERATURE STRUCTURE   Small-scale, in terms of vertical or horizontal dimensions, changes in temperature.

MIXING PROCESSES   Any process or condition that causes mixing of the seawater.

MOHOROVIČIĆ DISCONTINUITY   The sharp change in seismic velocity occurring at about 11 km depth in the ocean and 35 km depth under land that defines the top of the earth's mantle. This discontinuity, commonly called the Moho, may represent either a chemical or a phase change in the layering of the earth.

MONSOON   A term for seasonal winds, usually applied to the changing wind patterns in the Indian Ocean.

NANSEN BOTTLE   A common device used by oceanographers to obtain subsurface water samples.

NEAP TIDE   Weak tides that occur about every 2 weeks when the moon is in its quarter positions.

NEKTON   Animals that are able to swim independently of current action.

NERITIC   The part of the pelagic environment that extends from the nearshore zone out to a depth of about 200 m; in other words, the waters overlying the continental shelf.

NODE   A portion of a standing wave where the vertical motion is at its minimum but where horizontal velocity is highest.

NUTRIENTS   Compounds or ions that plants need for the production of organic matter.

OCEAN BASIN   That portion of the ocean seaward of the continental margin that includes the deep-sea floor.

OCEANOGRAPHIC STATION   A position at sea where oceanographic observations are made while the ship is stopped.

OOZE   Marine sediment that contains more than 30 percent of various microorganism shells.

OSMOSIS   The movement of dissolved ions or molecules through a semipermeable membrane. An osmotic pressure results when a difference in concentration exists on either side of the membrane. The greater the difference, the higher the osmotic pressure; the flow is toward the more concentrated solution.

OUTCROP   An exposure of rocks at the earth's surface.

OXIDATION   The process whereby an element or compound combines with oxygen or whereby electrons are removed from an ion or atom.

OXYGEN-MINIMUM ZONE   A layer below the surface where the oxygen content is very low or zero.

PALEOMAGNETISM   The study of variations in the earth's magnetic field as recorded in ancient rocks.

PANGAEA   Large continent that split apart about 200 million years ago to form the present continents.

PEAT   A brownish sediment composed of partially decomposed plant tissue.

PELAGIC   A division of the marine environment, including the entire mass of water. The pelagic environment can be divided into a neritic (water that overlies the continental shelf) province and an oceanic (the water of the deep sea) province.

PELAGIC SEDIMENTS   Sediments deposited in the deep sea that have little or no coarse-grained terrigenous material.

pH   A measure of the alkalinity or salinity of a solution. pH is the logarithm of the reciprocal of the hydrogen ion concentration, or $pH = \log 1/H^+$. A pH value of 7.0 indicates a neutral solution; lower than 7.0 is acidic; and higher is alkaline.

PHOTOSYNTHESIS   The production of organic matter by plants using water and carbon dioxide in the presence of chlorophyll and light; oxygen is released in the reaction.

PHOTOSYNTHETIC ZONE   That part of the ocean where photosynthesis is possible, usually defined by the availability of light. In the open ocean this zone usually extends from the surface to about 100 m.

PHYTOPLANKTON   Those plankton that are plants.

PINGER   An acoustic device used to determine distance of instruments above the bottom.

PLACER   Mineral deposits resulting from the reworking effect of waves or currents that remove the lighter density material, leaving the heavier (and often valuable) minerals behind.

PLANKTON   Floating or weakly swimming organisms that are carried by the ocean currents. The plankton range in size from microscopic plants to large jellyfish.

PLANKTON BLOOM   A large concentration of plankton within an area, due to a rapid growth of the organisms. The large numbers of plankton can color the water, causing a red tide in some instances.

PLEISTOCENE   A geological epoch that ended about 10,000 years ago and lasted about 1 to 2 million years. This epoch has been subdivided into four glacial stages and three interglacial stages. The last of the Pleistocene glacial stages is called the Wisconsin stage. The period we are now in, the Holocene Epoch

is not part of the Pleistocene and may be an interglacial stage.

POLYMORPHISM   Organisms that can occur in several different forms, independent of sexual differences.

PRESSURE   The force per unit area upon an object.

PRIMARY SHORELINE   Shoreline where the coastal region has been formed mainly by terrestrial agents. Examples are rivers, glaciers, deltas, volcanoes, folding, and faulting.

PRIMORDIAL   Pertaining to the beginning or initial times of the earth's history.

PRODUCTIVITY   The production of organic material.

PROTOPLANETS   The early planets that preceded and developed (according to the condensation theory on the origin of the planets) into the present planets.

PYCNOCLINE   A zone where the water density rapidly increases. The increase is greater than that in the water above or below it. The density change, or pycnocline, is due to changes in temperature and salinity.

RADIOACTIVE ELEMENTS   Those elements that are capable of changing into other elements by the emission of charged particles from their nuclei.

RARE GASES   Those gases (such as krypton, xenon, and argon) that are present in the earth's atmosphere in very small quantities.

RED TIDE   A brownish or reddish coloration of surface waters caused by large concentrations of microscopic dinoflagellates (sometimes no color is evident). Toxins produced by the dinoflagellates can cause large fish kills and be harmful to humans if they eat shellfish infected with these dinoflagellates.

RELICT SEDIMENTS   Sediments whose character does not represent present-day conditions but rather a past environment.

RESIDENCE TIME   The residence time of an element in seawater is defined as the total amount of the element in the ocean divided by the rate of introduction of the element or the rate of its precipitation to the sediments.

RESPIRATION   An oxidation process whereby organic matter is used by plants and animals and converted to energy. Oxygen is used in this process and carbon dioxide and water are liberated.

RIP CURRENT   A narrow seaward-flowing current that results from breaking of waves and subsequent accumulation of water in the nearshore zone.

ROCK WEATHERING   The chemical and physical processes that cause rocks to decay and eventually form soil.

SALINITY   The total amount of dissolved material in seawater. It is measured in parts per thousand by weight in 1 kg of seawater.

SALT DOME   A large cylindrical mass of salt that has risen through the surrounding sediment. It can form a trap for oil or gas.

SCUBA   Self-contained underwater breathing apparatus.

SEA   Waves within their area of generation; usually they are irregular without a definite pattern.

SEAMOUNTS   Isolated elevations, usually higher than 1,000 m, on the sea floor. They usually resemble an inverted cone in shape.

SECONDARY SHORELINE   Shoreline where the coastal region has been formed mainly by marine or biological agents. Examples are coral reefs, barrier beaches and marshes.

SEDIMENTARY ROCKS   Rocks that have formed from the accumulation of particles (sediment) in water or from the air.

SEICHE   See stationary waves.

SEISMICITY   Relates to earth movements or earthquakes; an area of high seismicity has numerous earthquakes.

SEXTANT   An instrument for measuring angles.

SHELF BREAK   The sharp break in slope that marks the edge of the continental shelf and beginning of the continental slope.

SHOAL   A submerged but shallow bank or ridge that can be dangerous to ships.

SHORELINE   The place where land and water meet.

SILL   A ridge separating one partially closed ocean basin from the ocean or another basin.

SLUMPING   The sliding or moving of sediments down a submarine slope.

SOLUBILITY   The degree to which a substance mixes with another substance.

SOUND CHANNEL   A zone where the sound velocity reaches a minimum value. Sound in this zone is refracted upward or downward back into the zone, with little energy loss. Thus the sound traveling in this channel can be transmitted over distances of many thousands of kilometers.

SOUNDING   The determination of the depth of the ocean either by lowering a line to the bottom or electronically by noting how long sound takes to travel to the bottom and return.

SPECIFIC HEAT   The number of calories needed to raise 1 g of a substance by 1°C.

SPONTANEOUS LIQUEFACTION   The movement or flow of an entire sediment mass, similar to an avalanche, that occurs when water-saturated sediments are subjected to a sudden shock, shear, or increase in pore water pressure and the internal grain-to-grain contacts within the sediment change.

SPRING TIDES   Strong tides that occur about every 2 weeks, when the moon is full or new.

STANDING CROP   See biomass.

STATIONARY WAVES   A type of wave in which the waveform does not move forward; however, the surface moves up and down. At certain fixed points, called nodes, the water surface remains stationary.

STORM SURGES   Abnormally high water levels due to strong winds blowing on the water surface.

SURF   Breaking waves in a coastal area.

SWELL   Waves that have traveled away from their generating area. These waves have a more regular pattern than sea waves.

SYMBIOSIS   A relationship existing between two organisms in which neither is harmed but one or both benefit.

SYNOPTIC MEASUREMENTS   Numerous measurements taken simultaneously over a large area.

TAXONOMIC CLASSIFICATION   A systematic method of classifying animals and plants.

TELEMETRY   A method, usually electronic, of measuring something and then transmitting the measurement to a receiving station.

TERRIGENOUS SEDIMENTS   Sediments composed of material derived from the land. Usually these deposits are found close to land.

THERMISTOR   A heat-sensitive device that can be used to measure temperature.

THERMOCLINE   A zone where the water temperature decreases more rapidly than the water above or below it. This zone usually starts from 10 to 500 m below the surface and can extend to over 1,500 m in depth.

THERMOHALINE CIRCULATION   A vertical circulation of seawater caused by differences in surface temperature or salinity that causes density differences and sinking of water.

TIDAL BORE   A large wave of tidal origin that travels up some rivers and estuaries.

TIDAL CURRENTS   Currents due to tides.

TIDES   The regular rising and falling of sea level, caused mainly by the gravitational attraction of the moon and sun on the earth.

TOPOGRAPHY   The study or description of the physical features of the earth's surface.

TSUNAMI   A long-period wave caused by a submarine earthquake, slumping, or volcanic eruptions. Tsunamis, sometimes called tidal waves, have heights of only a few centimeters in deep water, but can reach several tens of meters before they break on the beach.

TURBIDITE   A generally coarse-grained sediment that is deposited from a turbidity current. These sediments are generally found interbedded with the fine-grained muds typical of the deep sea.

TURBIDITY CURRENT   A turbid, relatively dense current composed of water and sediment that flows downslope through less dense seawater. The sediment eventually settles out, forming a turbidite.

TURBULENCE   A flow of water in which the motion of individual particles appears irregular and confused.

UPWELLING   The movement of water from depth to the surface.

VELOCITY   The rate of change of position (distance) in a given time, such as 50 km per hour.

VOLCANIC ROCKS   Rocks that have resulted from volcanic eruptions.

WAVE ATTENUATION   The decrease in the waveform or height with distance from a wave's origin.

WAVE CREST   The highest part of a wave.

WAVELENGTH   The horizontal distance between two wave crests (or similar points on the waveform) measured parallel to the direction of travel of the wave.

WAVE PERIOD   The time required for successive wave crests to pass by a fixed point.

WAVE REFRACTION   The change in direction of waves that occurs when one portion of the wave reaches shallow water and is slowed down while the other portion is in deep water and moving relatively fast.

WAVE TROUGH   The lowest part of a wave between two successive crests.

WAVE VELOCITY   The speed with which the waveform proceeds. It is equal to the wavelength divided by the wave period.

ZEOLITE   A particular type of authigenic mineral.

ZOOPLANKTON   Those plankton that are of the animal kingdom.